Sustainable Small-Scale Fisheries in the Sundarbans

This book discusses the sustainability challenges of small-scale fisheries in the Sundarbans, examining the socioecological dynamics and management approaches in this complex transboundary ecosystem shared between India and Bangladesh. Integrating oceanographic, ecological, and socioeconomic perspectives, it provides a multiscale analysis of artisanal fisheries in deltaic environments under anthropogenic climate stress.

Employing multidimensional vulnerability frameworks, the work analyzes ecological risks—climatic threats, oceanographic shifts, pollution dynamics—alongside the gender-disaggregated livelihood in fishing communities and the socioeconomic constraints creating poverty traps. The narrative moves beyond the Sundarbans to explore mangrove fisheries across Southeast Asia, Africa, and Latin America, identifying cross-regional sustainability patterns and innovative solutions. Resilience-based adaptive management frameworks are presented to help combat these issues, and the authors explore policy innovation and governance design frameworks for empowering small-scale fisheries at the intersection of ecological stewardship and socioeconomic equity.

This book is beneficial for postgraduates, undergraduates, professionals, researchers, fishery managers, policymakers, and academic scholars in fisheries science, environmental engineering, environmental management, sustainable development, and social ecology.

Anisa Mitra is an assistant professor of Zoology at Sundarban Hazi Desarat College, University of Calcutta, India. With over 15 years of teaching experience and 12 years in research, she specializes in fisheries and aquaculture. Dr Mitra earned her MPhil and PhD from the University of Calcutta and has published extensively in national and international journals. She authored a book on the biology and culture potential of *Chitala chitala* and serves as a regular reviewer for internationally reputed journals.

Prabal Barua holds a PhD in Environmental Sciences from Jahangirnagar University, Bangladesh, and is a visiting lecturer at Bangladesh Marine Fisheries Academy. He was awarded the MPhil in Fisheries Technology from the University of Calcutta and BSc and MSc degrees in Marine Sciences from the University of Chittagong. He has experience of visiting Bangladesh and Indian Sundarbans including coastal zones of both the nations. With 18 years of research experience and 14 years of professional experience, he has contributed over 100 publications to reputed journals and book chapters.

Sumit Homechaudhuri is Former Professor in the Department of Zoology at the University of Calcutta and the President of the Zoological Society of Kolkata. A gold medalist from the University of Calcutta with over 30 years of research experience, he has supervised 30 PhD and 24 MPhil students. He previously served as the Head of the Zoology Department and has published extensively in esteemed journals. He co-authored a book on *Chitala chitala* and regularly reviews for international journals.

Saeid Eslamian is Professor of Hydrology and Water Resources Sustainability at Isfahan University of Technology and Director of Excellence in Risk Management and Natural Hazards. His research focuses on statistical and environmental hydrology and climate change. A former visiting professor at Princeton University, ETH Zurich, and McGill University, he has contributed to over 1,000 publications, authored approximately 400 books and book chapters, and founded the *International Journal of Hydrology Science and Technology*. He serves on editorial boards of 140 Web of Science journals, including as associate editor for the *Journal of Hydrology* and several other prestigious publications.

Faezeh Eslamian holds a PhD in Bioresource Engineering from McGill University, where she developed a novel lime-based product to mitigate phosphorus loss from agricultural fields. She earned her bachelor's and master's degrees in Civil and Environmental Engineering from Isfahan University of Technology, researching natural absorbents for pollutant removal. An experienced multidisciplinary researcher with interests in soil and water quality, environmental remediation, water reuse, and drought management, Dr Eslamian currently works at GHD in Montreal, Canada.

Sustainable Small-Scale Fisheries in the Sundarbans

Challenges, Resilience, and Policy Pathways

Anisa Mitra
Prabal Barua
Sumit Homechaudhuri
Saeid Eslamian
Faezeh Eslamian

CRC Press
Taylor & Francis Group
Boca Raton London New York

CRC Press is an imprint of the
Taylor & Francis Group, an **informa** business

First edition published 2026
by CRC Press
2385 NW Executive Center Drive, Suite 320, Boca Raton FL 33431

and by CRC Press
4 Park Square, Milton Park, Abingdon, Oxon, OX14 4RN

CRC Press is an imprint of Taylor & Francis Group, LLC

ISBN: 978-1-041-21985-9 (hbk)
ISBN: 978-1-041-21986-6 (pbk)
ISBN: 978-1-003-72999-0 (ebk)

DOI: 10.1201/9781003729990

Typeset in Times
by SPi Technologies India Pvt Ltd (Straive)

Dedication

To the Sundarbans, where rivers whisper and mangroves stand tall, To the fisherfolk, whose courage dances with the tides, To every hand that casts a net, dreams a future, and nurtures the sea, May this work honor your resilience, celebrate your stories, and help the world see the beauty in your struggle.

Contents

About the Authors

Anisa Mitra holds a PhD in Zoology (Fisheries and Aquaculture Division) and an MPhil. (Fisheries Technology) from the University of Calcutta, West Bengal, India. She is currently an assistant professor of Zoology at Sundarban Hazi Desarat College, affiliated with the University of Calcutta, where she is actively engaged in teaching and research. Dr Mitra has published numerous scientific articles in reputed national and international journals and has authored a book on the biology and culture potential of *Chitala chitala*. She regularly serves as a reviewer for international journals. Dr Mitra has over 15 years of teaching experience and 12 years of research experience.

Prabal Barua is currently working as a program manager at Young Power in Social Action (YPSA), Bangladesh and has served as a visiting lecturer at the Bangladesh Marine Fisheries Academy in Chittagong, Bangladesh, since 2014. His teaching focuses on Biostatistics, Marine Fisheries Management, Coastal Resource Management, Maritime Environment, and Sustainability Management. Dr Barua holds a PhD in Environmental Sciences from Jahangirnagar University, Bangladesh, an MPhil in Fisheries Technology from the University of Calcutta, India and both a BSc and MSc in Marine Sciences from the University of Chittagong, Bangladesh. He has approximately 18 years of research experience and 14 years of professional experience. Dr Barua has published over 100 articles and book chapters in reputable national and international journals.

Sumit Homechadhuri holds a PhD in Fisheries Science and an MPhil. In Environmental Science from the University of Calcutta, where he was awarded a Gold Medal. He is a recipient of UGC Junior and Senior Research Fellowships as well as a CSIR Senior Research Fellowship. Prof. Homechaudhuri began his academic career as a lecturer in zoology at Darjeeling Government College, later serving as Assistant Professor and Reader at Presidency College. He currently serves as Professor in the Department of Zoology at the University of Calcutta and is the Honorary General Secretary of the Zoological Society of Kolkata. Prof. Homechaudhuri is a recognized PhD and MPhil supervisor and has previously acted as Head and PhD Convenor of the Zoology Department at the University of Calcutta. An active member of various academic committees, he has published extensively in national and international journals and co-authored a book on the biology and culture potential of *Chitala chitala*. Over his 30-year research career, he has supervised 24 MPhil. And 30 PhD students.

Saeid Eslamian is a full professor of hydrology and water resources sustainability at Isfahan University of Technology, Department of Water Engineering. His research focuses on statistical and environmental hydrology, climate change, and forecasting natural hazards such as floods, droughts, storms, and pollution to promote a sustainable environment. He currently serves as Director of Excellence in Risk

Management and Natural Hazards. Dr Eslamian has previously held visiting professorships at Princeton University (USA), ETH Zurich (Switzerland), and McGill University (Canada). He has contributed to over 1,000 publications, including journal articles, books, and technical reports and has authored or co-authored approximately 400 books and book chapters. Dr Eslamian is the founder and chief editor of the *International Journal of Hydrology Science and Technology* and serves on the editorial board or as a reviewer for about 140 Web of Science (ISI) journals. He has also held associate editor roles with several leading journals, including the *Journal of Hydrology* (Elsevier), *Ecohydrology and Hydrobiology* (Elsevier), *Journal of the Saudi Society of Agricultural Sciences* (Elsevier), *Water Reuse* (IWA), *Arabian Journal of Geosciences* (Springer), *International Journal of Climate Change Strategies and Management* (Emerald), and *Frontiers in Water and Frontiers in Climate* (MDPI).

Faezeh Eslamian holds a PhD in Bioresource Engineering from McGill University, where her research focused on developing a novel lime-based product to mitigate phosphorus loss from agricultural fields. She completed her bachelor's and master's degrees in civil and environmental engineering at Isfahan University of Technology, Iran, with research on natural and low-cost absorbents for removing pollutants such as textile dyes and heavy metals. Dr Eslamian has also conducted research on global water quality standards and wastewater reuse guidelines. Her multidisciplinary research interests include soil and water quality, environmental remediation, water reuse, and drought management. He is currently based at GHD in Montreal, Canada.

Preface

Fisheries have historically constituted a crucial worldwide enterprise, producing more than 100 million tons of fish and fisheries products per year. This sector sustains over 200 million livelihoods and delivers protein to more than a billion humans. Due to diminishing fish populations and mounting evidence of issues within fisheries, sustainability has emerged as a pivotal concern in fishery research and business. The escalating anxiety is primarily fueled by the reality that several fisheries are now in danger, with some needing immediate intervention. The advancement of sustainable fisheries depends on achieving fair enhancements in stakeholders' quality of life while preserving the ecosystem's long-term stability.

In scientific study and policy discourse, the industrial fishing industry has traditionally eclipsed small-scale fisheries (SSF). In recent years, initiatives by the worldwide scientific community and governments have started to emphasize their significance. These fisheries, often based on local resources and people, are essential in promoting resilience in poor nations, particularly in the context of climate change.

The Sundarbans, an extensive and intricate ecological and sociopolitical area shared by India and Bangladesh, offers a unique opportunity for cross-cultural collaboration to alleviate poverty and sustainably manage fish resources. Artisanal fisheries in this area account for over 60% of coastal fish production in eastern India and 80–90% of total marine fishery output in Bangladesh. These small-scale fisheries have encountered escalating obstacles over the last 20 years owing to climate change, anthropogenic activities, and sociopolitical upheavals. Although the Sundarbans constitute a unified ecosystem, they are sometimes seen as two distinct entities separated by the international boundary. A multitude of studies have been independently done in both nations to evaluate the state, exploitation, and management of fisheries. These studies have examined topics like fisheries exploitation in the Sundarbans of Bangladesh, the effects of climate change on small-scale fishers in India's Sundarbans, and the socioeconomic factors influencing fishing populations on both sides. This book represents the first transboundary initiative to comprehensively assess the socioecological changes affecting small-scale fishing in the Sundarbans. This research provides a thorough examination of the socioeconomic transformations throughout time and recognizes adaption patterns in both locations, highlighting the triumphs, problems, and prospects for the artisanal fisheries of the Sundarbans.

We, the authors aspire that this contribution will motivate future researchers, fishery managers, and policymakers to enhance their comprehension of the present condition of small-scale fisheries in the Sundarbans and strive for sustainable, cooperative solutions that advantage both ecosystems and the communities reliant on them.

FIGURE 0.1 Silence, salt air, and stories floating by.
(Photo credit: Sitesh Manna).

FIGURE 0.2 When the day melts into gold.
(Photo credit: Sitesh Manna).

FIGURE 0.3 A boat, a horizon, and endless calm.
(Photo credit: Sitesh Manna).

Acknowledgments

Like the intricate root systems of Sundarbans mangroves that create stability through interconnection, this book emerged from a vast network of relationships spanning fishing villages and university laboratories, government offices and women's cooperatives, ancient wisdom, and cutting-edge science. It would not have been possible without the extensive body of scholarship that preceded it and the generous contributions of researchers, institutions, and communities who facilitated access to both published literature and unpublished field reports. We are deeply indebted to the decades of foundational research conducted by marine biologists, anthropologists, economists, and policy analysts who have documented the complex socioecological dynamics of the world's largest delta system, from the pioneering ecological surveys of last few decades to the recent climate adaptation studies that inform our contemporary understanding. We are profoundly grateful to the fishing families who opened their homes and shared their knowledge across tidal creeks and seasonal cycles, to our academic colleagues worldwide who helped weave local insights into global understanding, to the NGO workers and community organizers who showed us pathways between conservation and livelihoods, to our families, whose unwavering encouragement, patience, and understanding provided the strength and stability needed throughout the process of writing this book, to the peer reviewers who pushed our thinking beyond comfortable boundaries, and provided critical feedback on our analytical framework, ensuring that this review accurately represents the evolution of thinking about sustainable fisheries management in mangrove ecosystems, to the meteorologists and oceanographers who helped us understand that the Sundarbans is not only a place but a process of constant becoming, to the students who asked the hardest questions, to the women fish processors whose economic innovations challenge traditional narratives, to the children who drew pictures of their changing world, and ultimately, we acknowledge our debt to the Sundarbans ecosystem itself—a living archive of human-environment interactions that continues to generate new knowledge even as rising seas and changing climate patterns, reshape both the questions we ask and the answers we seek where traditional knowledge and scientific method converge in the urgent work of imagining sustainable futures for both human communities and the largest mangrove ecosystem on Earth.

Dr Anisa Mitra
Dr Prabal Barua
Prof. Dr Sumit Homechadhuri
Prof. Dr Saeid Eslamian
Dr Faezeh Eslamian

1 Sustainability and Small-Scale Fishery

1.1 INTRODUCTION

"Nature–society relationships are conceptualized as a mutuality, rather than as a duality" (O'Brien et al. 2007). Nature sustains, society transforms—together, they define the trajectory of our world. The interface between natural and social systems is complex, unpredictable, and stochastic. Environmental and human subsystems encompass diverse dimensions, including ecological, genetic, biophysical, social, economic, and demographic factors (Porto-Gonçalves, 2013; Liberati et al., 2018). Researchers have described this interdependence through various frameworks, such as Coupled Human–Environment Systems (CHES) (Turner et al., 2003; Turner et al., 2003), Coupled Human And Natural Systems (CHANS) (Lutz et al., 2014), and the most widely used concept, Social–Ecological Systems (SES) (Berkes et al., 2003; Côrtes et al., 2015; Folke et al., 2005; Heikkila et al., 2018). These frameworks highlight the synergistic relationships that determine the condition, function, and response of, each subsystem or the system particularly in the face of disturbances, perturbations, or hazards. In essence, CHES depicts the trade-offs between environmental services and those services and human outcomes (Turner, 2010). In the 21st century, understanding CHANS as complex systems has gained widespread recognition. While current methods of complexity research can describe certain aspects of biological systems over limited parameters, they often fail to account for the initial emergence of complex systems, their evolution, state changes, and sudden transitions (Jaafar & Maideen, 2012).

As ecosystems become increasingly unstable, the theory of biocomplexity has emerged as crucial in predicting their future health (Barua et al., 2022; Mitra et al. 2023). Biocomplexity emerges from the "multiplicity of interconnected relationships and levels" among behavioral, biological, chemical, physical, and social interactions, emphasizing the connections between socioeconomic variables and environmental dynamics. Two key features of the biocomplexity concept are: (1) it emerges through three-dimensional complexity, which includes spatial heterogeneity, connectivity, and historical contingencies, enabling cross-disciplinary integration; and (2) each component becomes more complicated when attention moves from discrete units to the relationships between functional units, reflecting echoes, legacies, and the changing indirect impacts of previous and current states, which can result in unanticipated or emergent features. Sustainability principles can be derived from these intricate and sophisticated interactions (Masud & Kari, 2015; Paul et al., 2023; Sahoo & Bhaskaran, 2016). Each planet region possesses unique natural and human characteristics, with the biocomplexity of a given area determined by climatic, geological, hydrological, biological, and human-induced factors. These factors specify the features of zonal

DOI: 10.1201/9781003729990-1

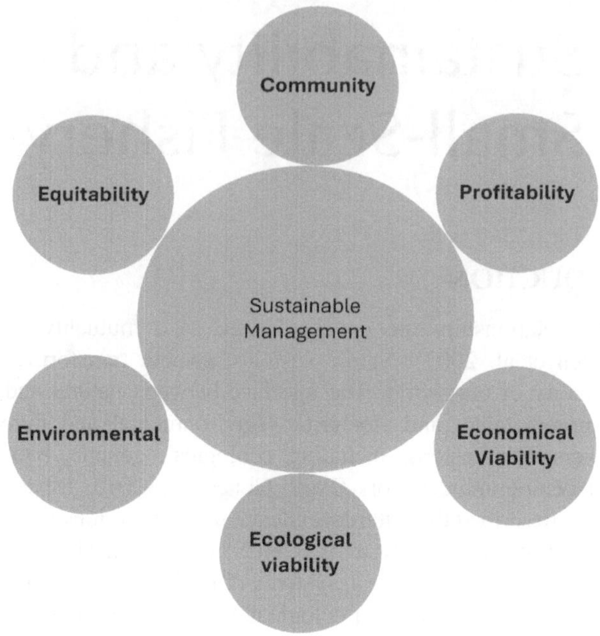

FIGURE 1.1 Sustainable natural resource management.

systems and the tension within them (Reddy et al., 2021). Designing economic, social, and natural resource policies that seek to advance sustainable health and quality of life in line with the trends of evolving landscapes requires the systematic collection and analysis of such indicators over time and space (Graziano et al., 2022; Paudel & Yuan, 2012) (Figure 1.1).

Understanding societal needs through the lens of biocomplexity is essential for crafting sustainable and resilient systems. The goal of meeting society's needs while protecting the planet's life support systems has greatly improved our understanding of how people and the environment interact, leading to the idea of "sustainability science." The subject of sustainability science has quickly evolved since its inception by Kates et al. (2001) as the scientific foundation for sustainable development (Fang et al., 2018; Wu, 2013). Sustainable development was established as a normative framework for balancing environmental stewardship and human welfare across temporal scales in the 1987 report "Our Common Future" by the Brundtland Commission. The operational conversion of this conceptual framework into quantifiable results, in particular socioecological contexts, is still a research challenge, though. Two key ideas are included in the Brundtland definition: (1) the importance of basic human needs, especially for underprivileged groups; and (2) the understanding of the constraints placed on the environment by social structure and technology (Brundtland 1987). With institutional arrangements and technological advancements, this framework suggests a dynamic equilibrium between ecological integrity and social welfare. Operationalizing this framework requires developing measurable indicators that capture both present-day well-being and future capacity for need satisfaction. Many ecologists and social scientists are now focusing on ecological sustainability, developing plans to ensure that resources and services will be available to a growing population (Goodland, 1995). Even though coping mechanisms and adaptation are crucial, there is increasing recognition that some social and economic factors may endanger social equality and long-term sustainability. To adapt to climate change, marginalized communities need to be reorganized and changed (Ali & Erenstein, 2020). In vulnerability research, system state desirability analysis highlights the unequal distribution of costs and benefits over time (Hossain et al., 2021b). Interdisciplinary scientific inquiry and decision-making must be enhanced if we are to deal with the susceptibility of integrated human–environment systems (Guillard-Gonçalves & Zêzere, 2018).

1.1.1 Sustainability: "Our Common Future"

The 2012 United Nations Conference on Sustainable Development (Rio+20) introduced the notion of the "blue economy" into discussions about oceans, coasts, and maritime issues, expanding upon the "green economy" framework established at the 1992 Rio Earth Summit. The blue economy seeks to integrate ecological sustainability with economic advantages, formulated within governance strategies and development models (Ayilu et al., 2022). The Sustainable Development Goals (SDGs) of the 2030 Agenda underscore the need for a comprehensive reform of the existing economic system. It is widely recognized that all SDGs are interconnected, and the achievement of one often facilitates progress in others. Partnerships at the global, regional, and national levels across sectors such as trade, finance, research, technology, and innovation are essential for the promotion of sustainable development. Industries are needed to play a key role in this process. To harmonize their trades with the objectives of sustainability and the environment, industries must be the voice of environmental issues, bring in activities that motivate the participation of the environment, and facilitate the development and utilization of environmentally friendly technologies (Rolib & Janasen, 2024) (Figure 1.2). Environmental sustainability, a fundamental pillar of sustainability, refers to the capacity of the natural environment to persistently produce and maintain human life over a prolonged duration. It requires the sustainable use of natural resources as both a waste repository and an economic resource (Mambet Doue et al., 2020) underline the correlation between carrying capacity and ecological integrity. They argue that the environmental effect of each of the three levels—geo-environment, eco-environment, and socioenvironment—are the main contributors to the total state of well-being (Hagihara & Hagihara, 2019; Tillema et al., 1991). Economic sustainability is making financially prudent and fair choices while considering other sustainability-related aspects (Zhai & Chang, 2019). Social sustainability is the main idea of the relationship between society and the environment (Giabardo, 2020). Social sustainability theory assumes that eradicating poverty should be done within the constraints of society's financial and environmental resources, thus not causing any unnecessary environmental degradation or economic instability (Kumar et al., 2014; Scopelliti et al., 2018). Grasping the development of social relations through a systemic view is very important for social sustainability (Mensah & Ricart Casadevall, 2019).

FIGURE 1.2 Steps to achieve sustainability.

The focus of using sustainability is the characteristics that make SES less susceptible or more resilient to the aggregate range of drives like stressors, perturbations, and disturbances. In a way, it is a balancing act between environmental services and human outcomes that try to reduce the unpredictability of nature while at the same time increasing the provisioning and regulating services, all linked to human well-being (Erős & Lowe, 2019; Gupta & Barua, 2017; Palomo-Campesino et al., 2018).

The concept of sustainability may be interpreted in both a limited and a wide context. A narrow interpretation emphasizes the conservation of capital or investments by protecting and preservation of natural resources. Conversely, a broader and meaningful view promotes the effective utilization and distribution of resources across time, emphasizing the welfare of society at large (Blanchard et al., 2017). Mitigating system vulnerability and enhancing system resilience are essential for attaining sustainability.

Vulnerability is defined as the conditions—whether physical, social, economic, or environmental—that increase a community's susceptibility to the impacts of hazards, which are typically external to the system or element at risk. Risk is often seen because of both the probability of hazards and the vulnerability of the exposed parts (Skilodimou et al., 2021). The fundamental definitions of risk components are tabulated in Table 1.1. Thus, vulnerability includes both susceptibility—systemic deficiencies—and adaptive strategies to manage external stressors. Adaptive capacity, defined as a system's ability to endure and react to stressors and disturbances, is included into the multidimensional vulnerability theory (Abid et al., 2016; Goniewicz et al., 2023; Hendriks, 2022). Vulnerability may be understood by combining biophysical factors with social, political, economic, and institutional dynamics (Jordan, 2015; Yousafzai et al., 2022). In other words, vulnerability denotes a society's fragility, defined as a

TABLE 1.1
Fundamental Definitions of Risk Components (Banerjee et al., 2023; Jha et al., 2022; Ebrahim et al., 2022)

Key Components	Characterizations
Risk	Outcomes that could occur when there is a valuable item at risk and the conclusion is unpredictable. The interplay of hazard, exposure, and susceptibility leads to risk.
Hazard	Any physical occurrence, trend, or impact—whether human-made or natural—that has the potential to cause harm, death, or other health repercussions in addition to the loss and destruction of ecosystems, environmental resources, infrastructure, livelihoods, services, and property.
Exposure	People, livelihoods, ecosystems or species, environmental services, resources, infrastructure, or economic, social, or cultural assets in potentially affected areas and habitats.
Vulnerability	The tendency or inclination to suffer negative consequences. A lack of ability to cope and adapt, as well as sensitivity or susceptibility to injury, are some of the notions and factors that make up vulnerability.
Sensitivity or susceptibility	The degree, to which climate-related stimuli affect a system, either favorably or unfavorably, is known as sensitivity. Indirect effects include damages from coastal flooding caused by sea level rise, whereas direct effects include changes in crop yield caused by temperature differences.
Adaptive capacity	The capacity of individuals, organizations, systems, and other organisms to adjust to potential threats, capitalize on opportunities, or respond to results.
Coping capacity	The ability of individuals, institutions, organizations, and systems to confront, handles, and surmount adverse conditions in the short to medium term by leveraging their skills, values, beliefs, resources, and opportunities.
Hazard chains	A situation when a series of occurrences is created by one hazard leading to another.
Intermediate impacts	Not a risk factor in and of itself, but a useful tool for comprehending the chain of events that leads to the risk. They depend on both vulnerability and hazard variables.

function of the risks faced by individuals or groups (exposure to risks), the sensitivity of their livelihoods to these risks (sensitivity), and their adaptive capacity to manage or recover from external stressors (adaptive capacity) (Islam et al., 2019; Tasnuva et al., 2021). Eriksen et al. (2021) characterize vulnerability as a relationship-shaped condition; instead of using the term "vulnerable people," which tends to label groups like women as inherently vulnerable, the term "marginalized people" focuses on the sociopolitical relationships, like gender and race, that create socially differentiated vulnerability. In many instances, this marginalization is explicitly political, with public resources allocated to certain groups as a kind of patronage to get political support. Thomas and Warner (2019) highlight the tendency to "weaponize vulnerability," whereby marginalized communities are portrayed as security threats to justify laws designed to protect the affluent from these disadvantaged populations (Figure 1.3).

Vulnerability assessment is crucial for mitigating the societal repercussions of hazard risks (Balogun & Onokerhoraye, 2022). A Vulnerability Assessment (VA) identifies, measures, and prioritizes a system's vulnerabilities. Certain VAs concentrates on climatic causes, while others, such as economic and demographic factors, are

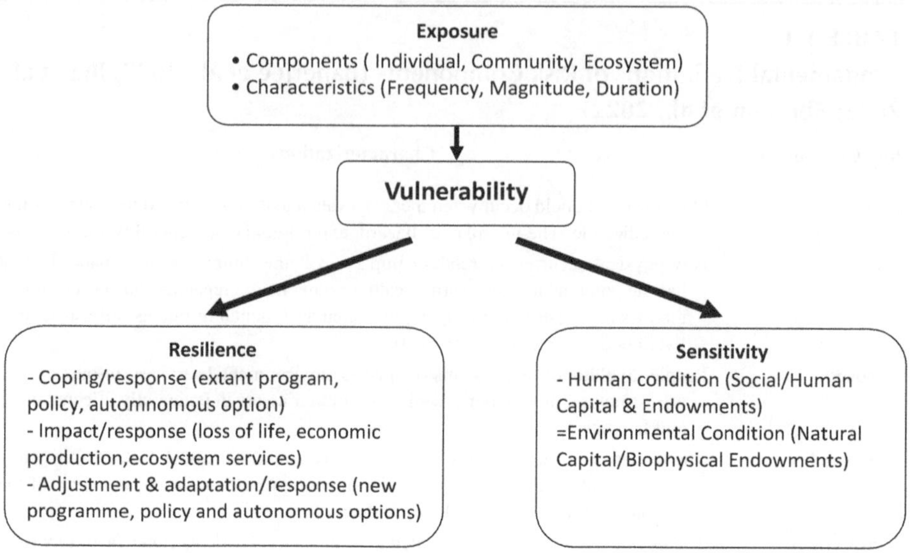

FIGURE 1.3 Social vulnerability influencing the community as well as coastal environment through exposure, sensitivity, and resilience.

classified as "second-generation non-climatic drivers" (Rakib et al., 2019). Two scales are often considered to measure vulnerability: the micro level (reductionist perspective) and the macro level (constraint perspective). The micro level pertains to localized occurrences that directly impact ecosystems or individuals within a small region. The macro level comprises wider, long-term processes that affect environmental conditions and livelihoods (Hossain et al., 2019, 2021a, 2021b; Xenarios et al., 2019). The inclusion of (1) exposure models that delineate factors identifying individuals and areas vulnerable to natural hazards (Penning-Rowsell et al., 2013); (2) social resilience metrics that assess the capacity to endure risks linked to vulnerability (Leal Filho et al., 2021); and (3) the integration of social resilience and potential exposures, focusing specific regions (Luo et al., 2015; Shakhawat et al., 2020) can act as the primary elements for vulnerability assessment. Hence a comprehensive analysis of multiscale processes is essential, since the systems and processes function over a wide range of geographical and temporal scales (James, 2008; Shah et al., 2020) to understand the vulnerability. Multiscale vulnerability assessments, examines structures and processes at many observational levels, are increasingly becoming the main subject of current research. Hierarchy theory regarded as a robust framework for investigating cross-scale interactions in interconnected human–environment systems, emphasizing their significant outcome at specific scales (Mafi-Gholami et al., 2021). Hence, by deconstructing complex systems into intricate systems across several geographical and temporal dimensions using diverse indicators, enable us to measure vulnerability in a more systematic way (Salik et al., 2020) (Figure 1.4). To compare different frameworks in a variety of contexts for developmental projects Vulnerability

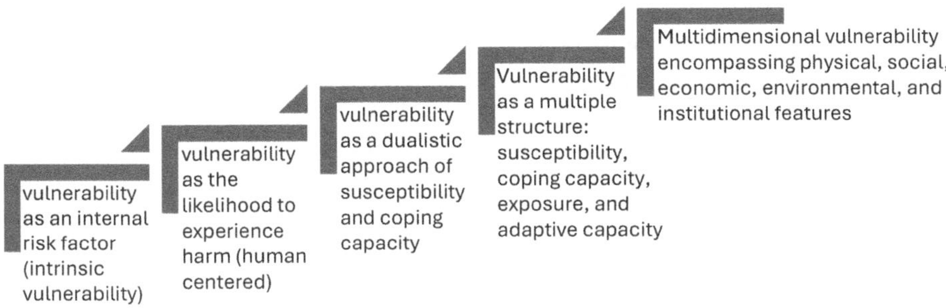

FIGURE 1.4 Multiscale approach to vulnerability assessment.

indices can serve as reference points. They help in directing the use of resources, which is especially necessary for adaptation and mitigation activities (Mallick et al., 2022; Niang et al., 2014).

Resilience in ecology refers to a system's capacity to function without change or in response to disturbances. Recent research has focused on persistence, adaptability, and transformability within SES, transcending the concept of "engineering" resilience. Contemporary resilience thinking encompasses learning, leadership, social capital, and institutions. There are two types of resilience in SES, namely, General resilience and Specified resilience. The concept of specified resilience is one where the ability of the system to cope with stress is studied, including threshold changes that cause sudden regime shifts or changes in the state of the system (Alemayehu & Bewket, 2020). The General resilience of the system, defined by Moser and Ekstrom (2020), as the capability to endure any type of shock regardless of nature and intensity. The dynamic concept of resilience, specifically in relation to socioecological systems, is often synonymous with the dilemma of retaining vs. changing systems, particularly in response to changing climate conditions. Understanding the balance between the twin goals is vital to promote sustainability and sustained system health.

Climate change, natural resources (also known as natural capital or bio-capacity), and the environment are deeply interconnected. Natural capital underpins the growth and development of humans and other living organisms. As the world population surpasses the Earth's carrying capacity, the demand for natural resources escalates, resulting in greater exploitation. Ecosystem services are deteriorating due to the significant and rapid impacts of human activities on ecosystems, climate, and the environment. Moreover, climate change exacerbates the ecological footprint in a cyclical manner. Hence, we must lower our ecological footprint to mitigate the worst impacts of climate change (Foley et al., 2005; Haberl et al., 2007; MEA, 2005). Effective governance can tackle climate change via the implementation of strong adaptation and mitigation strategies, alongside the SDGs, climate resilience initiatives, and nature-based solutions.

The landscape idea is among the most comprehensive approaches for achieving long-term sustainability. Landscape-scale sustainability operates at the systems level where the most impactful changes occur. This comprehensive methodology addresses interconnected challenges concurrently (Brundtland 1987). Landscape sustainability science integrates sustainability research, landscape ecology, and landscape planning and design to bridge the gap between the natural and social sciences (Opdam et al., 2018; Sorvari et al., 2009; Wu, 2013, 2019). Delta ecosystems are a perfect illustration of interconnectedness that defines landscape-scale contemplating. In landscape ecology, the "delta-sea mosaic" concept exemplifies complex social–ecological systems, where sediment flows, tidal dynamics, and human activity interact to shape vulnerability and resilience in spatial and temporal manner. Delta is a large land-locked complex river-floodplain ecosystem, with a diverse biota, and high environmental heterogeneity (Mosepele et al., 2022)—are vital for small-scale fisheries (SSFs). These areas provide essential habitats, support biodiversity, and sustain the livelihoods of millions. Sustainability in delta-sea mosaics is crucial to preserving their ecological richness, support livelihoods, and maintaining resilience against environmental change. In delta areas, future planning needs disaster risk reduction (DRR) techniques including ecosystem-based decision-making and risk-informed development plans. This will need spatially precise data on hazards, exposure, and vulnerabilities. This data must be integrated across SES, including both societal and ecological components. Comprehending the susceptibility of socioecological systems to tipping points[1] is essential for guaranteeing the sustained delivery of ecosystem services. The practical ramifications of these ecological transitions and their impact on socioeconomic systems (SE) are understudied (Milkoreit et al., 2018).

1.1.2 Sustainability and "A Sea of Small Boats"

A large proportion of the world's fish stocks are being fully exploited, over-exploited, or depleted causing potentially irreversible alterations to ecosystems due to unparalleled population growth and unsustainable practices (FAO, 2024; MEA 2005). In this context, sustainability of global fisheries is a paramount interest to achieve several SDGs, that is, SDG 1 (No Poverty), SDG 2 (Zero Hunger), and SDG 14 (Life Below Water) (Sampantamit et al., 2020). Capture fisheries being a common, open access resource extraction pool with low excludability and high subtract ability (McGoodwin, 1990; Pauly et al., 2002, 2005). This leads to "over-exploitation of a resource which is owned by many and not effectively managed by anyone and leads to reduction in catch and eventual poverty of users and others who would otherwise benefit from the harvests" (Béné et al., 2010) becoming the "Tragedy of the commons" (Hardin, 1968) unless those resources are privatized and/or regulated by the government (Costello et al., 2008; Feeny et al., 1990).

Under the broad spectrum of capture fisheries, Small-scale or Artisanal fisheries is crucial for alleviating poverty, ensuring food security, enhancing nutrition, and promoting sustainable livelihoods. Most of these operations occur in developing nations, particularly in Asia, and contribute around 50% of global fish catches (Das et al. 2026) (Figure 1.5). Fish as a vital component of local diets in many areas, strengthening access to small-scale fisheries could greatly improve food security (Ghosh and Dey 2025). According to FAO (2004)

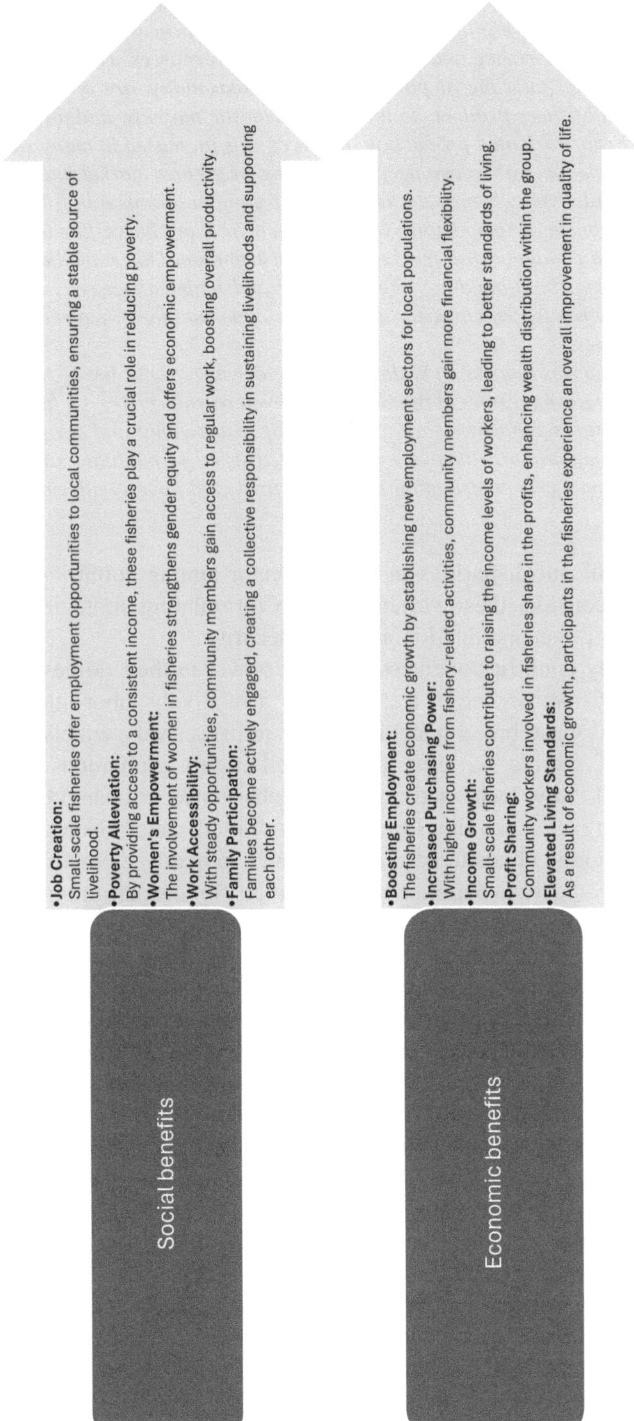

FIGURE 1.5 Benefits of small-scale fisheries.

"Small-scale fisheries can be broadly characterized as a dynamic and evolving sub-sector of fisheries employing labour-intensive harvesting, processing and distribution technologies to exploit marine and inland water fishery resources. The activities of this sub-sector conducted full-time or part-time, or just seasonally, are often targeted on supplying fish and fishery products to local and domestic markets, and for subsistence consumption. Export-oriented production, however, has increased in many small-scale fisheries during the last one to two decades because of greater market integration and globalization. While typically men are engaged in fishing and women in fish processing and marketing, women are also known to engage in near shore harvesting activities and men are known to engage in fish marketing and distribution. Other ancillary activities such as net-making, boat-building, engine repair and maintenance, etc. can provide additional fishery-related employment and income opportunities in marine and inland fishing communities.

Small-scale fisheries operate at widely differing organizational levels ranging from self-employed single operators through informal micro-enterprises to formal sector businesses. This sub-sector, therefore, is not homogenous within and across countries and regions and attention to this fact is warranted when formulating strategies and policies for enhancing its contribution to food security and poverty allevia"

Given the significant disparities in this subsector among countries and regions, it is important to consider these variances when formulating strategies and laws to enhance its role in poverty reduction and food security.

In general, many countries use vessel size or power in their domestic definitions of "small-scale" (Chuenpagdee et al., 2006) as a relatively labor-intensive fishing operations. Beyond these broad generalities, the term can vary widely with a synonym of "artisanal" fishing—a term that is itself highly ambiguous, but tends to describe traditional, "low tech" activity at the subsistence or near-subsistence level of economic activity level activity. In most cases, "small-scale" covers a much wider range of socioeconomic characteristics, all of which generally involve some form of commercial activity, be it via barter or cash exchange. So, the definition of small-scale and artisanal fishing can be unified by the following explanation: "fishing households [as opposed to commercial companies], using relatively small amount of capital and energy, relatively small fishing vessels [if any], making short fishing trips, close to shore" (FAO, 2015) and such fisheries frequently revolve around extraction of heterogeneous resources by fishers who employ diverse livelihood strategies (Allison & Ellis, 2001; Ostrom et al., 1994). Even individual fishing strategies are often presumed to follow one of several archetypical models of behavior, whether inherently ecologically and socially. This dominant imaginary of SSF is often spatialized, presumably limited to the tropical seas of the Third World, as opposed to the fully capitalist, industrial fisheries that inhabit the First World (Martin, 2005). From 1960 to 2015, several systematic and empirical research sought to define the artisanal fisheries sector, mostly via fleet size (65% of definitions), fishing gear (58%), and sociocultural characteristics (35%). Additional factors—including ecology and habitat, species, organization, labor, market integration, and onboard storage/preservation—were seldom referenced in the literature (Smith & Basurto, 2019) (Table 1.2).

Small-scale fisheries (SSFs) are rooted in complex SES where biophysical and social subsystems operate through interdependent feedback relationships under the

TABLE 1.2
Global Estimates of Small-Scale Fisheries Catch Per Capita, by Region (Average Annual Values, 2013–2017)

	Fish Catch Per Capita by Region and Type (kg/person/year)					
Region	Marine (Average)	Marine (Maximum)	Inland (Average)	Inland (Maximum)	Combined (Average)	Combined (Maximum)
Africa	~10	~10	~2.5	~10	~10	~10
Americas	~25	~50	~0.5	~1	~25	~50
Asia	~15	~20	~3	~33	~15	~20
Europe	~5	~5	~0.5	~5	~5	~5
Oceania	~50	~650	~1	~7	~50	~700

Source: After FAO (2023).
Note: Oceania dominates marine fish catch with maximum values reaching 650–700 kg/person/year.
Asia has the highest inland fish catch (3 kg average, 33 kg maximum).
Europe shows the lowest consumption across both categories.
America demonstrates significant variation between average (~25 kg) and maximum (~50 kg) for marine catch.

high levels of biological and ecological uncertainty (Cochrane et al. 2020; Ostrom 2009, Perry et al. 2010). Different country specific studies showed that the key factors that lead to either success or failure of these fisheries in developing countries are unique and inimitable elsewhere (Kosamu 2015). the perception of having SSF in good shape which is not universally true (Fernández et al. 2020). In last 25 years most studies of small-scale fisheries in developing countries emphasised small-scale fisherfolks' resource dependence and the open-access nature of fisheries that together lead to resource degradation, poverty and marginalisation (Defeo and Castilla 2012). This dependence on fisheries is acute at the regional and local level leading to the "the occupation of last resort" for "the poorest of the poor" (Be´ne´ et al., 2007; FAO, 2009). The toxic triad of underreporting catch, overfishing and condemning 'the environment' for the subsequent clutter, can easily be extended to toxic tetrad of fisheries (Pauly et al. 2014). Additionally, in the emerging countries, the poor access to resource, multi-species with multi-gear approach, single ownership with short run profit motives, weak governance, commoditization of fish and quest of economic growth through industrialization and market expansion (Asche et al. 2018; Botsford et al. 1997; Gordon n.d.; Pitcher and Lam 2015; Purcell and Pomeroy 2015; Smith 1969) make the resilience of SSFs is frequently low (Bueno and Basurto 2009; Pauly 2006) leading to their perturbation (Jentoft and Eide 2011).

There are two broad narratives that characterize the sustainable small-scale fisheries of the developing countries, that is, (1) economic benefits imposing high harvest levels and undermining ecological sustainability. Where virtual substitution of natural capital can be safely run down if enough human-made capital is built to in exchange and would not consider irreversible impacts in the ecological, socioeconomic or cultural domain (Neumayer, 1999). But many fundamental services provided by nature cannot be replaced at any level by human-made capital. According to the concept of Critical Natural Capital (Ekins et al., 2021), if a system aims to be sustainable, a minimum amount of this unique natural capital should be insured to maintain environmental functions overtime, otherwise these will irreversibly be lost over relevant time horizons. (2) the ecological integrity and social equity are non-negotiable, recognizing natural capital as irreplaceable and prioritizing collective social capital and long-term social-ecological resilience. (Brekke, 1997, Daly, 1999; Daly & Cobb, 1989; Garmendia et al., 2010; Neumayer, 2010; Pearce & Turner, 1990). These two narratives reflect the tension between weak and strong sustainability. The first prioritizes economic growth, assuming human-made capital can substitute natural capital, risking irreversible damage to ecosystems. The second treats natural capital as irreplaceable, emphasizing long-term social-ecological resilience. Sustainable small-scale fisheries must align with strong sustainability principles to ensure long-term viability.

1.2 CONCLUSION

In the small-scale fisheries sector, the intersection of sustainability, priorities with poverty alleviation, cultural preservation, and ecological resilience under constantly evolving environmental conditions. This translates to assessing current catch levels, income generation, and nutritional security alongside ecosystem health metrics, stock recruitment potential, and adaptive capacity under environmental change scenarios. The Brundtland concept of intergenerational equity has been developed by recent environmental justice research to consider distributional equality within and between generations into account (Gardiner, 2011). This includes the preservation of resource stocks and traditional knowledge systems for future generations, as well as issues of resource access, benefit distribution, and decision-making authority among present users in small-scale fisheries. Even though current fishing levels provide for immediate livelihood needs, they don't seem sustainable over the course of generations. Strict adherence to intergenerational equity principles may worsen current poverty among already marginalized communities, posing a classic "sustainability paradox." The Brundtland framework's focus on addressing "essential needs" becomes problematic when those needs clash with those of present and future generations due to resource scarcity and environmental degradation. This suggests the need for more advanced approaches to sustainability that explicitly addresses trade-offs between temporal and social equity objectives. Sustainable practices are grounded in the concept of carrying capacity, which guides the equitable and resilient use of natural resources. Carrying capacity involves significant unpredictability and is closely linked to resource availability, cultural values, and technical constraints. It delineates the maximum species population that a certain environment can support (Arrow et al., 1995). Numerous scholars have developed theories and methods for measuring fish resources from economic and social perspectives, grounded in the biological characteristics of fish. These contributions have laid the groundwork for policy and theoretical frameworks to support the sustainable management of fish resources (Zhang et al., 2022). In case of fisheries, as we deal with renewable resources, (1) a continued exploitation more than natural regeneration rates can turn potentially renewable

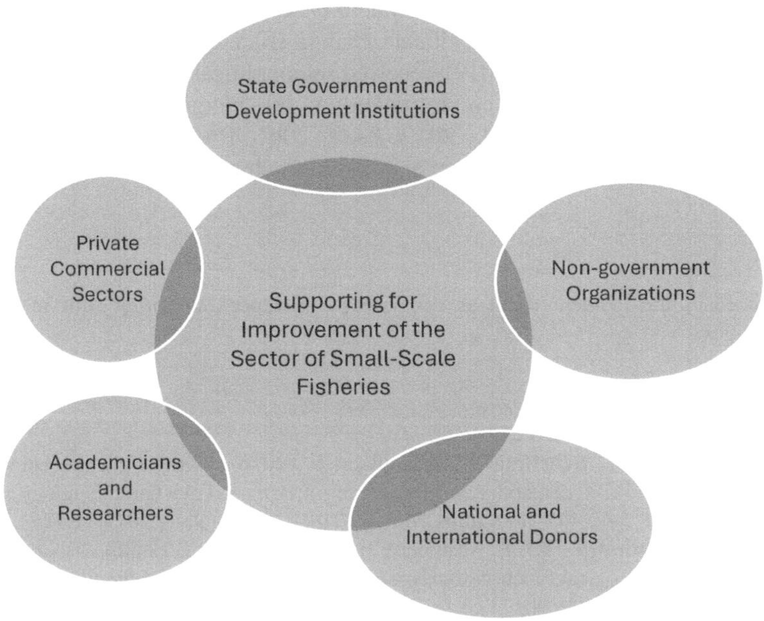

FIGURE 1.6 Supporting sectors for small-scale fisheries.

resources into non-renewable and lead the fishery to extinction. Therefore, an exhausted fish stock concedes certain socioeconomic standards and the integrity of the social functions including the diversity of formal and informal institutions that sustain the livelihood of fishing networks (2) As the small-scale fisheries are biologically and socially highly diverse, and make a diverse range of contributions to livelihoods. The universality of the problem of open access exists across all these fisheries is being questioned. Hence to identify the problems while implementing the policies that restrict this "open access" to a subset of fishers can reduce the effect of high harvest levels, but when implemented, they potentially compromise the achievement of wider social objectives. Given the importance and diversity of benefits of these complex socioecological systems where the fishers and other actors with a stake in the fishery benefit from a valuable food and income source, such interventions may create common value conflicts between the short-term social–economic goals and the longer-term goal of sustainability. (Garmendia et al. 2010). Small-scale fisheries exhibit ecological and social diversity, supporting a range of lifestyles. Although unrestricted access is a prevalent concern in these fisheries, limiting access to a select group of fishermen may mitigate the impacts of elevated harvest levels; however, it might undermine wider societal goals (Figure 1.6). The richness and variety of SSFs may lead to value conflicts between short-term socioeconomic objectives and long-term ecological sustainability in actions aimed at attaining sustainability. The poorly defined objectives and weaknesses at the institutional level, related to fishing rights and the prevalence of top-down management to name a few (Arthur, 2021; Martinez-Alier et al., 1998). Though in last two decades world has seen the widespread adoption of ecosystem-based approaches along with greater awareness of the implications of climate change (Barange et al., 2018; Bell et al., 2013; Hoegh-Guldberg et al., 2019), renewed sociopolitical commitments in fisheries to address and improve the circumstances of the marginalized small-scale and traditional fishers (Cohen et al., 2019; FAO, 2001, FAO, 2003; FAO, IFAD & WFP, 2015; Jentoft, 2017; Pikitch et al., 2004) to achieve the SDGs 2030.

NOTE

1 Defined as the moment when an ecosystem experiences an abrupt shift in regime or condition.

BIBLIOGRAPHY

Abid, M., Schilling, J., Scheffran, J., & Zulfiqar, F. (2016). Climate change vulnerability, adaptation and risk perceptions at farm level in Punjab, Pakistan. *Science of the Total Environment*, *547*, 447–460. https://doi.org/10.1016/j.scitotenv.2015.11.125

Alberti, M., & Marzluff, J. (2004). Resilience in urban ecosystems: Linking urban patterns to human and ecological functions. *Urban Ecosystems*, *7*, 241–265. https://doi.org/10.1023/B:UECO.0000044038.90173.c6

Alemayehu, A., & Bewket, W. (2020). Smallholder farmers' coping and adaptation strategies to climate change and variability in the central highlands of Ethiopia. *Local Environment*, *22*(7), 825–839. https://doi.org/10.1080/13549839.2017.1290058

Ali, A., & Erenstein, O. (2020). Assessing farmer use of climate change adaptation practices and impacts on food security and poverty in Pakistan. *Climate Risk Management*, *19*, 183–194. https://doi.org/10.1016/j.crm.2016.05.002

Allison, E. H., & Ellis, F. (2001). The livelihoods approach and management of small-scale fisheries. *Marine Policy*, *25*, 377–388. http://doi.org/10.1016/S0308-597X(01)00023-9

Allison, E. H. (2013). Book review: Poverty mosaics: Realities and prospects in small-scale fisheries. *International Journal of Maritime History*. https://doi.org/10.1177/084387141302500153

Arrow, K., Bolin, B., Costanza, R., Dasgupta, P., Folke, C., Holling, C. S., Jansson, B.-O., Levin, S., Maler, K. L., Perrings, C., & Pimentel, D. (1995). Economic growth, carrying capacity and the environment. *Science*, *268*, 89–90. https://doi.org/10.1126/science.268.5210.520

Arthur, W. B. (2021). Foundations of complexity economics. *Nature Reviews Physics*, *3*(2), 136–145.

Asche, F., Garlock, T. M., Anderson, J. L., Bush, S. R., Smith, M. D., Anderson, C. M., Chu, J., Garrett, K. A., Lem, A., Lorenzen, K., Oglend, A., Tveteras, S., & Vannuccini, S. (2018). Three pillars of sustainability in fisheries. *Proceedings. National Academy of Sciences. United States of America*, *115*(44), 11221–11225. https://doi.org/10.1073/pnas.1807677115

Ayilu, R. K., Fabinyi, M., & Barclay, K. (2022). Small-scale fisheries in the blue economy: Review of scholarly papers and multilateral documents. *Ocean & Coastal Management*, *216*, Article 105982. https://doi.org/10.1016/j.ocecoaman.2021.105982

Balogun, V. S., & Onokerhoraye, A. G. (2022). Climate change vulnerability mapping across ecological zones in Delta state, Niger Delta region of Nigeria. *Climate Services*, *27*, Article 100310. https://doi.org/10.1016/j.cliser.2022.100310

Banerjee, S., Chanda, A., Ghosh, T., Cremin, E., & Renaud, F. G. (2023). A qualitative assessment of natural and anthropogenic drivers of risk to sustainable livelihoods in the Indian Sundarban. *Sustainability*, *15*, 6146. https://doi.org/10.3390/su15076146

Barange, M., Bahri, T., Beveridge, M. C. M., Cochrane, K. L., Funge-Smith, S., & Poulain, F. (Eds.). (2018). *Impacts of climate change on fisheries and aquaculture: Synthesis of current knowledge, adaptation and mitigation options*. FAO Fisheries and Aquaculture Technical Paper No. 627. Rome, FAO.

Barua, P., Eslamian, S., & Mita, A. (2022). Sustainable management for conservation of urban aquatic forms in Bangladesh. In Eslamian, et al. (Eds.), *Handbook of Eurasian forecasting* (pp. 247–264). New York: Nova Science Publishers, Inc.

Bell, D. (2013). Climate change and human rights. *Wiley Interdisciplinary Reviews: Climate Change*, *4*(3), 159–170.

Bell, J. D., Ganachaud, A., Gehrke, P. C., Griffiths, S. P., Hobday, A. J., Hoegh-Guldberg, O., . . . & Waycott, M. (2013). Mixed responses of tropical Pacific fisheries and aquaculture to climate change. *Nature Climate Change*, *3*(6), 591–599.

Béné, C., Hersoug, B., & Allison, E. H. (2010). Not by rent alone: Analysing the pro-poor functions of small-scale fisheries in developing countries. *Development and Policy Review*, *28*, 325–358.

Béné C., Macfadyen, G., & Allison, E. H. (2007). *Increasing the contribution of small-scale fisheries to poverty alleviation and food security* (Vol. 481). Food & Agriculture Org

Berkes, F., Colding, J., & Folke, C. (2003). Navigating social-ecological systems: Building resilience for complexity and change. *Ecology and Society*, *9*(1, 1).

Blanchard, J. L., Watson, R. A., Fulton, E. A., Cottrell, R. S., Nash, K. L., Bryndum-Buchholz, A., Buchner, M., Carozza, D. A., & Cheung, W. W. L. (2017). Linked sustainability challenges and trade-offs among fisheries, aquaculture and agriculture. *Nature Ecology & Evolution*, *1*, 1240–1249. https://doi.org/10.1038/s41559-017-0258-8

Botsford, L. W., Castilla, J. C., & Peterson, C. H. (1997). The management of fisheries and marine ecosystems. *Science*, *277*, 509–515. https://doi.org/10.1126/science.277.5325.509

Brekke, K. A. (1997). *Economic growth and the environment*. Edward Elgar Publishing.

Brundtland, G.H. (1987) *Our common future: Report of the world commission on environment and development*. Geneva, UN-Dokument A/42/427.

Bueno, N., & Basurto, X. (2009). Resilience and collapse of artisanal fisheries: a system dynamics analysis of a shellfish fishery in the Gulf of California, Mexico. *Sustainability Science*, *4*(2), 139–149.

Chuenpagdee, R., & Jentoft, S. (2019). *Transdisciplinarity for small-scale fisheries governance, MARE publication series*. Springer International Publishing. https://doi.org/10.1007/978-3-319-94938-3

Chuenpagdee, R., Liguori, L., Palomares, M. L. D., & Pauly, D. (2006). Bottom-up, global estimates of small-scale marine fisheries catches.

Cochrane, K. L., Ortega-Cisneros, K., Iitembu, J. A., dos Santos, C. I., & Sauer, W. H. H. (2020). Application of a general methodology to understand vulnerability and adaptability of the fisheries for small pelagic species in the Benguela countries: Angola, Namibia and South Africa. *African Journal of Marine Science*, *42*(4), 473–493.

Cohen, P. J., Allison, E. H., Andrew, N. L., Cinner, J., Evans, L. S., Fabinyi, M., Garces, L. R., Hall, S. J., Hicks, C. C., Hughes, T. P., Jentoft, S., Mills, D. J., Masu, R., Mbaru, E. K., & Ratner, B. D. (2019). Securing a just space for small-scale fisheries in the blue economy. *Frontiers in Marine Science*, *6*, 434657. https://doi.org/10.3389/fmars.2019.00171

Côrtes, P. L., Torrente, M., Pinto Alves Pinto, A., Ruiz, M. S., Dias, A. J. G., & Rodrigues, R. (2015). Crise de abastecimento de água em São Paulo e falta de planejamento estratégico. *EstudosAvançados*, *29*(84), 7–26. https://doi.org/10.1590/S0103-40142015000200002

Costello, C., Gaines, S. D., Lynham, J. (2008). Can catch shares prevent fisheries collapse? *Science*, *321*(5896), 1678–1681. doi: 10.1126/science.1159478

Cudney-Bueno, R., & Basurto, X. (2009). Lack of cross-scale linkages reduces robustness of community-based fisheries management. *PLoS One*, *4*(7), e6253. https://doi.org/10.1371/journal.pone.0006253

Daly, H. (1999). Globalization versus internationalization—Some implications. *Ecological Economics, 31*, 31–37. 10.1016/S0921-8009(99)00087-7

Daly, H.E. and Cobb Jr., J.B. (1989) *For the common good: Redirecting the economy toward community, the environment, and a sustainable future*. Boston: Beacon Press.

Das, B. K., DebRoy, P., & Yateesh, D. C. (2026). Small-Scale Inland Fisheries of India: Trends, transitions, and the road ahead. In *Small-scale Inland Fisheries in India* (pp. 1–18). Singapore: Springer Nature Singapore.

Defeo, O., & Castilla, J. C. (2012). Governance and governability of coastal shellfisheries in Latin America and the Caribbean: Multi-scale emerging models and effects of globalization and climate change. *Current Opinion in Environmental Sustainability, 4*(3), 344–350.

Ebrahim, A. A., Miheretu, B. A., & Alemayehu, A. (2022). Vulnerability of smallholder farmers to climate variability and change across different agro-ecological zones in Oromo nationality administration (ONA), north East Ethiopia. *PLoS One, 17*(6), e0268094. https://doi.org/10.1371/journal.pone.0268094

Eriksen, S. H., Nightingale, A. J., & Eakin, H. (2021). Reframing adaptation: The political nature of climate change adaptation. *Global Environmental Change, 35*, 523–533. https://doi.org/10.1016/j.gloenvcha.2015.09.014

Erős, T., & Lowe, W. H. (2019). The landscape ecology of rivers: From patch-based to spatial network analyses. *Current Landscape Ecology Reports, 4*, 103–112. https://doi.org/10.1007/s40823-019-00044-6

Fang, X., Zhou, B., Tu, X., Ma, Q., & Wu, J. (2018). *What kind of a science is sustainability science? An evidence-based reexamination Sustainability, 10*(5), 1478.

FAO. (2001). *The state of food insecurity in the world*. Rome.

FAO. (2003). *Food energy—methods of analysis and conversion factors: FAO Food and Nutrition Paper 77*. Rome: FAO.

FAO. (2009). *The state of food and agriculture 2009: Livestock in the balance*. Rome, Italy. ISBN 978-92-5-106215-9.

FAO. (2015). *Voluntary guidelines for securing sustainable small-scale fisheries in the context of 632 food security and poverty eradication*. Rome: FAO.

FAO. (2024). *The state of world fisheries and aquaculture 2024 – Blue transformation in action*. Rome.

FAO Fisheries Report. No. 735. (2004). Rome: FAO, 21p.

FAO, IFAD & WFP. (2015). *The State of food insecurity in the world 2015: Meeting the 2015 international hunger targets: taking stock of uneven progress*. Rome: FAO.

FAO, World Fish & Duke University. (2023). Illuminating hidden harvests – The contributions of small-scale fisheries to sustainable development. *Rome*. https://doi.org/10.4060/cc4576en

Feeny, D., Berkes, F., McCay, B. J., et al. (1990). The tragedy of the commons: Twenty-two years later. *Human Ecology, 18*, 1–19. https://doi.org/10.1007/BF00889070

Fernández, M., Kriegl, M., Garmendia, V., Aguilar, A., & Subida, M. D. (2020). Evidence of illegal catch in the benthic artisanal fisheries of central Chile: patterns across species and management regimes. *Latin American Journal of Aquatic Research, 48*(2), 287–303.

Foley, J. A., Defries, R., Asner, G. P., Barford, C., Bonan, G., Carpenter, S. R., Chapin, F. S., Coe, M. T., Daily, G. C., Gibbs, H. K., Helkowski, J. H., Holloway, T., Howard, E. A., Kucharik, C. J., Monfreda, C., Patz, J. A., Prentice, I. C., Ramankutty, N., Snyder, P. K.. (2005). Global consequences of land use. *Science, 309*(5734), 570–574. doi: 10.1126/science.1111772

Folke, C., et al. (2005). Adaptive governance of social-ecological systems. *Annual Review of Environment and Resources, 30*, 441–473. https://doi.org/10.1146/annurev.energy.30.050504.144511

Food and Agriculture Organization. (2020). *Legislating for sustainable small-scale fisheries.* FAO.

Food and Agriculture Organization (FAO). (2003). *World agriculture: Towards 2015/2030.*

Gardiner, S. M. (2011). *A perfect moral storm: Climate change and intergenerational ethics.* Oxford University Press.

Garmendia, E., Gamboa, G., Franco, J., Garmendia, J. M., Liria, P., & Olazabal, M. (2010). Social multi-criteria evaluation as a decision support tool for integrated coastal zone management. *Ocean & Coastal Management, 53*(7), 385–403.

Ghosh, A., & Dey, S. (2025). Pronouncing the role of fisheries and aquaculture in enhancing food and nutritional security. In J. K. Sundaray, M. A. Rather, I. Ahmad, & A. Amin (Eds.), *Food security, nutrition and sustainability through aquaculture technologies.* Cham: Springer. https://doi.org/10.1007/978-3-031-75830-0_26

Giabardo, C. V. (2020). Climate change litigation and tort law: Regulation through litigation? http://dx.doi.org/10.2139/ssrn.3858956

Goniewicz, K., Sarker, M. N. I., & Schoch-Spana, M. (2023). Reimagining natural hazards and disaster preparedness: Charting a new course for the future. *BMC Public Health, 23*(1), 581. https://doi.org/10.1186/s12889-023-15471-8

Goodland, R. (1995). The concept of environmental sustainability. *Annual Review of Ecology and Systematics, 26,* 1–24. https://doi.org/10.1146/annurev.es.26.110195.000245

Gordon, H. S. The economic theory of a common-property resource: The fishery. *Journal of Political Economy, 62,* 124–166.

Graziano, M. P., Deguire, A. K., & Surasinghe, T. D. (2022). Riparian buffers as a critical landscape feature: Insights for riverscape conservation and policy renovations. *Diversity, 14*(3), 172. https://doi.org/10.3390/d14030172

Guillard-Gonçalves, C., & Zêzere, J. L. (2018). Combining social vulnerability and physical vulnerability to analyse landslide risk at the municipal scale. *Geosciences, 8*(2), 294–310. https://doi.org/10.3390/geosciences8080294

Gupta, H., & Barua, M. K. (2017). The relationship of transformational leadership with job satisfaction and organizational commitment: A review and discussion. *International Journal of Organizational Analysis, 25*(5), 735–756. https://doi.org/10.1108/IJOA-01-2017-1117

Gutierrez, N. L., Funge-Smith, S., Gorelli, G., Mancha-Cisneros, M. M., Defeo, O., Johnson, A. F., & Melnychuk, M. C. (2023). Production and environmental interactions of small-scale fisheries. In *FAO, Duke University & WorldFish, illuminating hidden harvests: The contributions of small-scale fisheries to sustainable development.* FAO; Duke University; WorldFish.

Haberl, H., Erb, K. H., Krausmann, F., Gaube, V., Bondeau, A., Plutzar, C., Gingrich, S., Lucht, W., & Fischer-Kowalski, M. (2007). Quantifying and mapping the human appropriation of net primary production in earth's terrestrial ecosystems. *Proceedings. National Academy of Sciences. United States of America, 104*(31), 12942–12947. https://doi.org/10.1073/pnas.0704243104

Hagihara, K., & Hagihara, Y. (2019). Sustainability and vulnerability: Well-being in the geo-,eco- and socio-environment. In C. Asahi (Ed.), *Building resilient regions, New Frontiers in regional science: Asian perspectives* (Vol. 35). Springer.

Hardin, G. (1968). The tragedy of the commons. *Science, 162,* 1243–1248. http://dx.doi.org/10.1126/science.162.3859.1243

Heikkila, T., Villamayor-Tomas, S., & Garrick, D. (2018). Bringing polycentric systems into focus for environmental governance. *Environmental Policy and Governance, 28,* 207–211. https://doi.org/10.1002/eet.1809

Hendriks, S. L. (2022). Sustainable small-scale fisheries can help people and the planet. *Nature, 606*(2), 650–652. https://doi.org/10.1038/d41586-022-01363-1

Hoegh-Guldberg, O., Pendleton, L., & Kaup, A. (2019). People and the changing nature of coral reefs. *Regional Studies in Marine Science*, *30*, 100699. https://doi.org/10.1016/j.rsma.2019.100699

Hossain, B., Ajiang, C., & Ryakitimbo, C. M. (2019). Responses to flood disaster: Use of indigenous knowledge and adaptation strategies in Char Village, Bangladesh. *Environmental Management and Sustainable Development*, *8*(4), 46–65.

Hossain, B., Ryakitimbo, C. M., & Sohel, M. S. (2021a). Climate change induced human displacement in Bangladesh: A case study of flood in 2017 in char in Gaibandha District. *Asian Research Journal of Arts & Social Sciences*, *11*(1), 47–60.

Hossain, B., Shi, G., Ajiang, C., Sarker, M. N. I., Sohel, M. S., & Sun, Z. (2021b). Impact of climate change on human health: Evidence from riverine island dwellers of Bangladesh. *International Journal of Environmental Health Research*, 1–17.

Hylton, K. N. (2019). When should we prefer tort law to environmental regulation? *Washburn Law Journal*, *41*, 515–534.

Islam, M. A., Hoque, M. A., Ahmed, K. M., & Butler, A. P. (2019). Impact of climate change and land use on groundwater salinization in southern Bangladesh-implications for other Asian deltas. *Environmental Management*, *64*(5), 640–649. https://doi.org/10.1007/s00267-019-01220-4

Jaafar, M., & Maideen, S. A. (2012). Ecotourism-related products and activities, and the economic sustainability of small and medium island chalets. *Tourism Management*, *33*(3), 683–691. https://doi.org/10.1016/j.tourman.2011.07.011

James, C. S. (2008). Hydraulics and the river environment - challenges of interdisciplinarity and complexity. http://search.informit.com.au/documentSummary;dn=610925896446758;res=IELENG;subject=Agriculture

Jentoft, S. (2017). Small-scale fisheries within maritime spatial planning: Knowledge integration and power. *Journal of Environmental Policy & Planning*, *19*(3), 266–278 https://doi.org/10.1080/1523908X.2017.1304210

Jentoft, S., & Eide, A. (Eds.). (2011). *Poverty mosaics: realities and prospects in small-scale fisheries*. Springer Science & Business Media.

Jha, S., Goyal, M. K., Gupta, B. B., Hsu, C. H., Gilleland, E., & Das, J. (2022). A methodological framework for extreme climate risk assessment integrating satellite and location based data sets in intelligent systems. *International Journal of Intelligent Systems*, *37*(12), 10268–10288. https://doi.org/10.1002/int.22475

Jordan, J. C. (2015). Swimming alone? The role of social capital in enhancing local resilience to climate stress: A case study from Bangladesh. *Climate and Development*, *7*(2), 110–123. https://doi.org/10.1080/17565529.2014.934771

Kates, R. W., Clark, W. C., Corell, R., Hall, J. M., Jaeger, C. C., Lowe, I., . . . Svedin, U. (2001). Sustainability science. *Science*, *292*(5517), 641–642. 10.1126/science.1059386

Kosamu, I. B. (2015). Conditions for sustainability of small-scale fisheries in developing countries. *Fisheries Research*, *161*, 365–373.

Kumar, A., Jain, V., & Sameer, K. (2014). A comprehensive environment friendly approach for supplier selection. *Omega*, *42*(1), 109–123. https://doi.org/10.1016/j.omega. 2013.04.003

Leal Filho, W., Totin, E., Franke, J. A., Andrew, S. M., Abubakar, I. R., Azadi, H., & Simpson, N. P. (2021). Understanding responses to climate-related water scarcity in Africa. *Science of the Total Environment*, *806*(1), 150420. https://doi.org/10.1016/j.scitotenv.2021.150420

Liberati, D., Sconocchia, P., Ricci, A., Gigliotti, G., Tacconi, C., Grenni, P., Tariciotti, S., & Caracciolo, A. B. (2018). Phyto-dehydration of confined sludge: A sustainable approach for the management of polluted ponds. *Journal of Soils and Sediments*, *18*, 2292–2304. https://doi.org/10.1007/s11368-018-1976-5

Luo, T., Maddocks, A., Iceland, C., Ward, P., & Winsemius, H. (2015). *World's 15 countries with the most people exposed to river floods*. World Resources Institute.

Lutz, A. F., Immerzeel, W. W., Shrestha, A. B., & Bierkens, M. F. P. (2014). Consistent increase in high Asia's runoff due to increasing glacier melt and precipitation. *Nature Climate Change*, *4*, 587–592. https://doi.org/10.1038/nclimate2237

Mafi-Gholami, D., Pirasteh, S., Ellison, J. C., & Jaafari, A. (2021). Fuzzy-based vulnerability assessment of coupled social-ecological systems to multiple environmental hazards and climate change. *Journal of Environmental Management*, *299*, 113573. https://doi.org/10.1016/j.jenvman.2021.113573

Mallick, B., Rogers, K. G., & Sultana, Z. (2022). In harm's way: Nonmigration decisions of people at risk of slow-onset coastal hazards in Bangladesh. *Ambio*, *51*, 114–134. https://doi.org/10.1007/s13280-021-01552-8

Mambet Doue, C., Navarro Carrascal, O., Restrepo, D., Krien, N., Rommel, D., Lemee, C., & Fleury-Bahi, G. (2020). The social riverine island dwellers for livelihood resilience. *Land Use Policy*, *94*, 400–420. https://doi.org/10.1016/j.landusepol.2020.104561

Martin, K. S. (2005). Mapping economic diversity in the first world: The case of fisheries. *Environment and Planning A: Economy and Space*, *37*(6), 959–979. https://doi.org/10.1068/a36296

Martinez-Alier, J., Munda, G., & O'Neill, J. (1998). Weak comparability of values as a foundation for ecological economics. *Ecological Economics*, *26*(3), 277–286. 10.1016/S0921-8009(97)00120-1

Massacci, A., Lagomarsino, A., & De Angelis, P. (2018). Phyto-dehydration of confined sludge: A sustainable approach for the management of polluted ponds. *Journal of Soils and Sediments*, *18*, 2292–2304. https://doi.org/10.1007/s11368-018-1976-5

Masud, M. M., & Kari, F. B. (2015). Community attitudes towards environmental conservation behaviour: An empirical investigation within MPAs, Malaysia. *Marine Policy*, *52*, 138–144. https://doi.org/10.1016/j.marpol.2014.10.015

McGoodwin, J. R. (1990). *Crisis in the world's fisheries, people problems and policies*. California: Stanford University Press.

Mensah, J., & Ricart Casadevall, S. (2019). Sustainable development: Meaning, history, principles, pillars, and implications for human action: Literature review. *Cogent Social Sciences*, *5*(1), 1653531. https://doi.org/10.1080/23311886.2019.1653531

Milkoreit, M., Hodbod, J., Baggio, J., Benessaiah, K., Calderón-Contreras, R., Donges, J. F., Mathias, J.-D., Rocha, J. C., Schoon, M., & Werners, S. E. (2018). Defining tipping points for social-ecological systems scholarship—An interdisciplinary literature review. *Environmental Research Letters*, *13*(3), 330–350. https://doi.org/10.1088/1748-9326/aaaa75

Millennium Ecosystem Assessment. (2005) *A report of the millennium ecosystem assessment*. Ecosystems and Human Well-Being. Washington, DC: Island Press.

Mitra, A., Abdel-Gawad, F. K., Bassem, S., Barua, P., Assisi, L., Parisi, C., Temraz, T. A., Vangone, R., Kajbaf, K., Kumar, V., & Guerriero, G. (2023). Climate change and reproductive biocomplexity in fishes: Innovative management approaches towards sustainability of fisheries and aquaculture. *Water*, *15*(4), 725. https://doi.org/10.3390/w15040725

Mosepele, K., Kolding, J., Bokhutlo, T., Mosepele, B. Q., & Molefe, M. (2022). The Okavango Delta: Fisheries in a fluctuating floodplain system. *Frontiers in Environmental Science*, *10*, 854835.

Moser, S. C., & Ekstrom, J. A. (2020). A framework to diagnose barriers to climate change adaptation. *Proceedings of the National Academy of Sciences*, *115*(5), 22026–22031. https://doi.org/10.1073/pnas.1007887107

Neumayer, E. (1998). Preserving natural capital in a world of uncertainty and scarce financial resources. *International Journal of Sustainable Development & World Ecology*, *5*(1), 27–42. https://doi.org/10.1080/13504509809469967

Neumayer, E. (1999). The ISEW: not an index of sustainable economic welfare. *Social Indicators Research*, *48*(1), 77–101.

Neumayer, E. (2010). Weak versus strong sustainability: exploring the limits of two opposing paradigms. In *Weak versus strong sustainability*. Edward Elgar Publishing.

Niang, I., Ruppel, O. S., Abdrabo, M. A., Essel, A., Lennard, C., Padgham, J., & Urquhart, P. (2014). *Climate change 2014: Impacts, adaptation and vulnerability*. Cambridge University Press.

O'Brien, K., Eriksen, S., Nygaard, L., & Schjolden, A. (2007). Why different interpretations of vulnerability matter in climate change discourses. *Climate Policy*, *7*, 73–88.

Opdam, P., Luque, S., Nassauer, J., Verburg, P. H., & Wu, J. (2018). How can landscape ecology contribute to sustainability science? *Landscape Ecology*, *33*(1). 10.1007/s10980-018-0610-7

Ostrom, E. (1994). 6. Constituting social capital and collective action. *Journal of Theoretical Politics*, *6*(4), 527–562. https://doi.org/10.1177/0951692894006004006

Ostrom, E. (2009). A general framework for Analyzing sustainability of social-ecological systems. *Science*, *325*, 419–422. 10.1126/science.1172133

Palomo-Campesino, S., González, J. A., & García-Llorente, M. (2018). Exploring the connections between agroecological practices and ecosystem services: A systematic literature review. *Sustainability*, *10*(9), 4339–4345. https://doi.org/10.3390/su10124339

Pascual-Fernández, J. J., Pita, C., & Bavinck, M. (Eds.). (2020). *Small-scale fisheries in Europe: Status, resilience and governance* (Vol. 23). Springer. https://doi.org/10.1007/978-3-030-37371-9

Paudel, S., & Yuan, F. (2012). Assessing landscape changes and dynamics using patch analysis and GIS modeling. *International Journal of Applied Earth Observation and Geoinformation*, *16*, 66–76. https://doi.org/10.1016/j.jag.2011.12.003

Paul, A. K., Paul, A., & Sardar, J. (2023). Susceptibility of the climate resilient landforms of the coastal tract of Odisha and West Bengal, India. *Journal of the Geological Society of India*, *99*(6), 827–839.

Pauly, D. (2006). Major trends in small-scale marine fisheries, with emphasis on developing countries, and some implications for the social sciences. *Maritime Studies*, (2).

Pauly, D., Belhabib, D., Blomeyer, R., Cheung, W. W., & Cisneros-Montemayor, A. M., Copeland, D., . . . & Zeller, D. (2014). China's distant-water fisheries in the 21st century. *Fish and Fisheries*, *15*(3), 474–488.

Pauly, D., Christensen, V., Guénette, S., et al. (2002). Towards sustainability in world fisheries. *Nature*, *418*, 689–695. https://doi.org/10.1038/nature01017

Pauly, D., Watson, R., & Alder, J. (2005). Global trends in world fisheries: Impacts on marine ecosystems and food security. *Philosophical Transactions of the Royal Society of London. Series B, Biological Sciences*, *360*(1453), 5–12. doi: 10.1098/rstb.2004.1574

Pearce, D., & Turner, R. K. (1990). *Economics of natural resources and the environment*. Baltimore: Johns Hopkins University Press.

Penning-Rowsell, E. C., Sultana, P., & Thompson, P. M. (2013). The 'last resort'? Population movement in response to climate-related hazards in Bangladesh. *Environment Science & Policy*, *27*, 44–59. https://doi.org/10.1016/j.envsci.2012.10.008

Perry, R. I., Barange, M., & Ommer, R. E. (2010). Global changes in marine systems: A social–ecological approach. *Progress in Oceanography*, *87*(1–4), 331–337.

Pikitch, E. K., Santora, C., Babcock, E. A., Bakun, A., Bonfil, R., Conover, D. O., Dayton, P., Doukakis, P., Fluharty, D., Heneman, B., Houde, E. D., Link, J., Livingston, P. A., Mangel, M., McAllister, M. K., Pope, J., & Sainsbury, K. J. Ecology. Ecosystem-based fishery management. Science, *305*(5682), 346–347. doi: 10.1126/science.1098222

Pikitch, E. K., Santora, C., Babcock, E. A., Bakun, A., Bonfil, R., Conover, D. O., . . . & Sainsbury, K. J. (2004). Ecosystem-based fishery management. *Science*, *305*(5682), 346–347.

Pitcher, T. J., & Lam, M. E. (2015). Fish commoditization and the historical origins of catching fish for profit. *Maritime Studies*, *14*(1), 2.
Porto-Gonçalves, C. W. (2013). *A Globalização da Natureza e a Natureza da Globalização* ((5th ed.). Civilização Brasileira.
Purcell, S. W., & Pomeroy, R. S. (2015). Driving small-scale fisheries in developing countries. *Frontiers in Marine Science*, *2*, 44.
Rakib, M. A., Sasaki, J., Pal, S., Newaz, A., & Bhuiyan, M. A. H. (2019). An investigation of coastal vulnerability and internal consistency of local perceptions under climate change risk in the southwest part of Bangladesh. *Journal of Environmental Management*, *231*, 419–428. https://doi.org/10.1016/j.jenvman.2018.10.054
Reddy, P. J., Sriram, D., Gunthe, S. S., & Balaji, C. (2021). Impact of climate change on intense bay of Bengal tropical cyclones of the post-monsoon season: A pseudo global warming approach. *Climate Dynamics*, *56*(9), 2855–2879. https://doi.org/10.1007/s00382-020-05622-7
Rolib, S., & Janasen, S. (2024). Regulations, policies, and governance in achieve sustainable development goals. *Indonesian Journal of Interdisciplinary Research in Science and Technology*, *2*(5), 893–906.
Sahoo, B., & Bhaskaran, P. K. (2016). Assessment on historical cyclone tracks in the bay of Bengal, east coast of India. *International Journal of Climatology*, *36*(1), 95–109. https://doi.org/10.1002/joc.4331
Salik, K. M., Jahangir, S., Zahdi, W. u. Z., & Hasson, S. (2020). Climate change vulnerability and adaptation options for the coastal communities of Pakistan. *Ocean & Coastal Management*, *122*, 61–73. https://doi.org/10.1016/j.ocecoaman.2016.10.003
Sampantamit, T., Ho, L. T., Lachat, C., Sutummawong, N., Sorgeloos, P., & Goethals, P. (2020). Aquaculture production and its environmental sustainability in Thailand : Challenges and potential solutions. *Sustainability*, *12*(5). https://doi.org/10.3390/su12052010
Saner, R., Yiu, L., & Nguyen, M. (2019). Monitoring the SDGs: Digital and social technologies to ensure citizen participation, inclusiveness and transparency. *Development Policy Review*, *40*(3), 56–65. https://doi.org/10.1111/dpr.12414
Scopelliti, M., Molinario, E., Bonaiuto, F., Bonnes, M., Cicero, L., De Dominicis, S., Bonaiuto, M., Admiraal, J., Beringer, A., Dedeurwaerdere, T., de Groot, W., Hiedanpää, J., Knights, P., Knippenberg, L., Polajnar Horvat, K., Popa, F., Porras-Gomez, C., Smrekar, A., & van den Born, R. (2018). What makes you a 'hero' for nature? Socio-psychological profiling of leaders committed to nature and biodiversity protection across seven EU countries. *Journal of Environmental Planning and Management*, *61*(5–6), 970–993. https://doi.org/10.1080/09640568.2017.1342612
Seidl, I., & Tisdell, C. A. (1999). Carrying capacity reconsidered: From Malthus' population theory to cultural carrying capacity. *Ecological Economics*, *31*, 395–408. https://doi.org/10.1016/S0921-8009(99)00063-4
Shah, K. U., Dulal, H. B., Johnson, C., & Baptiste, A. (2020). Understanding livelihood vulnerability to climate change: Applying the livelihood vulnerability index in Trinidad and Tobago. *Geoforum*, *47*, 125–137. https://doi.org/10.1016/j.geoforum.2013.04.004
Shakhawat, H. M., Arshad, M., Qian, L., Kächele, H., Khan, I., Islam, M. D. I., & Golam Mahboob, M. (2020). Climate change impacts on farmland value in Bangladesh. *Ecological Indicators*, *112*, 106181. https://doi.org/10.1016/j.ecolind.2020.106181
Skilodimou, H., Bathrellos, G., & Alexakis, D. (2021). Flood hazard assessment mapping in burned and urban areas. *Sustainability*, *13*(8), 4455. https://doi.org/10.3390/su13084455
Smith, H., & Basurto, X. (2019). Defining small-scale fisheries and examining the role of science in shaping perceptions of who and what counts: A systematic review. *Frontiers in Marine Science*, *6*, 443327. https://doi.org/10.3389/fmars.2019.00236

Smith, V. L. (1969). On models of commercial fishing. *Journal of Political Economy*, *77*(2), 181–198.

Sorvari, S., Tikka, P. M., Niemelä, J., Raivio, K., & Korhonen-Kurki, K. (2009). Breaking the boundaries: Multidisciplinary environmental research at the University of Helsinki. Retrieved from https://core.ac.uk/download/157587309.pdf

Tasnuva, A., Hossain, R., & Salam, R. (2021). Employing social vulnerability index to assess household social vulnerability of natural hazards: An evidence from southwest coastal Bangladesh. *Environment, Development and Sustainability*, *23*(7), 1021–1125. https://doi.org/10.1007/s10668-020-00769-z

Thomas, K. A., & Warner, B. P. (2019). Weaponizing vulnerability to climate change. *Global Environmental Change*, *57*(2), 56–75. https://doi.org/10.1016/j.gloenvcha.2019.101928

Tillema, H., Schoute, A., & Wijbrans, K. (1991). An operating environment for control systems on transputer networks. https://doi.org/10.1109/EMWRT.1991.144106

Turner, B. L., II, Kasperson, R. E., Matson, P., McCarthy, J. J., Corell, R. W., Christensen, L., Eckley, N., Kasperson, J. X., Luers, A., Martello, M. L., Polsky, C., Pulsipher, A., & Schiller, A. (2003). Framework for vulnerability analysis in sustainability science. *Proceedings of the National Academy of Sciences of the United States of America*, *100*, 8074–8079. https://doi.org/10.1073/pnas.1231335100

Turner, B. L., II, & Robbins, P. (2008). Land change science and political ecology: Similarities, differences, and implications for sustainability science. *Annual Review of Environment and Resources*, *33*, 6.1–6.22. https://doi.org/10.1146/annurev.environ.33.022207.104943

Turner, J. R. (2010). Responsibilities for sustainable development in project and program management. In H. Knoepfel (Ed.), *Survival and sustainability as challenges for projects*. International Project Management Association.

World Bank. (2017). Atlas of sustainable development goals 2017. *World Development Indicators*.

Wu, J. (2013). Landscape sustainability science: ecosystem services and human well-being in changing landscapes. *Landscape Ecology*, *28*(6), 999–1023.

Wu, J. (2019). Linking landscape, land system and design approaches to achieve sustainability. *Journal of Land Use Science*, *14*(2), 173–189. https://doi.org/10.1080/1747423X.2019.1602677

Xenarios, S., Nemes, A., & Wahed, G. (2019). Assessing vulnerability to climate change: Are communities in flood-prone areas in Bangladesh more vulnerable than those in drought-prone areas? *Water Resources and Rural Development*, *7*(2), 1–19. https://doi.org/10.1016/j.wrr.2016.03.001

Yousafzai, M. T., Shah, T., Khan, S., Ullah, S., Nawaz, M., & Han, H. (2022). Assessing socioeconomic risks of climate change on tenant farmers in Pakistan. *Frontiers in Psychology*, *13*(1), 87–97.

Zhai, T. T., & Chang, Y. C. (2019). Standing of environmental public-interest litigants in China: Evolution, obstacles and solutions. *Journal of Environmental Law*, *3*, 369–397. https://doi.org/10.1093/jel/eqz014

Zhang, G., Lu, Y., Fang, Z., Yang, H., & Wei, Z. (2022). Analysis of the ecological carrying capacity of fish resources in Shengjin Lake, Anhui Province, China. *International Journal of Environmental Research and Public Health*, *19*(13), 8177. https://doi.org/10.3390/ijerph19138177

2 The Perspective of Small-Scale Fishery in the Land of Distress

2.1 INTRODUCTION

Deltas are low-lying social, political, and environmental systems shaped by river–land–ocean interactions and human water management policies, hosting populations with 10 times the world average. Deltas account for only 1% of global land area, yet they are home to more than half a billion people, or around 7% of the world population (Ericson et al., 2006). The average coastline population density in developing countries is about 80 persons per km^2, which is twice the global average (Shamsuzzaman et al., 2017). Deltas are economic and environmental hotspots, food baskets for many nations, and hosts of exceptional biodiversity and rich ecosystems. Delta regions are vital for small-scale fisheries, providing livelihoods and food security to millions (Zaky et al., 2011). The Ganga Brahmaputra Meghna Delta (GBD), commonly called the Bengal Delta, is the largest in the world, shared between Bangladesh and West Bengal in India and hosting 250 million people. Ghosh (2005) vividly notes,

> "here, in the tide country, transformation is the rule of life: rivers stray from week to week, and islands are made and unmade in days. In other places forests take centuries, even millennia, to regenerate; but mangroves can recolonize a denuded island in ten to fifteen years. Could it be the very rhythms of the earth were quickened here so that they unfolded at an accelerated pace?"

This reflects the delta's intense sediment flux, strong tidal dynamics, and the exceptional adaptive capacity of mangroves, producing rapid geomorphological change and ecological renewal that define the Sundarbans ecosystem. This ecosystem also hosts a high diversity of fish and supports extensive fisheries upon which the economy and livelihoods of millions depend (Vass et al., 2011). The fisheries sector is vital for populations that live in this delta, as a source of animal protein (in Bangladesh around 50–60% of animal protein is supplied by fish, whereas in India this is about 12%) through subsistence fishing and as a source of employment, and for the wider economy (Lauria et al., 2018). Yet deltas are disappearing and deteriorating at an alarming rate as "victims" of anthropogenic activities, climatic impacts, and local exploration and have become "land of distress". In most cases, people who live in delta areas experience extreme poverty. Physical characteristics (such as low elevation and increased flood risk, significant land erosion and accretion, and dependence on sediment and fluvial water inputs) and socioeconomic factors (such as high population density, high rates of poverty, and limited socioeconomic advancement) combine to contribute to

DOI: 10.1201/9781003729990-2

this situation. Adaptation strategies include hard-engineering measures, nature-based solutions, spatial planning, accommodation measures, and managed retreat, though implementation depends on state wealth, governance modes, and civil society power. Current approaches privilege top-down technical solutions over locally informed vulnerability assessments, failing to address environmental and social problems affecting coastal communities. The proposed Global Delta Sustainability Initiative aims to integrate scientific knowledge on deltas as coupled socioecological systems, develop science-based risk assessment and decision support frameworks, build international data repositories, and implement demonstrations in selected deltas. Key components include Delta-SRES (Sustainability Research on Socio-ecological Systems), Delta-RADS (Risk Assessment and Decision Support Tools), Delta-DAT (Delta Data Repository), and Delta-GDVI (Global Delta Vulnerability Indices) (Fanchette, 2022; Foufoula-Georgiou et al., 2011).

2.2 THE SUNDARBANS: THE WORLD'S LARGEST AND MOST DYNAMIC DELTA

The Sundarbans landscape, spanning over 10,000 km^2 and situated between 21°32′ to 22°40′N and 88°05′ to 89°51′E at the confluence of Ganga, Brahmaputra, and Meghna rivers, represents the largest delta and mangrove forest ecosystem globally, acknowledged for its considerable ecological and economic importance, covering an area of 120,000 km^2 (Froese et al., 2017). Population size and population density are greatest in the GBD where the total population exceeded 108 million, and the population density in the study area is approximately 1280 people per km^2 (Ericson et al., 2006). This unique environment is a delicate balance between the saline waters of the Bay of Bengal and the freshwater currents of the tributaries and distributaries of the Ganga–Brahmaputra riverine system. The most important rivers of Bangladesh and Indian Sundarbans have been listed in Table 2.1. The river continuum culminates in this ecosystem as rivers meet the Bay of Bengal. The Sundarbans ecosystem provide a distinctive framework for comprehending the ongoing and dynamic interactions between rivers and their adjacent habitats which can be described by the River Continuum Concept (RCC) (Vannote et al., 1980). This ecological model describes river systems' transition from headwaters to mouth, outlining the changes in physical and biological attributes as they flow from headwaters to the mouth:

Headwaters (Upper Course): These are typically narrow, turbulent parts with calm waters. The primary energy source comes from allochthonous material, such as leaves and organic debris from surrounding forests. This area supports species adapted to these strong currents and low-temperature conditions.

Middle Course: The river becomes wider and slower, and the primary energy source shifts to autochthonous production, such as algae and aquatic plants. The biodiversity increases as the river flow stabilizes, and the temperature rises.

Lower Course: This part is more expansive, slower, and often characterized by floodplains. Here, sediments and nutrients accumulate, providing a rich habitat for various aquatic species. Detritus from upstream provides food for the organisms here, and the river starts interacting more with surrounding ecosystems like wetlands and deltas.

TABLE 2.1
Most Important Rivers of Bangladesh and Indian Sundarbans

Name of Estuary	Characteristics
Hooghly	It is the primary river of West Bengal and a direct extension of the Ganga River, delineating the western boundary of the Indian Sundarbans. The western side of this river hosts many of West Bengal's coastal operations.
Muriganga	It is the branch of the Hooghly River. Sagar Island, the largest island in the deltaic complex, is passed on its way east. Distinctive mangrove flora along its bank.
Saptamukhi	It was first established at Sultanpur. The Hatania-Duania Canal connects to the Muriganga (Bartala) branch of the Hooghly River.
Thakuran	It begins at 24 Parganas (South) near Jayanagar and has many connections to the Saptamukhi. Previously, it was linked to the Calcutta Canal by the now-dead Piyali and Kultali rivers.
Matla	This river originates at the meeting point of the Bidyadhari, Khuratya, and Rampur Khal in 24 Parganas (South), close to the town of Canning. Matla flows into the Bay of Bengal and is linked to Bidya. The freshwater discharge and connection of this river have lately disappeared.
Bidyadhari	Once a flourishing tributary of the Bhagirathi throughout the fifteenth and sixteenth centuries, it is now mostly used as a drainage outlet for Kolkata's surplus rainfall and sewage. The riverbed has completely silted and is on the brink of collapse.
Gosaba	Through a network of canals, the waters of Matla and Harinbhanga (Raimangal) combine to create the estuary. The reserve woodlands are intersected by the estuary and its many streams.
Harinbhanga	This river is the easternmost in the Indian Sundarbans deltaic complex. The Harinbhanga, also known as Ichamati and Raimangal, forms a natural boundary between India and Bangladesh.
Malancha	Near the Kadamtala station, the Malancha River, also known as the Chunar River, merges with the Ichamati River right before it enters the Sundarbans Forest. The Firingi River links Malancha and Jamuna. The Arpongachia and Bal Rivers are linked to the Kobadak and Kolpetua Rivers. The Arpongachia and Bal rivers join the Malancha River farther downstream.
Gorai	The Gorai is a significant river in Bangladesh. The Ganga River serves as its origin. Consequently, this river shapes the Sundarbans ecosystem. The river is 87 km in length. The average width is 300 m. During the monsoon, there is an increased flow; however, during lean periods, the flow diminishes. The tides have no effect. The river has more energy downstream. The flow of the Ganga River is intricately linked to that of the Gorai River during the dry season. The Ganga River often contributes sand to the system. Moreover, the Farakka Barrage is essential for regulating the flow of the Gorai River. The Gorai's flow during the dry season has markedly diminished since the Farakka barrier.
Shibsa	The Shibsa converges downstream after meandering alongside the Pasur River. It has a vital part in the riverine flow, extending 85 km in length. The typical value is 1500 m, and its breadth expands from upstream to downstream. It is a tidal river, and the tides use a mechanism known as tidal pumping to maintain the integrity of the tidal plain. Moreover, the river is linked to other rivers around the country. The river also provides a huge amount of flow to the Pasur River, which finally flows into the Bay of Bengal. Moreover, the river traverses the Sundarbans.

(Continued)

Table 2.1 (Continued)

Name of Estuary	Characteristics
Pasur	The Pasur is one of the most dynamic rivers in southwestern Bangladesh which is situated in Bagerhat district. It receives fluvial flow from upstream via Gorai and mostly constitutes the downstream end of the Gorai River. The river is 104 km in length. The typical value is 1100 m, and its breadth increases from upstream to downstream. The dynamics of this tidal river and the aquatic life that inhabits it are actively maintained by the tides. The river is integrated into the nation's whole hydrological system. The Pasur River's capacity to convey goods, cargo, and boats is vital to the nation's economic growth. The Pasur River traverses the Sundarbans Forest for most of its course, hence sustaining the forest's environment. Moreover, the Pasur River is critically significant for fish, trees, wildlife, vegetation, and fauna. Individuals earn their livelihood by fishing in this river, mostly for prawns. They can sell them at the local market.
Raimongol	This is a tidal estuary river located in the Sundarbans, spanning the South 24 Parganas district of West Bengal, India, and the Satkhira District of Bangladesh. The Ichamati, Kalindi, and Raimongol are designations for the same river, varying by location. Until entering the Sundarbans, it is known as the Raimongol. It then travels through the western part of the Satkhira district before emptying into the Bay of Bengal. The river mouth in the estuary resembles a funnel. The Raimongol River has tidal influences, resulting in saline water.
Kunga	The Mardat River, originating from the Pasur and Shibsa rivers, discharges into the Kunga River, which thereafter flows into the Hansharaj and Kunga at Hiron Point. A multiplicity of rivers originating in the catchment regions converge with the Pasur and Shibsa outside the Sundarbans, which flow into the estuary. Eventually, the upstream water flows pour into the Kunga estuary. It might aid in the creation of a healthy mangrove wetland ecosystem along the coastline.
Arpangasia	One of the most important rivers in the Sundarbans, the Arpangasia traverses the Satkhira and Khulna districts of Bangladesh. It discharges immediately into the sea. The river is 65 km in length. Due to receiving input from an additional channel upstream, its breadth expands at the downstream end. The average width is 1400 m. The river's morphology is mostly shaped by tidal effects. The tidal plain of this river is influenced by the tide. This river is influenced by tides, which actively sustain it.
Baleswar	The Baleswar River is one of the principal rivers of the Sundarbans. The Baleswar River now conveys freshwater into the eastern Sundarbans from upstream via the Gorai–Madhumati Rivers and the lower Meghna River. The Baleswar River contribute to the conservation of mangrove habitats and coastal landscapes.
Bangra river	The Pasur, Kanga, Barashiala, Selagang, and Bhola are the most prominent of the many tributaries that round the Bangra River's estuary. A complex network of rivers and canals connects these Bangra tributaries to Baleswar in the east and Pasur in the west. The Selagang and Bhola Rivers are the two main tributaries of the Ganga, which formerly carried freshwater from the Ganga via the Bhairab system and Jewdhara, as well as the Dhansagar Khal from the Baleswar-Madhumati system at Morelganj.

River ecologists have long explored the links between riparian zones, floodplains, and river ecosystems, with initial frameworks like the RCC, nutrient spiralling,[1] and the Serial Discontinuity Concept[2] providing linear models[3] of these systems (Elwood et al., 1983; Stanford & Ward, 1983; Vannote et al., 1980). Later, Junk et al. (1989) developed the "flood pulse concept" (FPC), which emphasized the importance of lateral connectivity between rivers and floodplains. The FPC suggests that flood periods promote nutrient availability and productivity in floodplain habitats, helping juvenile fish survival and growth. Seasonal flooding synchronizes with fish life cycles, enabling them to use consistent resources, whereas unpredictable floods demand adaptive strategies (Welcomme, 1985; Winemiller, 2004).

Energy flow in river systems can be understood through a combination of the RCC (downstream transport), the FPC (lateral nutrient exchange), and the "riverine productivity model"[4] focusing on in-stream production (Thorpe & Delong, 1994). Hydrological variability, as noted by Walker (1995), is essential at multiple temporal scales: the flood pulse (days to weeks), flow history (weeks to years), and flow regime (decades). This variability highlights ecosystem health, influencing long-term ecological dynamics (Poff et al., 1997).

Riparian zones affect fish communities by affecting light, water quality, and habitat diversity, which are vital for river ecosystem management (Naiman & Decamps, 1997; Pusey & Arthington, 2003). However, changes to water flows, like damming or inter-basin transfers, disrupt biodiversity by disconnecting rivers from floodplains, with greater impacts in tropical and seasonal rivers (Arrington & Winemiller, 2004; Halls et al., 1998). The "natural flows paradigm" proposed by Poff et al. (1997) suggests that preserving natural flow patterns is key to maintaining ecological integrity in river systems.

The intertidal zone in the Bengal delta continuum undergoes dynamic alterations due to erosion and accretion. Bengal delta continuum has a characteristic low topography (less than 5 m above mean sea level [MSL]) and a vast network of distributaries. Large-scale tidal channel infilling has occurred over the continuum during the last decades (Wilson et al., 2017). The continuous differences of properties within the river are dependent primarily on the specific composition of the organisms in different sections of the water. Throughout the river's continuum, the relative abundance of the four primary trophic groups, namely, shredders, collectors, grazers (scrapers), and predators, vary. According to the RCC, the position of a stream reach in its river network, measured as stream order (Strahler, 1952, 1957), is considered a proxy of multiple environmental conditions, including substrate composition and size, river slope, and canopy cover, which in turn influence water temperature, chemistry, and light penetration. All these factors influence food availability through primary production and particulate organic matter (POM), acting as environmental filters for aquatic animals according to their feeding patterns. The RCC links the trophic structure of benthic macroinvertebrates to changes in environmental conditions and energy inputs along the longitudinal gradient (Doretto et al., 2020). Both allochthonous (such as riverine loadings) and autochthonous (such as coastal wetlands, seagrass ecosystems, mangrove litter, benthic vegetation, freshwater, and marine particle organic matter) sources contribute organic matter to coastal habitats (Jennerjahn, 2012; Kristensen et al., 2008). The saline intertidal hydrological system acts as the largest nursery and breeding ground for fish and shellfish, supporting the livelihoods of approximately 0.1% of the global population (Batista et al., 2014; Fernandes et al., 2015; Hoq et al., 2013; Mozumder et al., 2018). Located at the northern edge of the Bay of Bengal, the GBD is contributed by a combined catchment area of 1.6×10^6 km^2, noted for the highest annual sediment discharge worldwide (~10^9 t a^{-1}), which is conveyed by various small- to medium-sized tributaries originating from the nearby highlands, alongside the Ganga and the Brahmaputra. The genesis of the Bengal basin, the origin of the GBD, commenced in the Jurassic period with the initiation of Pangaea's rifting and concluded in the Miocene with the docking of eastern India and the Burma plate (Islam & Shamsuddoha, 2018).

The sedimentation of the basin occurred over the three geotectonic provinces—the continental shelf, deeper basin areas, and the Chittagong–Tripura Fold Belt—and started nearly concurrently with rifting. The Rajmahal Garo gap, which formed during the Plio-Pleistocene epoch, marked the beginning of the GBD. The Holocene history of the delta was characterized by fluctuations in monsoon-related sediment discharge, land- and seaward migrations of depocentres, and at least five switching during the Brahmaputra (Bandyopadhyay, 2019). Four Holocene sedimentary facies, reaching a depth of approximately 90 m, may be recognized inside the delta itself. The coastal region of the delta is not older than 5–2.5 thousand years (ka) and was developed by formation of four overlapping and eastward-younging deltaic lobes. Unlike other large deltas, the significant prevalence of silts and sands over clays significantly restricts auto compaction in the GBD sediments (Chowdhury et al., 2012; Pethick & Orford, 2013). It is important to note that although evidence of land subsidence is common in the Sundarbans (Bandyopadhyay et al., 2021; Bomer et al., 2020; Hale et al., 2019), the bathymetry and topography of the coastline considerably affect nearshore water dynamics (Krien et al., 2016). The Bengal delta is subsiding at an uncertain and potentially fluctuating rate (Becker et al., 2020). Aside from the natural sediment compaction (Krien et al., 2019), increased groundwater extraction and general restriction of the sediment deposition are few of the anthropogenic reasons behind the diverse subsidence rate. The sediment deficiency within the polder caused subsidence of the interior due to compaction (Auerbach et al., 2015). This process occasionally lowers the ground beneath MSL. This exacerbates waterlogging within the polder, increases soil salinity, and amplifies the risk of extended tidal or storm surge flooding (Khan et al., 2022; Prasad et al., 2017). By 11.5 ka BP, the intensity of the southwest monsoon significantly escalated, leading to sediment discharge increasing to a minimum of 2.5 times its present level. The incised river valleys commenced the accumulation of sand (Becker et al., 2020). In contrast, a considerable sedimentation rate at the Bengal fan suggests that the emerging GBD could not sustain a large percentage of the load it encountered. Delta-building bio-tidal processes, at the southern edge of the GBD, was responsible for 302 phased accretion of all sea-front islands of the Sundarbans 5–2 ka BP, from which 304 supratidal landmass was developed. Riverine sediments undergo rapid transformation as they enter the estuaries. The majority exist as colloids or small charged particles that display mutual repulsion. This scenario is characterized by the electrolytes and organic substances in the seawater, resulting in the flocculation and sedimentation of the particles. In the quiescent phase at the apex of a tidal cycle, single or aggregated flocs settle into the substrate of a stream or mudflat and progressively undergo consolidation during the subsequent slack water period. During the forthcoming tidal cycle, the speed of the mid-tide currents may be insufficient to erode all previously deposited substances. In the continuous cycles of tidal deposition and erosion, accretion only occurs when net deposition exceeds erosion (Chatterjee et al., 2013; Dasgupta et al., 2020). Bain et al. (2019) determined the sediment flow rate of the Ganga–Brahmaputra to be 1133×10^6 m^3 annually and suggested that, at this rate, the delta should have advanced several kilometers into the sea. He recognized the Swatch of No Ground underwater canyon, which crosses the GBD shelf and stretches 30 km from the Kunga estuary, as the principal silt-trapper, obstructing delta extension. Fergusson (1863) concluded that whereas the eastern segment of the GBD is undergoing "rapid development," there has been "little or no modification or extension seaward" in the Sundarbans region over the past century. This pattern, recognized a century before, is apparent in numerous recent research and evaluations of coastal changes linked to the Sundarbans (Akhter et al., 2018; Auerbach et al., 2015; Chakraborty & Satpati, 2019; Gupta et al., 2017; Mitra, 2019; Nishat, 2019; Wilson et al., 2017). Though, none of these studies incorporate the entire Sundarbans region beyond a few kilometers north of the surface and have predominantly been confined to the western (Indian) or eastern (Bangladeshi) sides of the Raimangal–Hariabhanga. When the drainage and high water lines from Survey of India's 1901–1923 "inch" maps of the Sundarbans are overlaid on present-day satellite images, five key findings emerge: The erosion rate peaks in the west-central area between the Saptamukhi and Gosaba estuaries, reaching up to 40 m year, then gradually decreases westward and eastward, nearly approaching zero on the west bank of the Baleswar. Many interior streams and estuaries are undergoing siltation, primarily in the reclaimed western area. Outside this region, the banks of major estuaries are undergoing erosion; the situation is more diverse for the smaller streams of the forested islands. Substantial areas of the western Sundarbans were restored in 24 Parganas. The gradual and irreversible transformation of intra-island streams into stagnant water bodies, their degradation, and integration into agricultural zones are notable postreclamation changes occurring over approximately 50 years at the island level. A substantial segment of the former agricultural areas has been converted into shrimp farms, primarily in the reclaimed eastern region of the province. This started to acquire prominence in the 1980s. The stability of deltaic beaches is frequently influenced by upstream river inflows and 625 erosional wave and tidal forces. The Sundarbans requires sediment input from tidal sources to provide long-term stability against rising MSLs, as it is part of the fluvially abandoned western GBD.

An average of 109 ton of sediment is added to the Meghna estuary each year in the GBD's shelf region (Mitra et al., 2011; Yadav & Majumdar, 2020), with 95% of that amount being pumped during the monsoon season (Sinha et al., 2020). During the high-energy conditions that take place from May to September, some of this silt moves west and is intercepted by the Swatch of No Ground undersea canyon, after which it escapes southward into the deep ocean (Agarwal et al., 2019; Hazra, Ghosh, Das Gupta, & Sen, 2002; Jana et al., 2014). Ongoing erosion in the southern Sundarbans for the past 200 years, particularly in the west, signifies insufficient sediment replenishment in the external delta and channel banks, which are adversely affected by monsoon waves and tropical cyclones. Accreted by bio-tidal processes, the Sundarbans are shaped by marine and atmospheric agents whose intensity varies over time, and they are fluvially abandoned. Landforms of the region can be classified into the broad morphotypes of beach, dune, estuary bank, swamps, and reclaimed plains apart from process- and feature-based subtypes. Interior waterways, especially in the west, are accumulating silt despite the erosion of the Sundarbans' estuary edges and surface—up to 40 m annually—for many decades. The Sundarbans' low coastline zone makes them more vulnerable to climate change-related storm surges, tropical cyclones (Aipira et al., 2017), and rising sea levels. The shifting coastline has rendered residents in the ecologically sensitive deltaic regions of the Sundarbans increasingly susceptible to risk and exposure (Akhter et al., 2018; Barua et al., 2010; Hossain et al., 2015; Sahana et al., 2019). The Thakuran River, located in the western Indian Sundarbans, possesses the greatest number of fish species and demonstrated a much greater diversity of fish species compared to other rivers and streams (Chakraborty, 2011). Elevated water turbidity enhances the survival probability of fish juveniles by diminishing predator perception distance; augmented primary and secondary productivity results from nutrient retention, tidal mixing, and freshwater influx; and the habitat's physical and structural complexity, along with niche segregation, reduces competition (Mandal et al., 2013), contributing its status as a nursery with optimal potential (Mitra et al., 2009).

2.3 THE INTERCONNECTED DYNAMICS OF DELTAS AND ESTUARIES: ECOLOGICAL IMPORTANCE

Deltas and estuaries are significant geographical features at the interface of rivers and oceans, each playing a crucial role in coastal regions' environmental and ecological dynamics. Understanding their relationship involves examining their formation, characteristics, ecological importance, and interplay. The relationship between deltas and estuaries can be seen in their interconnectedness within the more extensive river–ocean system. In some regions, deltas and estuaries coexist, forming a continuum of environments from the river's mouth to the open sea. The delta's sediment deposition can influence the physical and chemical characteristics of the estuary, such as its sediment load, salinity gradients, and water quality. Conversely, estuaries can influence delta formation. The tidal actions in estuaries can affect sediments' deposition patterns, shaping the delta's growth and morphology. The mixing of fresh and saltwater in estuaries can also affect the sediment characteristics and nutrient availability in delta regions, influencing the types of vegetation and animal species that thrive there (Azad et al., 2019; Barua et al., 2010;Bergstrand et al., 2015; Chowdhury et al., 2012).

The estuarine habitat undergoes cyclical and periodic changes, allowing various fish species to exploit the niche at different intervals. The environment is not harshened by the tidal system. It facilitates the appropriate gradation and amalgamation of freshwater with saltwater, eventually creating a habitat for various species with differing salinity tolerance levels. The tidal flow and other potential disruptions inhibit any competing species from dominating a population. Both vertical and horizontal stratification are present in estuaries. The vertical stratification of the water column, influenced by sunlight penetration and ambient temperature, facilitates niche differentiation, reduces competition, and enhances species diversity in the estuary. The horizontal stratification emphasizes concerns such as ecotonal communities and edge species, signifying increased species variety (Allison et al., 2019; Ghosh et al., 2022; Panja et al., 2020).

2.4 PHYTOPLANKTON POWERHOUSE: THE FOUNDATION OF ESTUARINE FOOD WEBS IN THE SUNDARBANS

The estuarine food webs are spun with the energy of plankton. Phytoplanktons form the foundation stone of world fishery. Phytoplankton serve as the principal producers within the pelagic ecosystem. The pelagic system in the coastal zone is very productive due to the significant concentration of nutrients from human sources, runoff from nearby landmasses, trash, and mangrove detritus. Zooplankton contribute to the secondary production of the estuary, serving as a crucial connection between primary and tertiary crops.

In the Indian Sundarbans, 102 species of marine and estuarine phytoplankton have been recorded. However, previous research has documented that the eastern Sundarbans area is home to around 50 species of diatoms, 30 species of green algae, and 18 species of blue-green algae (Chakravarty, 2011). In the Bangladeshi Sundarbans, 98 species of phytoplankton, including nano and picoplankton communities of the Sundarbans, are available. Among these are 35 recorded benthic algae, including weeds (Aziz et al., 2022). *Catenella, Caloglossa, Bostrychia*, and *Cladophorella* are the most spectacular species growing on pneumatophores. The Sundarbans offer a unique environment that helps the faunal community survive. Nearly every species has evolved to withstand extreme salinity and needs brackish water for its life cycle. Every inhabitant of aquatic mammal, reptile, and amphibian species are adept swimmers who have developed a condition to fulfill dietary needs partly from marine sources.

Hussain and Acharya (1994) reported the presence of 53 species of pelagic fish across 27 families, 124 species of demersal fish from 49 families, 24 species of prawns, 7 species of crabs, 2 species of gastropods, 6 species of pelecypods, 8 species of locust lobsters, and 3 species of turtles in the Sundarbans of Bangladesh. Conversely, Rahaman and Asaduzzamn (2020) identified 20 prawn species, 8 lobster species, 7 crab species, 42 mollusk species, and 225 fish species, including 120 economically significant fish species, in the Bangladeshi Sundarbans. *Scylla serrata, Portunus pelagicus, Portunus cyguilontus, Baruna litrerata*, and *Sartori Ana spinizera* are the crab species that are edible and economically significant.

The oceanic upwelling zones have the greatest primary productivity owing to nutrient availability. The primary production in marine and estuarine environments is regulated by light, nutrients, temperature, and ocean currents. The Indian Sundarbans exhibit significant geographical and temporal variability in phytoplankton, mostly due to drastic regional variations in water parameters (Ganguly et al., 2006; Mitra, 2013; Mitra et al., 2009). Zooplankton and mangroves are integral components of coastal ecosystems, each playing crucial roles in maintaining ecological balance and productivity. Their relationship is vital for the health and sustainability of coastal ecosystems (Hossain et al., 2025). The symbiotic relationship with mangroves provides habitat and nutrients, while zooplankton contributes to the health of the mangrove ecosystem through their ecological functions. Zooplankton contributes significantly to mangrove ecosystems by forming a crucial link in the aquatic food web. They serve as primary consumers, feeding on phytoplankton and organic matter. In turn, they prey on higher trophic-level organisms (fish and crustaceans) associated with mangrove habitats. The presence of zooplankton in mangrove ecosystems aids in nutrient cycling and energy flow. Zooplankton controls algal blooms and maintains water quality by grazing on phytoplankton. In addition, their waste products contribute to recycling of nutrients, supporting the productivity of mangrove ecosystems. According to Sarkar et al. (1986), copepods make up between 61.0% and 96.4% of the zooplankton in the Sundarbans. Also the presence of clupeid fish, or *Hilsa*, in the Hooghly estuary indicate the abundance of the diatom species *Fragilaria oceanic* and *Hemidiscus hardmannianus* (Mitra et al., 2009).

2.5 MANGROVES AS BIOREACTORS: THE DYNAMIC ECOSYSTEMS OF THE SUNDARBANS AND THEIR ROLE IN OCEAN BIOGEOCHEMISTRY

Mangroves are characterized by their intricate and dense root systems, including aerial roots and pneumatophores, which create unique and productive habitats. The dense root systems effectively trap and stabilize sediments from tidal waters. As the roots slow down the water flow, the suspended particles settle and accumulate around the roots. Moreover, the roots attenuate the energy of the wave and the velocity of the tidal currents. This energy reduction allows finer sediments to settle rather than being resuspended and carried away. The calmer water conditions within mangrove stands facilitate continuous sediment deposition, contributing to a stable and elevated intertidal zone and creating ideal conditions for breeding, spawning, and the early stages of life in various fish species, ensuring a steady supply of fish for local fisheries. Over time, this leads to increased sediment deposition, raising the level of the land and contributing to tidal accretion for maintaining coastal stability and ecosystem health. These roots provide shelter and breeding grounds for various aquatic organisms, including zooplankton. The complex root structures protect zooplankton from predators and offer a stable environment for reproduction.

In addition, mangroves contribute to the nutrient dynamics of coastal waters. They contribute organic matter through leaf litter, decomposing roots, and other plant materials. Leaf litter and other organic matter from mangroves decompose and release nutrients into the water, supporting phytoplankton growth. Zooplankton, which feed on phytoplankton and detritus, benefit from the increased food availability. The nutrient-rich environment built by mangroves thus enhances the productivity of zooplankton populations. This organic matter mixes with the trapped sediments, enhancing the overall sediment volume and promoting soil formation. The buildup of organic-rich sediments further supports mangrove forest development and growth, generating a positive feedback loop. The mangrove forests are considered as one of the most productive ecosystems on the planet, producing much more organic carbon than the system requires and therefore becoming a major contributor to the global carbon cycle. Mitra et al. (2009) also suggested that the mangrove forest becomes resort complex that extends the residence period, which aids in the absorption of inorganic nutrients and the retention of suspended particle matter. The mangrove habitat was identified as the most promising nursery in the world and recognized to improve the survival rates of juvenile fish (Mumby et al., 2004). In addition to their critical ecological services, mangroves also generate economic benefits that support local communities and thus become essential to the Sundarbans' small-scale fishing. Fish catch is maximized in regions of elevated human population density, providing advantages for both fishermen and markets for the catch. Undoubtedly, mangroves situated near inhabited places are likely to face more threats than those in less populous regions, resulting in degradation, water pollution, or overfishing, hence diminishing their productivity. When mangroves are protected by effective management practices and their fisheries are properly regulated, they are likely to provide the most benefit. According to Hutchison et al. (2015), conservation and restoration initiatives in these regions near populated areas will probably provide the highest return on investment.

The ecosystem services provided by Mangrove are depicted in Table 2.2.

Studies indicated that pioneer 315 mangrove species like *Porteresia coarctata* colonize muddy tidal flats when tidal deposition rises above the low-tide line, and bio-tidal accretion takes over. According to the "classical successional view of mangrove dynamics" (Snedaker, 1982), mangroves "prepare the ground" for the next

TABLE 2.2

The Sundarbans Mangroves Forest's Ecosystem Services

Regulating service	Oxygen production, heat regulation, carbon sequestration, climate stabilization, hydrological and nutrient cycling, pollution control, and sediment management. For example, they alleviate the impacts of cyclones by protecting human life, securing properties, and preserving biodiversity from intense winds, waves, and seawater flooding during cyclonic occurrences.
Supporting service	Biodiversity preservation and conservation on land and in the ocean. Reproductive habitat for aquatic, benthic, and terrestrial organisms.
	Nursery, feeding, and breeding areas for commercially important fish and shellfish.
	Nypha fruticans leaves are widely used for thatching, but *Phoenix paludosa* timber is used in construction.
Provisioning service	Edible fish and shellfish, wax and honey from *Apis dorsata*, and fruits such as Keora (*Sonneratia apetala*) and Ora (*Sonneratia caseolaris*).
	Natural resources including timber for paper and fuel (species encompass *Ceriops decandra, Cynometra ramiflora*, and *Amoora cuculatta*).
	Multiple biocides, food additives such as alginates, and the foliage of *Phoenix paludosa* are used in the management of ringworm and pharyngitis.
Cultural service	Accessible comprehension of biodiversity mechanisms for educational and research objectives. Spiritual values relate to the environment and ceremonies within religious communities.
	Creating job opportunities through a rising influx of tourists engaging in cultural events, river cruises, and jungle adventures.

Source: Modified after Hilmi et al. (2025).

community as they raise the level of the fish population by inducing sedimentation and thereby reducing the tidal inundation time of a particular locality. At the end of the succession, the climax—largely nonmangrove—vegetation community evolves as the land is raised sufficiently above the tidal limits (Chapman, 1976; Thom, 1984). The primary productivity of this ecosystem is very rich, causing fish species to use the feeds without competition. Fish productivity from mangroves will be highest where mangrove productivity is high. It will increase with an increase in the total area of mangroves, notably with the length of the mangrove margin because, generally, it is the fringes of mangroves where fish populations are enhanced. Where there is high freshwater input from rivers and rainfall is where mangroves are in good condition. Fish production will be significantly enhanced by mangroves exhibiting more physical complexity, characterized by intricate patterns of channels, pools, and lagoons, as well as complex root structures that provide essential habitats for the refuge and development of certain bivalves (Table 2.2). Mangroves do not only provide physical habitat but also actively fuel estuarine food webs by supporting algae that serve as food for commercially important and resident fish species. *Coelathrum* sp. and *Ulva* sp. are very abundant in the pneumatophores and trunks of mangrove plants. These two algae have been identified in the stomachs of several fish species, including *P. pangasius, Pampus chinensis, Mugil tade*, and mudskippers *Perioptholmus* and *Boleopthalmus* (Chakravarty, 2011; Mitra & Homchaudhuri, 2013). *Plotinus canius* and *Pangasius pangasius* have been seen to voraciously consume the fruits of Keora (*Sonneratia apetala*) and Baen (*Avicennia* sp.). During August, September, and October, tons and tons of fruits fall on water and the fish congregate close the shore and consume them. The growth of fish and fat content suggests that these vegetable foodstuffs contribute to fat accumulation and development. The Baen leaves are significantly plagued by insects and pests (Mandal et al., 2013). Defoliation is expected to affect flower and fruit production, significantly influencing the migration of *P. pangasius* and *P. canius* (Chakravarty, 2011). This clearly indicates the trophic linkage between mangrove and the estuarine fish communities.

Mangroves often function as purifiers or biogeochemical sinks. The bioaccumulation of heavy metals such as Zn, Cu, Pb, and Hg by certain mangrove species (e.g., *Avicennia alba, Avicennia marina, Excoecaria agallocaha*, and *Acanthus ilicifolius*) is distinctive and serves to mitigate heavy metal-contaminated waste, thus purifying estuarine water (Mitra et al., 2009). Their complex root systems facilitate sediment deposition, microbial mediation, and metal sequestration, preventing bioavailability to aquatic fauna and protecting adjacent ecosystems. This phytoremediation capacity highlights their role in mitigating industrial and urban pollution, enhancing water quality and sustaining coastal resilience (Alongi, 2002; MacFarlane & Burchett, 2000; Marchand et al., 2016; Tam & Wong, 1995).

2.6 TREASURES OF THE TIDAL REALM: EXPLORING THE FISH AND SHELLFISH DIVERSITY OF THE SUNDARBANS

A multitude of research has been undertaken about the taxonomic diversity of fish species in the lower Gangetic delta. Pillay (1967) identified the presence of more than 120 species. Jhingran (1982) documented 172 species, noting that the high-salinity

TABLE 2.3
List of Fish Species Available in the Sundarbans

No.	Order (Family)	English Name	Scientific Name	Max Length	Distribution	Habitat	IUCN Red List Status	Commercial Importance
1	Orectolobiformes (Hemiscylliidae)	Grey Bamboo Shark	*Chiloscyllium griseum*	77.0 cm TL male/unsexed	Indo-West Pacific	M, B	NT	Commercial
2	Carcharhiniformes (Carcharhinidae)	Dog Shark	*Scoliodon laticaudus*	100.0 cm TL male/unsexed	Indo-West Pacific	M, B	NT	Commercial
3		Shark	*Glyphis gangeticus*	204 cm TL male/unsexed	Indo-West Pacific	M, F, B	EN	Of no interest
4			*Rhizoprion odonacutus*	175 cm TL male/unsexed	Eastern Atlantic	M, F, B	NE	Commercial; gamefish: yes
5		Blacktip Reef Shark	*Carcharhinus melanopterus*	200 cm TL male/unsexed	Indo-Pacific	M, B	NT	Fisheries: commercial; gamefish: yes; aquarium: public aquariums
6	Carcharhiniformes (Sphyrnidae)	Scalloped Hammerhead Shark	*Sphyrna lewini*	430 cm TL male/unsexed	Circumglobal in coastal warm temperate and tropical seas	M, B	EN	Commercial; gamefish: yes
7		Hammer head Shark	*Eusphyra blochii*	186 cm TL male/unsexed	Indo-West Pacific	M, B	EN	Commercial
8	Carcharhiniformes (Triakidae)	Hardnose Smoothhound	*Mustelus mosis*	150 cm TL male/unsexed	Indian Ocean	M	DD	Fisheries: commercial; gamefish: yes
9	Torpediniformes (Narkidae)	Brown Spotted Numbfish	*Narcine brunnea*	22.0 cm TL male/unsexed	Indo-West Pacific	M	NE	DD

(Continued)

Table 2.3 (Continued)

No.	Order (Family)	English Name	Scientific Name	Max Length	Distribution	Habitat	IUCN Red List Status	Commercial Importance
10	Torpediniformes (Narcinidae)	Electric Ray	*Narcine timlei*	38.0 cm TL male/unsexed	Indo-West Pacific	M	DD	Fisheries: commercial; aquarium: commercial
11	Rajiformes (Rhinobatidae)	Spottail Sleeper Ray	*Narke dipterygia*	18.0 cm TL male/unsexed	Indo-West Pacific	M	DD	Minor commercial
12		Gulter Fish	*Rhynchobatus djiddensis*	310 cm TL male/unsexed	Western Indian Ocean	VU	M, B	Commercial; gamefish: yes; aquarium: public aquariums
13		Sharpnose Guiterfish, Shovelnose	*Glaucostegus granulatus*	280 cm SL male/unsexed	Indo-Pacific	M	VU	Commercial
14	Myliobatiformes (Dasyatidae)	Scaly Whipray	*Brevitrygon imbricata*	25.5 cm WD male/unsexed	Indian Ocean	M, F, B	DD	Minor commercial
15		Dwarf Whipray	*Brevitrygon walga*	20.3 cm WD male/unsexed	Indian Ocean	M	NT	Commercial
16		Stingray	*Himantura undulata*	410 cm TL male/unsexed	Indo-West Pacific	M	VU	Minor commercial
17		Leopard Stingray, Reticulate Whipray, Honeycomb Stingray	*Himantura uarnak*	200 cm WD male/unsexed	Indo-Pacific	M, B	VU	Minor commercial; gamefish: yes; aquarium: commercial

18		Cowtail Stingray	*Pastinachus sephen*	183 cm WD male/unsexed	Northwestern Indian Ocean	M, F, B	NE	Minor commercial; aquarium: commercial
19		Bleeker's Whipray	*Pateobatis uarnacoides*	119 cm WD male/unsexed	Indo-Malay Arch	M	VU	Bycatch
20		White Spotted Stingray	*Maculabatis gerrardi*	200 cm TL male/unsexed	Indo-West Pacific	M, B	VU	Fisheries: commercial; gamefish: yes; aquarium: commercial
21		Cowtail Stingray	*Pastinachus sephen*	183 cm WD male/unsexed	Northwestern Indian Ocean	M, F, B	NT	Minor commercial; aquarium: commercial
22		Sharp Snout Stingray	*Telatrygon zugei*	29.0 cm WD male/unsexed	Northwest Pacific	M, B	NT	Minor commercial
23	Myliobatiformes (Gymnuridae)	Longtail Butterfly Ray, Butterfly Ray	*Gymnura poecilura*	250 cm WD male/unsexed	Indo-Pacific	M	NT	Commercial
24	Elopiformes (Elopidae)	Tenpounder, Tarpon	*Elops machnata*	118 cm FL male/unsexed	Indo-West Pacific	M, B	LC	Commercial; gamefish: yes
25	Elopiformes (Megalopidae)	Indo-Pacific Tarpon	*Megalops cyprinoides*	150 cm TL male/unsexed	Indo-Pacific	M, F, B	DD	Minor commercial; aquaculture: commercial; gamefish: yes
26	Anguilliformes (Muraenidae)	Red Sea White-Spotted Moray	*Gymnothorax punctatus*	90.0 cm TL male/unsexed	Western Indian Ocean	M	NE	Minor commercial
27		Moray Eel	*Gymnothorax tile*	60.0 cm TL male/unsexed	Indo-West Pacific	M, F, B	NE	Minor commercial
28		Moray Eel	*Gymnothorax sp.*	400 cm TL male/unsexed	Indo-West Pacific	M, B	NE	Minor commercial; gamefish: yes

(*Continued*)

Table 2.3 (Continued)

No.	Order (Family)	English Name	Scientific Name	Max Length	Distribution	Habitat	IUCN Red List Status	Commercial Importance
29		Slender Giant Moray	*Strophidon sathete*	400 cm TL male/unsexed	Indo-West Pacific	M, B	NE	Minor commercial; gamefish: yes
30	Anguilliformes (Muraenesocidae)	Yellow Pike Conger	*Congresox talabonoides*	250 cm TL male/unsexed	Indo-West Pacific	M, B	NE	Commercial
31		Eel	*Congresox talabonoides*	250 cm TL male/unsexed	Indo-West Pacific	M, B	NE	Commercial
32		Daggertooth Pike Conger	*Muraenesox cinereus*	220 cm TL male/unsexed	Indo-West Pacific	M, F, B	NE	Highly commercial; aquaculture: commercial; gamefish: yes
33	Anguilliformes (Ophichthidae)	Boro Snake Eel	*Pisodonophis boro*	100.0 cm TL male/unsexed	Indo-West Pacific	M, F, B	LC	Minor commercial
34	Anguilliformes (Anguillidae)	Purple Spaghetti Eel	*Moringua raitaborua*	44.0 cm SL male/unsexed	Eastern Indian Ocean	F, B	NE	No interest
35		Giant Mottled Eel	*Aquilla bengalensis*	200 cm TL male/unsexed	Asia	M, F, B	NT	Commercial; aquaculture: likely future use; gamefish: yes
36	Clupeiformes (Clupeidae)	Chacunda Gizzard Shad, Shortnose Gizard Shad	*Anodontostoma chacunda*	22.0 cm SL male/unsexed	Indo-West Pacific	M, F, B	NE	Commercial
37		Indian River Shad	*Gudusia chapra*	20.0 cm TL male/unsexed	Asia: rivers of India and Bangladesh affluent to the Bay of Bengal	F, B	LC	Subsistence fisheries

38		Kelee Shad	*Hilsa kelee*	35.0 cm TL male/unsexed	Indo-West Pacific	M, F, B	NE	Highly commercial
39		Bloch's Gizzard Shad, Long-finned Gizzard Shad	*Nematalosa nasus*	27.2 cm TL male/unsexed	Indo-West Pacific	M, F, B	LC	Minor commercial
40		White Sardine	*Escualosa thoracata*	10.0 cm SL male/unsexed	Indo-West Pacific	M, F, B	NE	Commercial
41		Gold Stripe Sardine	*Sardinella gibbosa*	29.6 cm TL male/unsexed	Indo-West Pacific	M	NE	Highly commercial
42		Sardine	*Sardinella fimbriata*	19.0 cm TL male/unsexed	Indo-West Pacific	M, B	NE	Commercial
43		Indian Oil Sardine	*Sardinella longiceps*	23.0 cm SL male/unsexed	Indian Ocean	M	NE	Highly commercial
44		Blacktip Sardinella	*Sardinella melanura*	20.3 cm SL male/unsexed	Indo-West Pacific	M	NE	Minor commercial
45	Clupeiformes (Engraulidae)	River Shad, Hilsa Shad	*Tenualosa ilisha*	60.0 cm SL male/unsexed	Indian Ocean	M, F, B	LC	Minor commercial; aquaculture: experimental
46		Toli Shad, Shad	*Tenualosa toli*	60.0 cm TL male/unsexed	Indo-West Pacific	M, F, B	NE	Highly commercial
47		Gold-spotted Grenadier Anchovy	*Coilia dussumieri*	20.0 cm SL male/unsexed	Indian Ocean	M, F, B	NE	Commercial
48	Clupeiformes Chirocentridae	Neglected Grenadier Anchovy	*Coilia neglecta*	17.0 cm SL male/unsexed	Indian Ocean	M, B	LC	Subsistence fisheries
49		Ramcarat Grenadier Anchovy	*Coilia ramcarati*	25.0 cm TL male/unsexed	Indian Ocean	M, B	NE	Subsistence fisheries

(Continued)

Table 2.3 (Continued)

No.	Order (Family)	English Name	Scientific Name	Max Length	Distribution	Habitat	IUCN Red List Status	Commercial Importance
50		Gangetic Hairfin Anchovy	*Setipinna phasa*	40.0 cm TL male/unsexed	Asia: Indian fresh and brackish waters	F, B	NE	Minor commercial
51		Scaly Hairfin Anchovy	*Setipinna taty*	15.3 cm TL male/unsexed	Indo-West Pacific	M, B	NE	Minor commercial
52		Indian Anchovy	*Stolephorus indicus*	15.5 cm SL male/unsexed	Indian Ocean	M, B	NE	Commercial
53		Common Hairfin Anchovy	*Setipinna tenuifilis*	22.0 cm SL male/unsexed	Eastern Indian Ocean	M, B	NE	Minor commercial
54		Spined Anchovy	*Stolephorus tri*	9.5 cm SL male/unsexed	Western Pacific	NE	M, B	Commercial
55		Anchovy	*Thryssa dussumieri*	11.0 cm SL male/unsexed	Indo-Pacific	M, B	LC	Commercial
56		Hamilton's Thryssa	*Thryssa hamiltonii*	27.0 cm SL male/unsexed	Indo-Pacific	M, B	NE	Commercial
57		Oblique Jaw Thryssa, Gangetic Anchovy	*Thryssa purava*	15.5 cm SL male/unsexed	Indian Ocean	M, B	NE	Minor commercial
58		Wolf Herring	*Chirocentrus nudus*	100.0 cm SL male/unsexed	Indo-West Pacific	M	LC	Highly commercial
59	Clupeiformes (Dussumieriidae)	Dorab WolfHerring	*Chirocentrus dorab*	100.0 cm SL male/unsexed	Indo-Pacific	M, B	NE	Commercial; gamefish: yes
60		Rainbow Sardine	*Dussumieria acuta*	20.0 cm SL male/unsexed	Indo-West Pacific	M, F, B	LC	Highly commercial

61	Clupeiformes (Pristigasteridae)	Smooth Back Herring	*Raconda russeliana*	19.0 cm SL male/unsexed	Indo-Pacific	M, B	NE	Minor commercial
62		Indian Pellona Herring	*Pellona ditchela*	16.0 cm SL male/unsexed	Indo-West Pacific	M, F, B	LC	Commercial
63		Indian Ilisha	*Ilisha melastoma*	22.0 cm SL male/unsexed	Indo-Pacific	M, B	LC	Minor commercial
64		Bigeye Herring	*Ilisha megaloptera*	36.5 cm SL male/unsexed	Indo-Pacific	M, F, B	LC	Commercial
65		Coromandal Ilisha	*Ilisha filigera*	22.0 cm SL male/unsexed	Indo-Pacific	M, F, B	DD	Subsistence fisheries
66		Long-Finned Herring	*Opisthopterus tardoore*	23.0 cm TL male/unsexed	Indo-West Pacific	M, B	NE	Minor commercial
67	Cypriniformes (Cobitidae)	Guntea Loach	*Lepidocephalichthys guntea*	15.0 cm TL male/unsexed	Asia	F, B	LC	Of no interest; aquarium: commercial
68	Cypriniformes (Cyprinidae)	Swamp Barb	*Puntius chola*	15.0 cm TL male/unsexed	Asia	F	LC	Commercial; aquarium: commercial
69		Barb	*Puntius terio*	10.0 cm TL male/unsexed	Asia	F	LC	Commercial; aquarium: commercial
70		Gangetic Scissortail Rasbora	*Rasbora*	13.0 cm TL male/unsexed	Asia	F, B	LC	Commercial; aquarium: commercial
71	Siluriformes (Plotosidae)	Canine Catfish Eel	*Plotosus canius*	111 cm TL male/unsexed	Indo-West Pacific	M, F, B	NE	Commercial
72		Striped Ell Tail Catfish	*Plotosus lineatus*	35.2 cm TL male/unsexed	Indo-Pacific	M, B	NE	Commercial; aquarium: commercial

(*Continued*)

Table 2.3 (Continued)

No.	Order (Family)	English Name	Scientific Name	Max Length	Distribution	Habitat	IUCN Red List Status	Commercial Importance
73	Siluriformes (Schilbeidae)	Gagra	*Clupisoma garua*	60.9 cm SL male/unsexed	Asia	F, B	LC	Commercial; gamefish: yes
74	Siluriformes (Ailiidae)	Gangetic Ailia	*Ailia coila*	30.0 cm TL male/unsexed	Asia	F, B	NT	Commercial
75		Silond Catfish, Silondia Vacha	*Silonia silondia*	183 cm TL male/unsexed	Asia	F, B	LC	Commercial; gamefish: yes
76	Siluriformes (Pangasiidae)	Fatty Catfish	*Pangasius pangasius*	300 cm SL male/unsexed	Asia	F, B	LC	Commercial; aquaculture: commercial; gamefish: yes
77	Siluriformes (Bagridae)	Tengara Catfish	*Mystus tengara*	18.0 cm TL male/unsexed	Asia	F, B	LC	Subsistence fisheries
78		Catfish	*Mystus bleekeri*	17.7 cm TL male/unsexed	Asia	F, B	LC	Minor commercial; aquarium: commercial
79		Long-Whiskered Catfish	*Mystus gulio*	46.0 cm TL male/unsexed	Asia	F, B	LC	Commercial
80	Siluriformes (Clariidae)	Walking Catfish	*Clarias batrachus*	47.0 cm TL male/unsexed	Asia	F, B	LC	Commercial; aquaculture: commercial; gamefish: yes; aquarium: commercial
81	Siluriformes (Sisoridae)	Gangetic Goonch, Devil Catfish	*Bagarius bagarius*	230 cm TL male/unsexed	Asia	F, B	LC	Minor commercial; gamefish: yes

82	Siluriformes (Ariidae)	Gagora Catfish	*Arius gagora*	91.4 cm SL male/unsexed	Indian Ocean	M, F, B	NT	Commercial
83	Siluriformes (Heteropneustidae)	Spotted Catfish, Sea Catfish	*Arius maculatus*	80.0 cm TL male/unsexed	Indo-West Pacific	M, F, B	NE	Commercial
84		Yellow Sea Catfish, Marine Catfish	*Arius venosus*	30.0 cm TL male/unsexed	Indo-West Pacific	M, B	NE	Commercial
85		Threadfin Sea Catfish	*Arius arius*	40.0 cm SL male/unsexed	Indo-West Pacific	M, B	LC	Commercial
86		Blacktip Sea Catfish	*Plicofollis dussumieri*	85.0 cm TL male/unsexed	Indo-West Pacific	M, F, B	LC	Bycatch
87		Flatmouth Sea Catfish	*Plicofollis platystomus*	31.0 cm TL male/unsexed	Indian Ocean	M, B	LC	Commercial
88		Dusky Catfish, Sona Sea Catfish	*Sciades sona*	92.0 cm TL male/unsexed	Indo-West Pacific	M, B	NE	Commercial
89		Engraved Catfish	*Nemapteryx nenga*	30.0 cm TL male/unsexed	Indo-West Pacific	M, B	NE	DD
90		Engraved Catfish	*Nemapteryx caelata*	59.0 cm FL male/unsexed	Indo-Pacific	M, B	NE	Commercial
91		Giant Sea Catfish	*Netuma thalassina*	185 cm TL male/unsexed	Indo-West Pacific	M, F, B	NE	Commercial; gamefish: yes
92		Bronze Catfish	*Netuma bilineata*	90.0 cm TL male/unsexed	Indo-West Pacific	M, F, B	NE	Minor commercial
93		Stinging catfish	*Heteropneustes fossilis*	41.2 cm TL male/unsexed	Asia	F, B	LC	Highly commercial; aquaculture: commercial; aquarium: commercial

(Continued)

Table 2.3 (Continued)

No.	Order (Family)	English Name	Scientific Name	Max Length	Distribution	Habitat	IUCN Red List Status	Commercial Importance
94	Aulopiformes (Synodontidae)	Bombay Duck	*Harpadon nehereus*	61.0 cm TL male/unsexed	Indo-West Pacific	M, B	NE	Highly commercial
95		Greater Lizard Fish	*Saurida tumbil*	60.0 cm FL male/unsexed	Indo-West Pacific	M	LC	Commercial
96	Gadiformes (Bregmacerotidae)	Unicorn Cod	*Bregmaceros mcclellandi*	9.6 cm SL male/unsexed	Indian Ocean	M, B	NE	Commercial
97	Lophiiformes (Antennariidae)	Shaggy Angler, Zebra Frogfish	*Antennarius hispidus*	Indo-West Pacific	Indo-West Pacific	M	NE	DD
98	Beloniformes (Belonidae)	Banded Needle Fish, Square Tail Alligator Gar	*Strongylura leiura*	100.0 cm TL male/unsexed	Indo-West Pacific	M, B	NE	Commercial; gamefish: yes
99		Spottail Needle Fish	*Strongylura strongylura*	40.0 cm SL male/unsexed	Indo-West Pacific	M, B	NE	Commercial
100		Needle Fish	*Tylosurus crocodilus*	150 cm TL male/unsexed	Indo-West Pacific	M, B	NE	Commercial; gamefish: yes
101		Silver Needle Fish	*Xenentodon cancila*	40.0 cm TL male/unsexed	Asia	M, F, B	LC	Minor commercial; aquarium: potential
102	Beloniformes (Hemiramphidae)	Congaturi Halfbeak	*Hyporhamphus limbatus*	35.0 cm TL male/unsexed	Indo-West Pacific	M, F, B	LC	Minor commercial
103		Georges Halfbeak, Longbilled Halfbeak	*Rhynchorhamphus georgii*	31.0 cm male/unsexed	Indo-West Pacific	M, F, B	NE	Minor commercial

104	Beloniformes (Zenarchopteridae)	Buffon's Halfbeak, Buffon's Garfish	*Zenarchopterus buffonis*	23.0 cm TL male/unsexe	Indo-West Pacific	M, B	NE	Subsistence fisheries
105		Ectuntio Halfbeak	*Zenarchopterus ectuntio*	18.0 cm TL male/unsexed	Asia	F, B	NE	Commercial
106	Beloniformes (Exocoetidae)	Tropical Two-Winged Flying Fish	*Exocoetus volitans*	30.0 cm TL male/unsexed	Widespread in tropical and subtropical zones of all oceans	M	LC	Commercial
107	Syngnathiformes (Fistulariidae)	Red Cornetfish, Flute-Mouth	*Fistularia petimba*	200 cm TL male/unsexed	Indo-West Pacific, Atlantic	M, B	LC	Subsistence fisheries; gamefish: yes
108	Syngnathiformes (Syngnathidae)	Sea Horse, Smooth Seahorse	*Hippocampus kuda*	30.0 cm TL male/unsexed	Indo-Pacific	M, B	VU	Highly commercial; aquaculture: commercial; aquarium: commercial
109	Synbranchiformes (Synbranchidae)	Cuchia, Gangetic Mud Eel	*Monopterus cuchia*	79.7 cm TL male/unsexed	Asia	F, B	LC	Commercial
110	Synbranchiformes (Mastacembelidae)	Lesser Spiny Eel	*Macrognathus aculeatus*	38.0 cm TL male/unsexed	Asia	F, B	NE	Minor commercial; aquarium: commercial
111		Striped Spiny Eel	*Macrognathus pancalus*	18.9 cm TL male/unsexed	Asia	F, B	LC	Minor commercial
112	Scorpaeniformes (Scorpaenidae)	Plaintail Turkeyfish, Russell's Firefish	*Pterois russelii*	30.0 cm SL male/unsexed	Indo-Pacific	M, B	NE	Minor commercial; aquarium: public aquariums

(Continued)

Table 2.3 (Continued)

No.	Order (Family)	English Name	Scientific Name	Max Length	Distribution	Habitat	IUCN Red List Status	Commercial Importance
113	Scorpaeniformes (Synanceiidae)	Miles Lionfish	*Pterois miles*	43.0 cm TL male/unsexed	Indian Ocean	M	NE	Commercial; aquarium: commercial
114		Grey Stingfish	*Minous monodactylus*	15.0 cm TL male/unsexed	Indo-West Pacific	M	NE	Minor commercial
115	Scorpaeniformes (Platycephalidae)	Painted Stringer	*Minous pictus*	10.2 cm SL male/unsexed	Indo-West Pacific	M	NE	
116		Rough Flathead	*Grammoplites scaber*	30.0 cm TL male/unsexed	Indo-West Pacific	M, B	NE	Minor commercial
117	Perciformes (Epinephelidae)	Flathead	*Cociella punctata*	35.0 cm TL male/unsexed	Indo-West Pacific	M	LC	Minor commercial
118		Spiny Flathead	*Kumococius rodericensis*	25.0 cm TL male/unsexed	Indo-West Pacific	M	NE	Minor commercial
119		Spotted Flathead	*Cociella crocodilus*	50.0 cm TL male/unsexed	Indo-West Pacific	M, B	NE	DD
120		Thorny Flathead	*Rogadius asper*	20.5 cm TL male/unsexed	Indo-West Pacific	M	LC	Minor commercial
121		Flathead	*Rogadius pristiger*	50.0 cm TL male/unsexed	Indo-West Pacific	M	LC	DD
122		Bartail Flathead	*Platycephalus indicus*	100.0 cm TL male/unsexed	Indo-West Pacific	M, B	NE	Commercial; aquaculture: commercial; gamefish: yes

123		Cloudy Grouper, Cloudy Rock Cod	*Epinephelus erythrurus*	45.0 cm TL male/unsexed	Indo-West Pacific	M, B	VU	
124		Orange-spotted Grouper	*Epinephelus coioides*	120 cm TL male/unsexed	Indo-West Pacific	M, B	NT	Commercial; aquaculture: commercial; gamefish: yes
125		Blacktip Grouper	*Epinephelus fasciatus*	52.0 cm TL male/unsexed	Indo-Pacific	M, B	LC	Commercial; gamefish: yes; aquarium: commercial
126		Grouper	*Epinephelus tauvina*	Max length: 100.0 cm TL male/unsexed	Indo-Pacific	M	DD	Minor commercial; aquaculture: commercial; gamefish: yes
127		Vermillion Grouper	*Cephalopholis miniata*	50.0 cm TL male/unsexed	Indo-Pacific	M	LC	Commercial; gamefish: yes; aquarium: commercial
128		Giant Grouper	*Epinephelus lanceolatus*	270 cm TL male/unsexed	Indo-Pacific	M, B	VU	Subsistence fisheries; aquaculture: commercial; gamefish: yes; aquarium: commercial
129	Perciformes (Terapontidae)	Terapon Perch, Three-striped Tiger Fish	*Terapon jarbua*	Max length: 36.0 cm TL male/unsexed	Indo-Pacific	M, F, B	LC	Minor commercial; aquaculture: commercial
130		Big Eye	*Terapon theraps*	30.0 cm SL male/unsexed	Indo-West Pacific	M, F, B	LC	Minor commercial

(*Continued*)

Table 2.3 (Continued)

No.	Order (Family)	English Name	Scientific Name	Max Length	Distribution	Habitat	IUCN Red List Status	Commercial Importance
131	Perciformes (Priacanthidae)	Moontail Bulls Eye	*Priacanthus hamrur*	45.0 cm TL male/unsexed	Indo-Pacific	M	LC	Minor commercial; aquarium: commercial
132		Purple Spotted Big Eye	*Priacanthus tayenus*	35.0 cm TL male/unsexed	Indo-West Pacific	M	LC	Minor commercial
133	Perciformes (Apogonidae)	Broad-banded Cardinalfish	*Ostorhinchus fasciatus*	12.6 cm TL male/unsexed	Indo-West Pacific	M	NE	No interest
134		Three-Striped Cardinalfish	*Apogon septemstriatus*	5.3 cm SL male/unsexed	West Pacific	M	NE	DD
135	Perciformes (Sillaginidae)	Gangetic Sillago	*Sillaginopsis panijus*	44.0 cm TL male/unsexed	Western Indian Ocean	M, F, B	NE	Commercial
136		Silver Sillago	*Sillago sihama*	31.0 cm SL male/unsexed	Indo-West Pacific	M, B	LC	Commercial; aquaculture: commercial
137	Perciformes (Channidae)	Asiatic Snakehead	*Channa orientalis*	33.0 cm TL male/unsexed	Asia	F, B	NE	Minor commercial; aquarium: commercial
138		Striped Snakehead	*Channa striata*	100.0 cm SL male/unsexed	Asia	F, B	NE	Highly commercial; aquaculture: commercial; aquarium: public aquariums

139		Spotted Snakehead	*Channa punctata*	31.0 cm TL male/unsexed	Asia	F, B	LC	Commercial; aquaculture: commercial; aquarium: commercial
140	Perciformes (Rachycentridae)	Cobia, Black King Fish	*Rachycentron canadum*	200 cm TL male/unsexed	Worldwide in tropical and subtropical waters	M, B	LC	Minor commercial; aquaculture: commercial; gamefish: yes
141	Perciformes (Echeneidae)	Common Remora	*Remora remora*	86.4 cm TL male/unsexed	Cosmopolitan in warm waters	M	LC	Commercial; gamefish: yes
142	Perciformes (Carangidae)	Razorbelly Scad	*Alepes kleinii*	18.2 cm TL male/unsexed	Indo-West Pacific	M	LC	Minor commercial
143		Shrimped Scad	*Alepes djedaba*	40.0 cm TL male/unsexed	Indo-Pacific	M	LC	Commercial; gamefish: yes
144		Black-Fin Scad	*Alepes melanoptera*	25.0 cm TL male/unsexed	Indo-Pacific	M, B	LC	Minor commercial
145		Threadfin Trevally	*Alectis indica*	165 cm TL male/unsexed	Indo-Pacific	M, B	LC	Commercial; gamefish: yes; aquarium: public aquariums
146		Threadfin Trevally	*Alectis ciliaris*	150 cm TL male/unsexed	Circumglobal in tropical, subtropical and temperate seas	M	LC	Minor commercial; gamefish: yes; aquarium: commercial
147		Black-Fin Jack	*Atropus atropos*	31.6 cm TL male/unsexed	Indo-West Pacific.	M	LC	Minor commercial
148		Oxeye Scad	*Selar boops*	25.0 cm FL male/unsexed	Pacific Ocean	M	LC	Commercial
149		Bigeye Scad	*Selar crumenophthalmus*	70.0 cm TL male/unsexed	Circumtropical	M	NE	Highly commercial; gamefish: yes

(Continued)

Table 2.3 (Continued)

No.	Order (Family)	English Name	Scientific Name	Max Length	Distribution	Habitat	IUCN Red List Status	Commercial Importance
150		Longfin Trevally	*Carangids armatus*	120 cm TL male/unsexed	Indo-Pacific	M, B	NE	commercial; gamefish: yes
151		Bigeye Trevally	*Caranx sexfasciatus*	120 cm TL male/unsexed	Indo-Pacific	M, B	LC	Commercial; gamefish: yes
152		Bumpnose Trevally	*Carangoides hedlandensis*	32.0 cm TL male/unsexed	Indo-West Pacific	M	NE	Commercial; gamefish: yes
153		Trevally	*Carangoides malabaricus*	32.0 cm TL male/unsexed	Indo-West Pacific	M	LC	Commercial; gamefish: yes
154		Giant Trevally, Giant Kingfish	*Caranx ignobilis*	170 cm TL male/unsexed	Indo-Pacific	M, B	LC	Commercial; aquaculture: commercial; gamefish: yes; aquarium: public aquariums
155		Red-Tailed Mackerel Scad	*Decapterus kurroides*	45.0 cm TL male/unsexed	Indo-West Pacific	M	NT	Commercial
156		Torpedo Scad	*Megalaspis cordyla*	80.0 cm TL male/unsexed	Indo-West Pacific	M, B	LC	Highly commercial
157		Black Promfet	*Parastromateus niger*	75.0 cm TL male/unsexed	Indo-West Pacific	M, B	NE	Highly commercial
158		Double-Spotted Queenfish	*Scomberoides commersonnianus*	120 cm TL male/unsexed	Indo-West Pacific	M, B	LC	Minor commercial; gamefish: yes
159		Queen Fish	*Scomberoides tol*	60.0 cm TL male/unsexed	Indo-West Pacific	M, B	NE	60.0 cm TL male/unsexed

160		Black-Banded Trevally	*Seriolina nigrofasciata*	77.5 cm TL male/unsexed	Southeast Atlantic	M, B	LC	Minor commercial; gamefish: yes
161		Longrakered Trevally	*Ulua mentalis*	100.0 cm TL male/unsexed	Indo-West Pacific	M	LC	Commercial; gamefish: yes
162	Perciformes (Menidae)	Moon Fish	*Mene maculata*	30.0 cm TL male/unsexed	Indo-West Pacific	M, B	NE	Commercial
163	Perciformes (Leiognathidae)	Orangefin Ponyfish	*Photopectoralis bindus*	14.0 cm TL male/unsexed	Indo-West Pacific	M, B	NE	Minor commercial
164	Perciformes (Lactariidae)	Shortnose Ponyfish	*Leiognathus brevirostris*	18.2 cm FL male/unsexed	Indo-West Pacific	M, B	NE	Commercial
165		Striped Ponyfish	*Leiognathus fasciatus*	21.0 cm TL male/unsexed	Indo-West Pacific	M, B	LC	Minor commercial
166		Common Ponyfish	*Leiognathus equulus*	28.0 cm TL male/unsexed	Indo-West Pacific	M, F, B	NE	Minor commercial; aquaculture: commercial
167		Deep Pugnose Pony Fish	*Secutor ruconius*	*8.0 cm TL male/unsexed*	*Indo-West Pacific*	M, F, B	NE	Minor commercial
168		Pugnose Pony Fish	*Secutor insidiator*	11.3 cm SL male/unsexed	Indo-West Pacific	M, B	NE	Commercial
169		Tooth Pony	*Gazza minuta*	21.0 cm FL male/unsexed	Indo-Pacific	M, B	NE	Commercial
170		False Trovally	*Lactarius*	40.0 cm TL male/unsexed	Indo-West Pacific	M, B	NE	Commercial
171	Perciformes (Lutjanidae)	John's Snapper, Red Snapper	*Lutjanus johnii*	106 cm TL male/unsexed	Indo-West Pacific	M, B	LC	Commercial; aquaculture: commercial; gamefish: yes
172	Perciformes (Uranoscopidae)	Malabar Red Snapper	*Lutjanus malabaricus*	100.0 cm TL male/unsexed	Indo-West Pacific	M, B	NE	Highly commercial; gamefish: yes; aquarium: commercial

(Continued)

Table 2.3 (Continued)

No.	Order (Family)	English Name	Scientific Name	Max Length	Distribution	Habitat	IUCN Red List Status	Commercial Importance
173		Pinjalo Snapper	*Pinjalo pinjalo*	80.0 cm TL male/unsexed	Indo-West Pacific	M	LC	Minor commercial
174		Blood Snapper	*Lutjanus sanguineus*	100.0 cm TL male/unsexed	Western Indian Ocean	M	NE	DD
175		Stargazer	*Astroscopus guttatus*	59.0 cm TL male/unsexed	Indo-West Pacific	M	LC	Minor commercial
176	Perciformes (Datnioididae)	Dollfus' Stargrazer	*Uranoscopus guttatus*	29.8 cm TL male/unsexed	Indian Ocean	M	NE	DD
177		Stargazer	*Ichthyscopus inermis*	60.0 cm TL male/unsexed	Indo-West Pacific	M	NE	DD
178		Stargazer	*Ichthyscopus lebeck*	60.0 cm TL male/unsexed	Indo-West Pacific	M	NE	DD
179		Four-Barred Tigerfish	*Datnioides polota*	30.0 cm SL male/unsexed	Asia and Oceania	F, B	NE	Commercial
180	Perciformes (Gerreidae)	Whiptail Silverbiddy	*Gerres filamentosus*	39.0 cm TL male/unsexed	Indo-West Pacific	M, F, B	LC	Minor commercial
181		Silverbiddy	*Gerres sp.*	39.0 cm TL male/unsexed	Indo-West Pacific	M, B	NE	Minor commercial
182	Perciformes (Haemulidae)	Silver Grunt	*Pomadasys argenteus*	70.0 cm TL male/unsexed	Indo-West Pacific	M, F, B	LC	Commercial
183		Blotched Grunt	*Pomadasys maculatus*	59.3 cm TL male/unsexed	Indo-West Pacific	M, B	LC	Commercial
184	Perciformes (Sparidae)	Ongspine Seabream	*Argyrops spinifer*	80.0 cm TL male/unsexed	Indian Ocean	M	NE	Commercial

185		Yellow Seabream	*Acanthopagrus latus*	40.0 cm FL male/unsexed	Northwest Pacific	M, F, B	DD	Commercial; aquaculture: commercial; gamefish: yes
186		River Bream	*Acanthopagrus berda*	90.0 cm TL male/unsexed	Indian Ocean	M, F, B	LC	Commercial; gamefish: yes
187	Perciformes (Nemipteridae)	Double-Whip Threadfin Bream	*Nemipterus nematophorus*	20.0 cm SL male/unsexed	Indo-West Pacific	M	NE	Minor commercial
188		Pink Perch	*Nemipterus japonicus*	34.0 cm TL male/unsexed	Indo-Pacific	M	NE	Commercial
189	Perciformes (Sciaenidae)	Goatee Croaker	*Dendrophysa russelii*	38.3 cm TL male/unsexed	Indo-Pacific	M, F, B	NE	Minor commercial
190		Croaker	*Chrysochir aureus*	30.0 cm SL male/unsexed	Indo-Pacific	M, B	NE	Minor commercial
191		Blackmouth Croaker	*Atrobucca nibe*	45.0 cm SL male/un	Indo-West Pacific	M	NE	Commercial
192		Sharpnose Hammer Croaker	*Johnius borneensis*	34.8 cm TL male/unsexed	Indo-West Pacific	M, F, B	NE	Minor commercial
193		Silver Croaker	*Pennahia argentata*	40.0 cm SL male/unsexed	Indo-pacific	M	NE	DD
194		Large-Eye Croaker	*Johnius plagiostoma*	10.0 cm SL male/unsexed	Indo-West Pacific	M, F, B	NE	Of no interest
195		Coitor Croaker	*Johnius coitor*	16.0 cm SL male/unsexed	Indo-West Pacific	M, F, B	LC	Minor commercial
196		Large-Fined Croaker	*Johnius macropterus*	25.0 cm TL male/unsexed	Indo-West Pacific	M, F, B	NE	Commercial
197		Black Croaker	*Johnius dussumieri*	40.0 cm TL male/unsexed	Indian Ocean	M	NE	Minor commercial

(Continued)

Table 2.3 (Continued)

No.	Order (Family)	English Name	Scientific Name	Max Length	Distribution	Habitat	IUCN Red List Status	Commercial Importance
198		Gangetic Bola, Croaker	*Johnius gangeticus*	12.0 cm SL male/unsexed	Asia	M, B	NE	Minor commercial
199		Kathala Croaker	*Kathala axillaris*	27.0 cm TL male/unsexed	Indian Ocean	F, B	NE	Commercial
200		Cuja Croaker	*Macrospinosa cuja*	152 cm TL male/unsexed	Indian Ocean	M	NE	Commercial
201		Pama Croaker	*Otolithes parna*	160 cm TL male/unsexed	Indo-Pacific	M, F, B	NE	DD
202		Lesser Tiger-Toothed Croaker	*Otolithes cuvieri*	39.0 cm TL male/unsexed	Western Indian Ocean	M	NE	Commercial
203		Tiger-Toothed Croaker	*Otolithes ruber*	90.0 cm TL male/unsexed	Indo-West Pacific	M	NE	Commercial; gamefish: yes
204		Pama Croaker	*Otolithoides pama*	160 cm TL male/unsexed	Indo-Pacific	M, B	NE	Commercial
205		Bronze Croaker	*Otolithoides biauritus*	160 cm SL male/unsexed	Indo-West Pacific	M, B	NE	Minor commercial
206		Panna Croaker	*Panna microdon*	30.0 cm SL male/unsexed	Western Central Pacific	M, F, B	NE	Minor commercial
207		Donkey Croaker	*Pennahia anea*	30.0 cm SL male/unsexed	Indo-West Pacific	M, B	NE	Commercial
208		Spindle Croaker	*Pseudotolithus elongatus*	62.5 cm TL male/unsexed	Eastern Atlantic	M	LC	DD
209		Spotted Croaker	*Protonibea diacanthus*	165.0 cm TL (female)	Indo-West Pacific	M, F, B	NE	Minor commercial; aquaculture: commercial

210		Blotched Tiger-Toothed Croaker	*Pterotolithus maculatus*	45.0 cm SL male/unsexed	Indo-Pacific	M, B	LC	Minor commercial
211	Perciformes (Olynemidae)	Fourfinger Threadfin	*Eleutheronema tetradactylum*	200 cm TL male/unsexed	Indo-Pacific	M, B	NE	Highly commercial; aquaculture: commercial
212		Indian Threadfin	*Leptomelanosoma indicum*	142 cm TL male/unsexed	Indo-West Pacific	M, F, B	NE	Commercial; gamefish: yes
213		Blackspot Threadfin	*Polydactylus sextarius*	30.0 cm TL male/unsexed	Eastern Indian Ocean and Western Pacific	M, F, B	NE	DD
214		Golden Threadfin	*Polydactylus sexfilis*	61.0 cm TL male/unsexed	Indo-Pacific	M, B	NE	Commercial; aquaculture: experimental; gamefish: yes; aquarium: commercial
215		Paradise Threadfin	*Polynemus paradiseus*	30.0 cm NG male/unsexed	Eastern Indian Ocean and Western Pacific	M, F, B	NE	Commercial
216	Perciformes (Mullidae)	Red Sea Goatfish	*Parupeneus forsskali*	28.0 cm TL male/unsexed	Indian Ocean	M, F, B	NE	Minor commercial; aquarium: commercial
217		Goatfish	*Parupeneus heptacanthus*	36.0 cm TL male/unsexed	Indo-West Pacific	M, B	LC	Commercial
218		Goldband Goatfish	*Upeneus moluccensis*	22.5 cm TL male/unsexed	Indo-West Pacific	M, B	LC	Commercial
219		Sulphur Goatfish	*Upeneus sulphureus*	23.0 cm TL male/unsexed	Indo-West Pacific	M, B	LC	Minor commercial; aquarium: commercial

(Continued)

Table 2.3 (Continued)

No.	Order (Family)	English Name	Scientific Name	Max Length	Distribution	Habitat	IUCN Red List Status	Commercial Importance
220		Finstripe Goatfish	*Upeneus taeniopterus*	33.0 cm TL male/unsexed	Indo-Pacific	M, B	LC	Commercial
221	Perciformes (Toxotidae)	Largescale Archerfish	*Toxotes chatareus*	40.0 cm SL male/unsexed	Asia and Oceania	M, B	NE	Minor commercial; aquarium: commercial
222		Banded Archerfish	*Toxotes jaculatrix*	30.0 cm TL male/unsexed	Asia and Oceania	F, B	LC	Minor commercial; aquarium: commercial
223	Perciformes (Drepanidae)	Spotted Sickle Fish	*Drepane punctata*	50.0 cm TL male/unsexed	Indo-West Pacific	F, B	NE	Commercial; aquarium: commercial
224		Spadefish	*Ephippus orbis*	25.0 cm TL male/unsexed	Indo-West Pacific	M, F, B	NE	Commercial
225		Banded Drepane	*Drepane longimana*	50.0 cm TL male/unsexed	Indo-West Pacific	M, B	NE	Minor commercial; aquarium: commercial
226	Perciformes (Mugilidae)	Longarm Mullet	*Osteomugil cunnesius*	41.0 cm SL male/unsexed	Indo-West Pacific	M, B	NE	Commercial
227		Striped Mullet	*Mugil cephalus*	100.0 cm SL male/unsexed	Cosmopolitan in coastal waters of the tropical, subtropical and temperate zones	M, F, B	LC	Highly commercial; aquaculture: commercial; gamefish: yes
228		Greenback Mullet	*Planiliza subviridis*	40.0 cm SL male/unsexed	Indo-Pacific	M, F, B	NE	Commercial; aquaculture: commercial

229		Corsula Mullet	*Rhinomugil corsula*	45.0 cm TL male/unsexed	Asia	M, F, B	LC	Commercial; aquaculture: commercial
230		Mullet	*Valamugil speigleri*	35.0 cm TL male/unsexed	Indo-West Pacific	F, B	NE	Commercial
231		Yellowtail Mullet	*Sicamugil cascasia*	12.5 cm TL male/unsexed	Asia	M, F, B	LC	DD
232		Goldspot Mullet	*Liza parsia*	23.5 cm TL male/unsexed	Indian Ocean	F	NE	Commercial; aquaculture: commercial
233		Tade Grey Mullet	*Chelon planiceps*	70.0 cm TL male/unsexed	Indo-Pacific	M, F, B	NE	Commercial; aquaculture: commercial
234		Goldspot Mullet	*Chelon parsia*	23.5 cm TL male/unsexed	Indian Ocean	M, F, B	NE	Commercial; aquaculture: commercial
235	Perciformes (Uranoscopidae)	Stargazer	*Uranoscopus guttatus*	29.8 cm TL male/unsexed	Indian Ocean	M, F, B	NE	DD
236		Stargazer	*Ichthyscopus lebeck*	60.0 cm TL male/unsexed	Indo-West Pacific	M	NE	DD
237		Two-Spined Yellowtail Stargazer	*Uranoscopus cognatus*	16.5 cm TL male/unsexed	Eastern Indian Ocean	M	NE	Bycatch
238	Perciformes (Eleotridae)	Duckbill Sleeper	*Butis butis*	15.0 cm TL male/unsexed	Indo-Pacific	M	LC	Minor commercial
239		Black-Spot Sleeper Goby	*Butis humeralis*	14.2 cm SL male/unsexed	Indo-Pacific	M, F, B	NE	DD
240		Sleeper Goby	*Butis melanostigma*	18.0 cm TL male/unsexed	Indo-West Pacific	M, F, B	NE	Of no interest

(*Continued*)

Table 2.3 (Continued)

No.	Order (Family)	English Name	Scientific Name	Max Length	Distribution	Habitat	IUCN Red List Status	Commercial Importance
241		Dusky Sleeper	*Eleotris fusca*	26.0 cm TL male/unsexed	Indo-West Pacific	M, F, B	LC	Minor commercial; aquarium: commercial
242	Perciformes (Gobiidae)	Mudskipper	*Boleophthalmus boddarti*	22.0 cm TL male/unsexed	Indo-West Pacific	M, F, B	LC	Of no interest
243		Tank Goby	*Glossogobius giuris*	50.0 cm SL male/unsexed	Africa to Oceania	M, F, B	NE	Minor commercial; aquaculture: commercial; aquarium: commercial
244		Goby	*Apocryptes bato*	26.0 cm TL male/unsexed	Indian Ocean	M, F, B	NE	Minor commercial
245		Bearded Worm Goby	*Taenioides cirratus*	30.0 cm TL male/unsexed	Indo-Pacific	M, F, B	DD	Of no interest
246		Bumblebee Goby	*Brachygobius nunus*	2.5 cm SL male/unsexed	Asia	M, F, B	NE	Of no interest
247		Goby	*Zappa confluentus*	4.4 cm SL male/unsexed	Asia and Oceania	F, B	NE	DD
248		Goby	*Pogonogoibius planiformes*	27.0 cm TL male/unsexed	Indo-West Pacific	M, F, B	NE	Minor commercial
249		Mudskipper	*Periophthalmodon schlosseri*	27.0 cm TL male/unsexed	Indo-West Pacific	M, B	NE	Minor commercial
250		Mudskipper	*Periophthalmus barbarus*	Eastern Indian Ocean and Western Pacific	30.0 cm NG male/unsexed	M, F, B	LC	LC

251		Rubicundus Eelgoby	*Odontamblyopus rubicundus*	33.4 cm SL male/unsexed	Indo-West Pacific	M, F, B	NE	Minor commercial
252		Pointed-Tailed Goby	*Pseudapocryptes elongatus*	20.0 cm TL male/unsexed	Indo-Pacific	M, F, B	LC	Minor commercial
253		Walking Goby	*Scartelaos histophorus*	14.0 cm SL male/unsexed	Indo-West Pacific	F, B	NE	Minor commercial
254		Knight Goby	*Stigmatogobius sadanundio*	9.0 cm SL male/unsexed	Asia	M, B	NE	Of no interest; aquarium: commercial
255		Eel Goby	*Taenioides buchanani*	30.0 cm TL male/unsexed	Indian Ocean	F, B	NE	Of no interest
256		Burrowing Goby	*Trypauchen vagina*	22.0 cm TL male/unsexed	Indo-Pacific	M, B	NE	Minor commercial
257	Perciformes (Callionymidae)	Arrow Dragonet	*Callionymus sagitta*	11.0 cm TL male/unsexed	of no interest	M, B	NE	Of no interest
258	Perciformes (Ephippidae)	Spadefish	*Ephippus orbis*	25.0 cm TL male/unsexed	Indo-West Pacific	M	NE	Commercial
259	Perciformes (Ambassidae)	Elongate Glassy Perchlet	*Chanda nama*	11.0 cm TL male/unsexed	Asia	M	LC	Minor commercial; aquarium: public aquariums
260		Himalayan Glassy Perchlet	*Parambassis baculis*	7.0 cm TL male/unsexed	Asia	F, B	LC	Of no interest
261		Glassy Fish	*Parambassis ranga*	9.5 cm TL male/unsexed	Asia	F	LC	Subsistence fisheries; aquarium: commercial
262	Perciformes (Scatophagidae)	Spotted Scat, Spotted Butterfish	*Scatophagus argus*	45.0 cm TL male/unsexed	Indo-Pacific	F, B	LC	Minor commercial; aquaculture: commercial; aquarium: commercial

(Continued)

Table 2.3 (Continued)

No.	Order (Family)	English Name	Scientific Name	Max Length	Distribution	Habitat	IUCN Red List Status	Commercial Importance
263	Perciformes (Siganidae)	Streaked Rabbitfish	*Siganus javus*	53.0 cm TL male/unsexed	Indo-Pacific	M, F, B	LC	Commercial; aquarium: commercial
264		White-Spotted Spinefoot	*Siganus canaliculatus*	40.0 cm TL male/unsexed	Indo-West Pacific	M, B	LC	Commercial; aquaculture: commercial
265	Perciformes (Sphyraenidae)	Bigeye Barracuda	*Sphyraena forsteri*	92.3 cm FL male/unsexed	Indo-Pacific	M, B	NE	Commercial
266		Barracuda	*Sphyraena barracuda*	200 cm TL male/unsexed	Indo-Pacific	M	LC	Minor commercial; gamefish: yes; aquarium: public aquariums
267		Yellowstripe Barracuda	*Sphyraena chrysotaenia*	32.2 cm TL male/unsexed	Indo-Pacific	M	NE	DD
268		Banded Oor Indian Barracuda	*Sphyraena jello*	150 cm TL male/unsexed	Indo-West Pacific	M, B	NE	Commercial; gamefish: yes
269	Perciformes (Trichiuridae)	Small-Head Ribbon Fish	*Eupleurogrammus muticus*	87.0 cm TL male/unsexed	Indo-West Pacific	M, B	NE	Commercial
270		Large-Head Ribbon Fish	*Trichiurus lepturus*	234 cm TL male/unsexed	Circumtropical and Circumtemperate	M, B	NE	Highly commercial; gamefish: yes
271		Savalani Ribbon Fish	*Lepturacanthus savala*	100.0 cm SL male/unsexed	Indo-West Pacific	M, B	NE	Commercial

272	Perciformes (Scombridae)	Mackerel Tuna	*Euthynnus affinis*	100.0 cm FL male/unsexed	Indo-West Pacific	M, B	LC	Highly commercial; gamefish: yes
273		Island Mackerel	*Rastrelliger faughni*	20.0 cm FL male/unsexed	Indo-West Pacific	M	DD	Commercial; gamefish: yes
274		Indian Mackerel	*Rastrelliger kanagurta*	36.0 cm male/unsexed	Indo-West Pacific	M	LC	Highly commercial; gamefish: yes; bait: occasional
275		Indian Mackerel	*Rastrelliger brachysoma*	34.5 cm FL male/unsexed	Pacific Ocean	M	DD	Highly commercial; gamefish: yes
276		Striped Bonito	*Sarda orientalis*	102 cm FL male/unsexed	Indo-Pacific	M, B	LC	Minor commercial; gamefish: yes
277		Buulet Tuna	*Auxis rochei*	50.0 cm FL male/unsexed	Atlantic, Indian and Pacific	M	NE	Highly commercial; gamefish: yes
278		Seer Fish	*Scomberomorus lineolatus*	94.0 cm FL male/unsexed	Indo-West Pacific	M, B	LC	Commercial; gamefish: yes
279		Barred Mackerel	*Scomberomorus commerson*	240 cm FL male/unsexed	Indo-West Pacific	M	NT	Highly commercial; gamefish: yes
280		Indo-Pacific King Mackerel	*Scomberomorus guttatus*	81.5 cm FL male/unsexed	Indo-West Pacific	M	DD	Highly commercial; gamefish: yes
281	Perciformes (Stromateidae)	Silver Pomfret	*Pampus argenteus*	60.0 cm SL male/unsexed	Indo-West Pacific	M, B	NE	Highly commercial
282		Chinese Pomfret	*Pampus chinensis*	40.0 cm TL male/unsexed	Indo-West Pacific	M	NE	Minor commercial

(*Continued*)

Table 2.3 (Continued)

No.	Order (Family)	English Name	Scientific Name	Max Length	Distribution	Habitat	IUCN Red List Status	Commercial Importance
283	Perciformes (Lethrinidae)	Ornate Emperor	*Lethrinus ornatus*	45.0 cm TL male/unsexed	Indo-West Pacific	M, B	LC	Minor commercial
284	Perciformes (Lobotidae)	Tripletail	*Lobotes surinamensis*	110 cm TL male/unsexed	Tropical and subtropical waters of all oceans	M	LC	Commercial; gamefish: yes; Aquarium: commercial
285	Perciformes (Kurtidae)	Indian Lamphead	*Kurtus indicus*	12.6 cm TL male/unsexed	Asia	M, B	NE	Minor commercial
286	Perciformes (Latidae)	Barramundi	*Lates calcarifer*	200 cm TL male/unsexed	Indo-West Pacific	M, F, B	NE	Highly commercial; aquaculture: commercial; gamefish: yes; aquarium: public aquariums
287	Perciformes (Anabantidae)	Climbing perch	*Anabas testudineus*	25.0 cm TL male/unsexed	Asia	F, B	DD	Commercial; aquaculture: commercial; aquarium: commercial
288	Pleuronectiformes (Soleidae)	Javanese Flounder	*Pseudorhombus javanicus*	35.0 cm SL male/unsexed	Indo-West Pacific	M, F, B	NE	Commercial
289		Commerson`S Sole	*Synaptura commersonnii*	36.0 cm TL male/unsexed	Indo-West Pacific	M	NE	Commercial
290		Zebra Sole	*Zebrias altipinnis*	20.0 cm NG male/unsexed	Indo-West Pacific	M, B	NE	DD

291		Oriental Sole	*Brachirus orientalis*	38.0 cm TL male/unsexed	Indo-West Pacific	M, F, B	NE	Commercial
292		Sole	*Brachirus pan*	11.1 cm TL male/unsexed	Indo-West Pacific	M, F, B	LC	DD
293	Pleuronectiformes (Paralichthyidae)	Malayflounder	*Pseudorhombus malayanus*	35.0 cm SL male/unsexed	Indo-West Pacific	M, B	NE	Commercial
294		Large-Tooth Flounder	*Pseudorhombus arsius*	45.0 cm TL male/unsexed	Indo-West Pacific	M	NE	Commercial; gamefish: yes
295		Deep Flounder	*Pseudorhombus elevatus*	37.5 cm TL male/unsexed	Indian Ocean	M, B	NE	Commercial
296		Indian Hailbut	*Psettodes erumei*	64.0 cm TL male/unsexed	Indo-West Pacific	M	NE	Commercial
297	Pleuronectiformes (Cynoglossidae)	Halibut	*Psettodes belcheri*	80.0 cm TL male/unsexed	Eastern Atlantic	M	DD	Commercial
298		Fourlined Tonguesole	*Cynoglossus bilineatus*	44.0 cm SL male/unsexed	Indo-West Pacific	M, B	NE	Commercial; gamefish: yes
299		Gangetic Tonguesole	*Cynoglossus cynoglossus*	20.0 cm TL male/unsexed	Indo-West Pacific	M, B	NE	Commercial
300		Long Tonguesole	*Cynoglossus kopsii*	60.0 cm SL male/unsexed	Indo-West Pacific	M	NE	Of no interest
301		Tonguesole	*Symphurus trifasciatus*	11.6 cm SL male/unsexed	Indian Ocean	M	NE	of no interest
302			*Paraplagusia bilineata*	31.6 cm TL male/unsexed	Indo-Pacific	M	NE	Minor commercial
303		Double-Lined Tonguesole	*Cynoglossus lingua*	Indo-West Pacific	Indo-West Pacific	M, F, B	NE	Commercial
304		Speckled Tonguesole	*Cynoglossus puncticeps*	35.0 cm SL male/unsexed	Indo-West Pacific	M, F, B	NE	Commercial

(*Continued*)

Table 2.3 (Continued)

No.	Order (Family)	English Name	Scientific Name	Max Length	Distribution	Habitat	IUCN Red List Status	Commercial Importance
305	Pristiformes (Pristidae)	Largescale Tonguesole	*Cynoglossus arel*	Indo-West Pacific	Indo-West Pacific	M, B	NE	Commercial
306		Saw Shark	*Anoxypristis cuspidata*	470 cm TL male/unsexed	Indo-West Pacific	M, F, B	EN	Commercial; gamefish: yes
307	Tetraodontiformes (Triacanthidae)	Large-Tooth Saw Fish	*Pristis microdon*	700 cm TL male/unsexed	Indo-West Pacific	M, F, B	NE	Minor commercial; gamefish: yes
308		Short-Nosed Tripod Fish	*Triacanthus biaculeatus*	30.0 cm TL male/unsexed	Indo-West Pacific	M, F, B	NE	Minor commercial
309		Tripod Fish	*Pseudotriacanthus strigilifer*	25.0 cm TL male/unsexed	Indo-Pacific	M	NE	Minor commercial; aquarium: commercial
310	Tetraodontiformes (Balistidae)	Triggerfish	*Abalistes stellaris*	60.0 cm TL male/unsexed	Indo-West Pacific	M	NE	Commercial; aquarium: commercial
311	Tetraodontiformes (Ostraciidae)	Yellow Box Fish	*Ostracion cubicus*	45.0 cm TL male/unsexed	Indo-Pacific	M	NE	Minor commercial; aquarium: commercial
312	Tetraodontiformes (Tetraodontidae)	Gangetic Pufferfish	*Chelonodontops patoca*	38.0 cm SL male/unsexed	Indo-Pacific	M, F, B	LC	Minor commercial
313		Bengal Reticulated Puffer	*Chelonodontops bengalensis*	31.2 cm SL male/unsexed	Eastern Indian Ocean	M, B	NE	DD
314		Puffer Fish	*Arothron stellatus*	120 cm TL male/unsexed	Indo-Pacific	M, B	LC	Of potential interest
315		Puffer Fish	*Leiodon cutcutia*	15.0 cm TL male/unsexed	Asia	F, B	NE	Of no interest

316		Puffer Fish	*Carinotetraodon travancoricus*	3.5 cm TL male/unsexed	Asia	F	VU	Of no interest; aquarium: commercial
317		Diamond-Back Puffer	*Lagocephalus guentheri*	37.4 cm TL male/unsexed	Indo-West Pacific	M	LC	DD
318		Green Pufferfish	*Lagocephalus lunaris*	45.0 cm SL male/unsexed	Southeast Atlantic	M	LC	Fisheries: of no interest; aquarium: commercial
319		Lattice Blaasop	*Takifugu oblongus*	40.0 cm TL male/unsexed	Indo-West Pacific	M, B	LC	Fisheries: of potential interest
320		Green Pufferfish	*Dichotomyctere fluviatilis*	17.0 cm TL male/unsexed	Asia: India, Sri Lanka, Bangladesh, Myanmar and Borneo. Recorded from the Mekong delta	F, B	LC	Fisheries: of no interest; aquarium: commercial
321	Tetraodontiformes (Diodontidae)	Spotted Porcupine Fish	*Diodon hystrix*	91.0 cm TL male/unsexed	Circumtropical	M	LC	Fisheries: minor commercial; gamefish: yes; aquarium: commercial
322	Batrachoidiformes (Batrachoididae)	Grunting Toadfish	*Allenbatrachus grunniens*	30.0 cm TL male/unsexed	Indo-West Pacific	M, B	NE	Fisheries: of no interest

Source: Modified after Habib et al., 2020; Froese, R. & Pauly, D. (Eds.) (2025). FishBase. World Wide Web electronic publication. www.fishbase.org
Abbreviations: EN—Endangered | VU—Vulnerable | NT—Near Threatened | LC—Least Concern | DD—Data Deficient | NE—Not Evaluated | F—Freshwater | B—Brackish | M—Marine).

region of the Indian Sundarbans, marked by mangrove predominance, exhibits a considerably greater diversity. His estimates reveal that 99 species originate from marine or higher salinity environments, whereas 73 species derive from freshwater sources. In the aquatic subsystem of the Indian Sundarbans, Mandal and Nandi (1989) discovered 141 species spanning 100 taxa, whereas Chaudhuri and Choudhury (1994) recorded 250 species among 96 genera. Khan (2011) identified 107 species within the Sundarbans Biosphere Reserve; however, this count excludes species only located in the upper low-saline zone of the Hooghly–Matla estuary complex. Fish use estuarine corridors for migratory purposes. Certain marine fish species, including *Hilsa, Pama, Sillaginopsis*, and *Polynemus*, migrate upstream via estuaries to spawn in freshwater. In addition, there is the group of freshwater fish that come into the estuaries to reproduce, such as *Arius* sp., *Osteogeniosus* sp., *Polydactylus* sp., and *Eleutheronema* sp., among others. Fish species also recorded from estuary include the eel *Anguilla bicolour, Moringa raitaborua*, the partially migratory *Pisidonophis boro* (Borofish), *Congresox talabonoides*, and *Muraenesox cinereus* (Chaudhuri & Choudhury, 1994; Mandal et al., 2013; Mandal & Ghosh, 1989). In the estuarine waters of the Sundarbans, two categories of fish exist: transients (migrants) and permanent residents. Species that occupy every estuary zone throughout the year and during all phases of their life cycle are designated as resident species. The recognized species comprise *Mugil parsia, Mugil tade, Otolithoides biuritus, Lates calcarifer, Hilsa toli, Arius jella, Harpodon nehereus, Setipinna taty, Ilisha elongata, Setipinna phasa, Coilia ramcarati, Pama pama*, and *Sillaginopsis panijus*, among others. Temporary or migratory fish reside in the inshore areas of the Bay of Bengal and its related estuaries for a restricted period. Migrants can be categorized into three groups according to their migratory trajectory and patterns (Jhingran, 1982): (1) transient marine species, such as *Tenualosa ilisha, Polynemus paradiseus, Sillaginopsis panijus*, and *Pama pama*, which migrate downstream to spawn in freshwater regions of the estuary, characterized by near-zero salinity; (2) freshwater species that spawn in saline estuarine areas, represented by *Pangasius pangasius*; and (3) Marine species that spawn in estuarine waters with reduced salinity (Mukherjee et al., 2021), ranging from 5.0 to 10.0 psu, including *Arius jella, Osteogeneious militaris*, and *Polydactylus indicus*.

Conversely, additional research is necessary regarding the diversity of fish species in Bangladesh's Sundarbans. Hamilton (1822) presented the inaugural documentation of fish species in the Sundarbans. Fifty-one of the 71 fish species he identified from the Gangetic estuaries were endemic to the Sundarbans. Since Bangladesh attained independence in 1971, particularly after 1978, numerous authors and scientists have produced books, guides, project reports, conference papers, and other publications concerning the diversity of fish species. Seidensticker and Hai (1983), Rahman (1989), Acharya and Kamal (1994), Chantarasri (1994), Bernacsek (2001), Bernacsek and Haque (2001), Shah and Hossain (2006), and Rahman et al. (2019) constitute a limited selection of references regarding the availability of species in the Sundarbans waterways.

The Sundarbans is home to about 322 fish species (Habib et al., 2020) (Table 2.3), 20 prawn species, and 44 crab species, which including two edible varieties (Chandra & Sagar, 2003). The revised list of fish includes 322 species from 217 genera, 96 families, and 22 orders, based on direct observations and published data. Four fish species have also been recently discovered in the Bangladeshi waters (Habib et al., 2019): *Uranoscopus cognatus* (Cantor, 1849); *Lagocephalus guentheri* (de Miranda Ribeiro, 1915); *Carangoides hedlandensis* (Whitley, 1934); and *Mustelus mosi* (Hemprich & Ehrenberg, 1899). Extensive study has been conducted on all facets of fish biology in the Sundarbans. Significant publications (Annandale, 1922; Hamilton, 1822; Hora, 1934; Kemp, 1917; Pantulu, 1966; Shetty et al., 1961; Talwar & Joglekar, 1970) have advanced the research on the ichthyofauna of the Hooghly–Matla estuary. The fish and fishery resources of the Sundarbans mangrove-based estuary have been documented by Gopalakrishnan and Rao (1968), Gopalkrishnan (1971), Gopalakrishnan et al. (1973), Sen (1975), Naskar and Guhabakshi (1987), Mandal and Nandi (1989), Chaudhuri and Choudhury (1994), Naskar (2004). Sen (1975) documented 61 finfish species, Naskar and Guhabakshi (1987) identified 92 species, Mandal and Nandi (1989) recorded 141 species across 100 genera, Chaudhuri and Choudhury (1994) reported 150 species within 96 genera of class Pisces, Bose et al. (1999) noted 74 species, Sanyal (1999) recorded 134 species, Debnath (2001) documented 91 species, Khan (2011) reported 106 species, and Naskar (2004) identified 139 species of fish. A total of 267 fish species were documented and classified into 81 families and 17 orders (Mandal et al., 2013).

2.7 WHERE TRADITION MEETS SURVIVAL: SMALL-SCALE FISHERIES OF THE SUNDARBANS

The Bangladesh Delta region is among the most impoverished areas globally (FAO, 2006a). Since the 1980s, the coastal population of Bangladesh has doubled, exceeding 16 million, which is about 10% of the nation's overall population, with a considerable segment facing poverty ansd environmental susceptibility (Allison et al., 2009). The Indian segment of the GBM delta (Indian Sundarbans Delta, West Bengal) consists of 102 islands, 54 of which are populated. The population is around 4.6 million, increasing by 2% annually (Hazra et al., 2002). Alterations in coastal morphology resulting from erosion and accretion and human activities affect coastal ecosystems and their functionality. These alterations affect the socioeconomic welfare of the residents (Wahl et al., 2018). In Bangladesh, the artisanal sector accounts for 99% of the total volume of landings. Marine fishing operations are placed at shallow depths (up to 100 m), although deep-water resources remain unexploited by Bangladeshi fishermen. Reports indicate substantial illicit foreign fishing offshore, which remains unaddressed owing to insufficient monitoring efforts (De Young, 2006). Fisheries hold significant importance in Bangladesh, with catches exceeding 13.5 million tonnes from 1950 to 2010 (Ullah et al., 2014), because many individuals rely on this method to sustain their families. However, species with higher commercial value are not targeted for subsistence fishing; for instance, the low-value Bombay duck is the predominant subsistence species, accounting for over 12% of the total catch.

The coastal artisanal fishery of the Sundarbans is characterized by a diverse range of traditional, labor-intensive, low-technology, and low-investment crafts and gears (Tables 2.3 and 2.4). Fishers use various fishing crafts and gears, many unchanged for centuries, whereas others have been modified in response to changes in target species, environmental conditions, and socioeconomic factors. The fishing area, target species, economic conditions, and regulatory issues influence the selection of crafts and gears. Fishers may adapt or change their methods based on market competition and fish availability, with a rising trend toward modernization and

TABLE 2.4
Fishing Crafts Used in Sundarbans Small-Scale Fishery

Craft Name	Dimensions (m)	Material	Bottom	Crews (no.)	Carrying Capacity (ton)	Life Span (years)	Main Gear	Ecological Impact
Nonmechanized								
Kosha Nauka	L: 8–10, W: 2–4, H: 1–2	Wood, Bamboo	Flat	2–3	4–5	4–5	Monofilament gill net, Cast net	Minimal impact due to low catch capacity and limited range. However, overfishing of nearshore species is possible.
Kosha Nauka	L: 6–7, W: 2–3, H: 1–2	Wood, Bamboo	Flat	3–5	3–5	3–5	Chandi Jal	Low to moderate impact. Traditional nets may unintentionally catch juveniles and nontarget species.
Dinghi Nauka	L: 6–7, W: 2–3, H: 1–2	Wood, Bamboo, Iron	Rounded	4	0.6–0.8	4–5	Fixed purse net, Behundi Jal, Bata Jal	Moderate impact due to increased catch from mechanization. Some gears, like the purse net, may cause habitat damage.
Dinghi Nauka	L: 5–6, W: 2–3, H: 1–2	Wood, Bamboo	Rounded	5–6	4–5	4–5	ESBN, Chandi Jal	Low to moderate impact. Limited by small scale, but overuse of specific areas could harm local populations.
Dinghi Nauka	L: 5–8, W: 2–2.5, H: 2	Wood, Bamboo	Rounded	3–6	3–5	3–5	Chandi Jal	Moderate ecological pressure in areas with high fishing activity. Nonselective nets may catch nontarget species.

Mechanized Trawler	L: 16–18, W: 6–8, H: 3	Wood	Rounded	15	150–200	20	Gillnet, Fixed purse net (Behundi Jal, Bata Jal)	High ecological impact due to large catch size and potential for overfishing. Can also cause habitat disruption, especially with trawling gear. Noise pollution and high carbon emission.
Tempu Nauka	L: 8–10, W: 2–4, H: 2–3	Wood	Flat/Rounded	5–8	5–6.5	4–5	Gillnet, Fixed purse (Behundi Jal, Bata Jal)	Moderate to high impact. Mechanized boats increase catch, contributing to overfishing and habitat degradation in some areas. Noise pollution and high carbon emission.
Choto Tempu	L: 8–10, W: 2–3, H: 2–3	Wood, Iron	Rounded	5–8	4–5	4–5	Estuarine Set Bag Net (ESBN), Chandi Jal	Moderate impact. Selective pressure on certain species. Potential bycatch and habitat disruption. Noise pollution and high carbon emission.
Chandi Nauka	L: 10–15, W: 5–8, H: 2–4	Wood, Iron	Rounded	8–10	6–10	8–10	ESBN, Chandi Jal	High impact due to increased mechanization. Significant pressure on fish stocks and potential for bycatch of nontarget species. Noise pollution and high carbon emission.
Balam Nauka	L: 10–15, W: 5–8, H: 2–4	Wood, Iron	Rounded	10–15	8–10	8–10	ESBN, Chandi Jal	High impact. Larger size and higher capacity can lead to overfishing and more bycatch, affecting the marine ecosystem. Noise pollution and high carbon emission.

L-Length, W-Width, H-Height
(Modified after Mustafa et al., 2023)

mechanization. The crafts and gears used have significant ecological and socio-economic impacts. The Department of Fisheries (DoF) identifies the most significant artisanal fisheries, including mechanized gillnet, pots and traps, and estuarine set bag net (ESBN) fisheries. According to model projections, catch increases in Bangladesh are caused by increased fishing pressure brought on by an increase in the coastal population rather than an increase in marine productivity (Fernandes et al., 2015). It has been estimated that Hilsa shad is fished at a rate that is two to three times the maximum sustainable yield (MSY).[5]

The Sundarbans' small-scale fisheries are projected to yield an annual total catch of 1318.60 kg per boat, with hook-line gear contributing the highest annual catch of 172.80 kg per boat. The economic assessment indicates that the average income per fishing expedition ranges from BDT 1307 to 1562, with the hook-line gear exhibiting the greatest financial return (Shaha et al., 2014). In the GBM delta area, over 68% of all boats are nonmechanized, the majority being less than 20 m in total length. The primary gear types used in artisanal vessels are driftnets, boat seines, and gillnets, and they include plank-built boats and catamarans. The primary functions of mechanized boats include gillnetting, long-lining, purse-seining, and trawling (FAO, 2006b). Approximately 67% of the total fish produced in both the countries is consumed fresh; almost 6% is allocated for fish meal manufacturing; the remaining 27% is exported (Planning Commission, 2010). Understanding the evolution, diversity, and reasons behind adopting specific devices is essential for effective fisheries management, conservation, and development strategies in the Sundarbans (Mustafa et al., 2023). Although the diversity of fishing crafts and gears (Tables 2.4 and 2.5) enhances the catch potential, it raises concerns about sustainability and stock vulnerability, necessitating careful management to balance economic needs with ecological health.

2.8 CONCLUSION AND THE ROADMAP AHEAD

The shared Sundarbans represents India and Bangladesh's most complex ecological and sociopolitical landscapes and signifies a substantial opportunity for strategic, cross-cultural cooperation for poverty reduction and sustainable development of fish resources (IWA Annual Report 2019-2020). The Sundarbans' economy is based mostly on small-scale fishing and fish processing (Chandra & Sagar, 2003). Small-scale fisheries in this region comprise around 60% of the coastal fishery in eastern India and 80–90% of the total marine fisheries production in Bangladesh (DISHA, 2006).

However, over the past two decades, the small-scale fishery of this region has been facing challenges due to climate change and anthropological and sociopolitical interruptions (Chandra & Sagar, 2003). The Sundarbans ecosystem is often divided into two distinct ecosystems by the international boundary. A growing body of literature has emerged examining various facets of the small-scale fisheries in the Sundarbans, the world's largest mangrove ecosystem, straddling both Bangladesh and India. For instance, Hoq (2007) investigated fisheries exploitation and management practices in the Bangladeshi Sundarbans, offering insights into resource use patterns and institutional gaps. Salagrama (2012) focused on the perceptions of small-scale fishers in the Indian Sundarbans considering climate change, emphasizing the community's vulnerability to environmental variability. Chacraverti (2014) further explored the unstable livelihoods of Sundarbans fishers inhabiting a stressed mangrove estuarine system, highlighting structural disadvantages and ecological risks and documented traditional

TABLE 2.5
Different Nets Used by Small-Scale Fishers of the Sundarbans

Net Type	Description	Dimensions	Target Species	Mesh Size	CPUE (Catch Per Unit Effort)	Area Used	Fishing Season	Ecological Impact
Bag Net (Been Jal)	Conical-shaped net with a bag at the end, fixed in estuarine tidal zones		Juvenile shrimp, small fish	Small (varying)		Tidal rivers and estuaries	Premonsoon, postmonsoon	High bycatch of juvenile species, can disrupt breeding grounds of shrimp and fish, though limited to tidal areas.
Drag Net (Ber Jal)	Rectangular net dragged through shallow waters by a group of fishers	500–700 × 9–12	Hilsa, small carps, prawn	Varies 0.19–22	40–300	Shallow rivers, estuaries	Throughout the year	Disturbs benthic communities, can lead to overfishing of target species.
Seine Net (Current Jal)	Long rectangular net with small mesh, hauled or set in a circle around fish schools		Hilsa, smaller marine and estuarine fish	Fine mesh		Rivers and estuaries	Monsoon, postmonsoon	High potential for overfishing, captures juvenile fish; alters fish migration patterns and contributes to habitat degradation.
Gill Net (Fash Jal)	Net with varying mesh sizes, fish caught by their gills		Large carps, hilsa	Varying mesh sizes		Rivers, inlets	Postmonsoon, winter	Moderate ecological impact as selective fishing is possible; may still catch nontarget species such as juvenile fish.
Push Net (Thela Jal)	Manually pushed triangular net mounted on bamboo frames, used in shallow waters	Arms is 2–3 m and front is 1–1.5 m	Prawns, crabs, small fish	Fine mesh 0.2–6		Shallow waters, mudflats	Postmonsoon	High bycatch rate, disturbs shallow water ecosystems and bottom-dwelling species.

(Continued)

Table 2.5 (Continued)

Net Type	Description	Dimensions	Target Species	Mesh Size	CPUE (Catch Per Unit Effort)	Area Used	Fishing Season	Ecological Impact
Cast Net (Jhaki Jal)	Circular net thrown manually into shallow waters, designed to sink quickly	3.7–6 m(L) × 6.1–12.2 m (mouth diameter)	Small fish, prawns	0.2–0.6,0.5–3,0.625–1.25	1–2 or 1–5	Shallow rivers, inlets	Throughout the year	Less ecological impact due to manual, small-scale use, but may still catch nontarget juvenile species.
Longline Fishing	Baited hooks attached to a long line, set at different depths		Large fish like catfish, sharks	No net, only hooks	4	Deep river channels, estuaries	Monsoon, winter	Minimal ecological disturbance; risks catching endangered species such as sharks.
Stake Net (Khata Jal)	Fixed net attached to poles in tidal areas, traps fish moving with tides		Shrimps, smaller fish	Fine mesh		Tidal zones, river mouths	Premonsoon, postmonsoon	Traps nontarget species and juveniles, potentially causing population imbalance and habitat alteration in sensitive tidal zones.
Chandi Jal	Drag net used from river banks; rectangular in shape	$650–800 \times 10–14 m^2$	Mainly Hilsa, Poa, Taposhi	Small 1.57–4.5	8–150	Riverbanks	Year-round	Causes disturbance in benthic habitats, affects juvenile fish populations.
Bata Jal	Gill net mainly for catching bata fish; rectangular in shape	$6.5–33 \times 1.25–2\ m^2$	Bata or medium to large fish species	Medium 0.98–1.97	0.1–10	Rivers and estuaries	Monsoon	mesh allows juveniles of other fishes to escape but can cause overfishing if used excessively.
Coral Jal	Circular fishing net for small-scale coastal fisheries	$60–70 \times 150–200\ m^2$ $150–200 \times 10–15\ m^2$	Small coastal fishes	Small 7		Coastal areas	Postmonsoon to winter	Limited ecological impact due to selective fishing of small coastal species.

Poa Jal	Set gill net; rectangular in shape	60–70 × 3.6–4.5 m^2 600–700 × 3.5–4.5 m^2 200 feet × 3.66–4.57 feet	Medium to large fishes	Medium 1.38	5–100	Rivers and creeks	Postmonsoon	Causes bycatch of nontarget species, disrupts local fish populations.
Ghongra Jal	Specially designed for shrimp fishing; rectangular in shape		Shrimp, prawns	Fine mesh 10–15	100–200	Estuarine areas	Year-round	High bycatch of juvenile fish and nontarget species, affecting biodiversity.
Ilish Jal	Specific rectangular net for targeting hilsa during migration periods	35–40 × 12–150 m^2 1250–2500 × 10–25 m^2	Hilsa	Medium 4–6.5 3–3.5	300–400 250–300	Rivers and estuaries	Monsoon	Affects hilsa population due to overfishing during breeding season.
Lata Jal	Gill nets rectangular in shape used for general fish species		Various species of fish	Medium 0.5–1	50–100	Rivers	Winter	Minimal impact, depends on the size of the catch.
Duba Jal	Submerged gill net rectangular in shape used in deeper water	100–500 × 15–20m^2	Larger river fish species	Large 5–6		Deeper river sections	Year-round	Potential disruption of migration routes for fish species.
Moshari Jal	Fine-mesh net triangular in shape primarily for shrimp fishing	8–10 × 3–4 m^2	Shrimp, prawns	Very fine mesh 0.1–0.5	8–20	Shallow estuarine zones	Postmonsoon to winter	High bycatch rates affecting juvenile fish populations.
Bagdhara Jal	Bag net used for larger marine and estuarine species; triangular in shape	Width 0.1–0.5	Large estuarine fish species	Large 2–3		Estuarine and coastal	Year-round	Causes habitat disruption in estuarine and coastal areas, high impact on nontarget species.

(Continued)

Table 2.5 (Continued)

Net Type	Description	Dimensions	Target Species	Mesh Size	CPUE (Catch Per Unit Effort)	Area Used	Fishing Season	Ecological Impact
Atto Jal	Fixed gill net used in narrow river channels; triangular in shape	Arms in 2–3 m front is 1–1.5m	Mainly prawn	Medium	0.2–1	River channels	Year-round	Causes blockage in natural fish migration routes, affecting breeding patterns of migratory species.
Behundi Jal	Large fixed bag net used in estuarine and coastal fisheries	45–55 × 16–20m^2 12–15 × 11–12 m^2 15.24–60.96 × 18.79–47.72 m^2 8.5–41 × 2–7 m^2 12–15 × 11–12 m^2	All types of small fishes Large estuarine fish, shrimp, prawns	Medium	2–2.5 and 0.2–0.5 1.97–2.46 and 0.2–0.5 0.5–1, mouth 5.51–0.79 and cod end 8.7–0.2 1.97–2.46 and 0.19–0.49	Coastal, estuarine	Year-round	High ecological impact due to high bycatch rates, overexploitation of both target and nontarget species.
Marine Set Bag Net	Large-scale bag conical net used in marine areas to target multiple fish species	18–40,10–23 at mouth, 0.47–0.98 cod end 20–55,10–14 at mouth, 0.60–1.38 cod end 8–40,5–30 at mouth, 0.47–0.98 cod end	Mixed fish species, high-value commercial species (Ribbon fish, silver pomfret, Bombay duck, anchovy, rainbow shrimp Bombay Duck, shrimp, Lates	Large	9–35,60	Marine zones	Year-round	Significant impact on marine biodiversity, large bycatch, affects benthic communities, disrupts food chains.

knowledge systems and practices within the Sundarbans Tiger Reserve on the Indian side. In a series of studies, Hoque et al., 2018 examined the socioecological dynamics and economic characteristics of small-scale fisheries in Bangladesh.

Although these studies provide valuable insights into specific issues—ranging from governance and ecological stress to climate vulnerability and livelihood diversity—they remain largely compartmentalized in terms of disciplinary focus, geographic scope, or thematic emphasis. There exists a critical gap in understanding how these fragmented insights can be synthesized into a broader, more integrated narrative that captures the cross-border socioecological realities of the region's small-scale fisheries. In addition, existing literature rarely compares the institutional, ecological, and cultural dynamics of both the Indian and Bangladeshi Sundarbans in a comparative framework, thereby limiting the potential for shared learning and transboundary policy reflection. In the chapters that follow, we extend the transboundary lens to unravel the complexities of socioecological resilience and adaptive capacity in the small-scale fisheries of the Sundarbans bridging the theoretical insights and applied realities. The chapters address a specific dimension of our threefold objective:

1. To present a holistic analysis of the socioecological trends—spanning ecological degradation, livelihood adaptation, policy transitions, and climate-related challenges—that shape the small-scale fishery systems across both territories;
2. To document the transformation of the socioeconomic system over time, capturing critical inflection points in governance, livelihood structure, and demographic shifts; and
3. To identify and analyze areas of convergence (agreement) and divergence (incongruity) in adaptive responses, thereby drawing lessons on what has worked, what has not, and what opportunities lie ahead for encourgaing sustainability and resilience in the Sundarbans' small-scale fishery landscape comparing the small-scale fisheries across the globe.

NOTES

1 The interdependent processes of nutrient cycling and downstream transport.
2 Impoundments disrupt a river's longitudinal patterns, shifting biotic and abiotic processes and resources downstream, creating discontinuities in ecological gradients.
3 One-way flow of energy and nutrients along the river's course, from source to sea.
4 Highlights the importance of local organic production (autochthonous production) and direct inputs from the riparian zone (leaves, organic carbon) as significant sources of energy and material in large rivers, particularly those with constricted channels.
5 The largest catch or yield that can be taken from a species' stock over an indefinite period without causing the population to decline.

BIBLIOGRAPHY

Acharya, & Kamal, D. (1994). Fisheries. In Z. Hussain & G. Acharya (Eds.), *Mangrove of the Sundarbans* (Vol. 2, pp. 101–140). Bangladesh: IUCN.

Agarwal, S., Fazli, P., Zaman, S., Pramanick, P., & Mitra, A. (2019). Seasonal variability of acidification in major estuaries of Indian Sundarbans. *Global Journal of Engineering Science Research*, *6*(4), 493–498.

Aipira, C., Kidd, A., & Morioka, K. (2017). Climate change adaptation in pacific countries: Fostering resilience through gender equality. In W. Leal Filho (Ed.), *Climate change adaptation in Pacific countries: Fostering resilience and improving the quality of life* (pp. 225–239). Springer.

Akhter, M., Reza, M. S., Jamil, N. A. H. M., & Uddin, M. N. (2018). Assessment water quality and seasonal variations based on aquatic biodiversity of Sundarbans mangrove forest, Bangladesh. *IOSR Journal of Biotechnology and Biochemistry*, *4*(1), 6–15.

Allison, E. H., Perry, A. L., Badjeck, M. C., Adger, W. N., Brown, K., Conway, D., Halls, A. S., Pilling, G. M., Reynolds, J. D., Andrew, N. L., & Dulvy, N. K. (2009). Vulnerability of national economies to the impacts of climate change on fisheries. *Fish and Fisheries*, *10*(2), 173–196. https://doi.org/10.1111/j.1467-2979.2008.00310.x

Allison, M. A., Khan, S. R., Goodbred, S. L., Jr., & Kuehl, S. A. (2019). Stratigraphic evaluation of the late Holocene Ganges-Brahmaputra lower delta plain. *Sedimentary Geology*, *155*(2), 317–342.

Alongi, D. M. (2002). Present state and future of the world's mangrove forests. *Environmental Conservation*, *29*(3), 331–349.

Annandale, N. (1922). The marine elements in the fauna of Ganges. *Bijdragen tot de Dierkunde*, *22*, 143–154.

Annual report 2010–11. Governement of India. Planning Commission

Arrington, D. A., & Winemiller, K. O. (2004). Organization and maintenance of fish diversity in shallow waters of tropical floodplain rivers. In *Proceedings of the second international symposium on the management of large rivers for fisheries* (Vol. 2, pp. 25–36). FAO Regional Office for Asia and the Pacific Bangkok, Thailand.

Auerbach, L. W., Goodbred, S. L., Jr., Mondal, D. R., Wilson, C. A., Ahmed, K. R., Roy, K., & Ackerly, B. A. (2015). Flood risk of natural and embanked landscapes on the Ganges Brahmaputra tidal delta plain. *Nature Climate Change*, *5*(2), 153–157.

Azad, A. K., Jensen, K. R., & Lin, C. K. (2019). Coastal aquaculture development in Bangladesh: Unsustainable and sustainable experiences. *Environmental Management*, *44*, 800–809.

Aziz, A., Rahman, M., & Ahmed, A. (2022). Diversity, distribution and density of estuarine phytoplankton in the Sundarban mangrove forests. *Bangladesh Journal of Botany*, *51*(2), 87–95.

Bain, R. L., Hale, R. P., & Goodbred, S. L. (2019). Flow reorganization in an anthropogenically modified tidal channel network: An example from the southwestern Ganges-Brahmaputra-Meghna Delta. *Journal of Geophysical Research: Earth Surface*, *124*, 2141–2159. https://doi.org/10.1029/2018JF004996

Bandyopadhyay, S. (2019). *Sundarban: A review of evolution and geomorphology*. World Bank Group.

Bandyopadhyay, S., Dasgupta, S., Khan, Z. H., & Wheeler, D. (2021). Spatiotemporal analysis of tropical cyclone landfalls in Northern Bay of Bengal, India and Bangladesh. *Asia-Pacific Journal of Atmospheric Sciences*, *57*(4), 799–815. https://doi.org/10.1007/s13143-021-00227-4

Barua, P., Chowdhury, M. S. N., & Sarker, S. (2010). Climate change and its risk of mangrove ecosystem. *Bangladesh Journal of Research and Publication*, *20*(2), 45–65.

Batista, V. S., Fabré, N. N., Malhado, A. C., & Ladle, R. J. (2014). Tropical artisanal coastal fisheries: Challenges and future directions. *Reviews in Fisheries Science & Aquaculture*, *22*(1), 1–15.

Becker, M., Papa, F., Karpytchev, M., Delebecque, C., Krien, Y., Khan, J. U., Ballu, V., Durand, F., Cozannet, G. L., Islam, A. K. M. S., Calmet, S., & Shum, C. K. (2020). Water level changes, subsidence and sea level rise in the Ganges-Brahmaputra-Meghna delta. *Proceedings of the National Academy of Sciences*, *117*(4), 1867–1876.

Bergstrand, K., Mayer, B., Brumback, B., & Zhang, Y. (2015). Assessing the relationship between social vulnerability and community resilience to hazards. *Social Indicators Research*, *122*, 391–409.

Bernacsek, G., & Haque, E. (2001). *Fishing gears of the Sundarbans (draft): Internal notes.* Sundarbans Biodiversity Conservation Project, Aquatic Resources Program, Ministry of Environment and Forests.

Bernacsek, G. M. (2001). *Guide to the finfishes of Bangladesh Sundarbans* [Technical Report].

Biplab, S., Mahmudul, M. A., Islam, H. M., Lubna, A., Khan, K. U. A., &Mazlin, M. (2015). Assessment of external supports for harvesting artisanal Galda (Macrobrachium rosenbergii De man) fisheries from the Sundarbans mangrove ecosystem in Bangladesh. *Social Science ResearchNetwork*, *8*(1), 1–23.

Bomer, E. J., Wilson, C. A., Hale, R. P., Hossain, A. N. M., & Rahman, F. A. (2020). Surface elevation and sedimentation dynamics in the Ganges-Brahmaputra tidal delta plain, Bangladesh: Evidence for mangrove adaptation to human-induced tidal amplification. *Catena*, *187*, 104312.

Bose, T., Baguli, T. K., Ghosh, A. K., & Naskar, K. R. (1999). *A review of shell fish and fin fish species from the brackish water fisheries of the Sundarbans and their impact on the mangrove.*

Cantor, T. (1849). *Catalogue of Malayan fishes* (Vol. *12*). Baptist Mission Press.

Chacraverti, S. (2014). *The Sundarbans fishers: Coping in an overly stressed mangrove estuary*. International Collective in Support of Fishworkers.

Chakraborty, A., & Satpati, L. N. (2019). Vulnerability, resilience and quality of life: A micro-level study of Ghoramara Island in the Sundarban region of West Bengal, India. In B. R. K. Sinha (Ed.), *Multidimensional approach to quality of life issues* (pp. 241–252). Springer Nature.

Chakraborty, S. K. (2011). Mangrove ecosystem of Sundarbans, India: Biodiversity, ecology, threats and conservation. In J. N. Metras (Ed.), *Mangroves: Ecology, biology and taxonomy* (pp. 83–112). USA: NOVA Publisher.

Chandra, G., & Sagar, R. (2003). Fisheries in Sundarbans: Problems and prospects. *Journal of Fisheries*, *20*(2), 56–65.

Chantarasri, S. (1994). Fisheries resources management for the Sundarbans reserved forest. In *Integrated resource development of the Sundarbans reserved forest, Bangladesh (4th draft final report)*. FAO/UNDP.

Chapman, V. J. (1976). *Mangrove vegetation*.

Chatterjee, M., Shankar, D., Sen, G. K., Sanyal, P., Sundar, D., Michael, G. S., Chatterjee, A., Amol, P., Mukherjee, D., Suprit, K., Mukherjee, A., Vijith, V., Chatterjee, S., Basu, A., Das, M., Chakraborti, S., Kalla, A., Misra, S. K., Mukhopadhyay, S., & Sarkar, K. (2013). Tidal variations in the Sundarbans estuarine system, India. *Journal of Earth System Science*, *122*(4), 899–933.

Chaudhuri, A. B., & Choudhury, A. (1994). *Mangroves of the Sundarbans. Volume 1: India*. World Conservation Union.

Chaudhuri, P., Chaudhuri, S., & Ghosh, R. (2019). The role of mangroves in coastal and estuarine sedimentary accretion in Southeast Asia. In B. Dericks (Ed.), *Sedimentary processes-examples from Asia, Turkey and Nigeria*. IntechOpen.

Chowdhury, S. R., Hossain, M. S., Shamsuddoha, M., & Khan, S. M. M. H. (2012). *Coastal fishers' livelihood in peril: Sea surface temperature and tropical cyclones in Bangladesh*. CPRD.

Das, M. K. (1999). Fish and prawn disease epizoobiotics in aquatic ecosystems around Sundarban – An overview. In D. N. Guha Bakshi, P. Sanyal, & K. R. Naskar (Eds.), *Sundarbans Mangal* (pp. 485–492). Naya Prokash.

Dasgupta, H., Mustafa, S., & Huqu, H. (2021). Impact of climate change and aquatic salinization on fish habitats and poor communities in southwest coastal Bangladesh and Bangladesh Sundarbans. *Disaster*, *50*(3), 45–65.

Dasgupta, S., Wheeler, D., Sobhan, M. I., Bandyopadhyay, S., Nishat, A., & Paul, T. (2020). *Coping with climate change in the Sundarbans: Lessons from multidisciplinary studies*. World Bank Group. https://doi.org/10.1596/978-1-4648-1587-4

de Miranda Ribeiro, A. (1915). *Fauna brasiliense. Peixes. Tomo V. [Eleutherobranchios aspirophoros]. Physoclisti. Arquivos do Museu Nacional de Rio de Janeiro, 17*, 1–679.

De Young, C. (ed.). (2006). Review of the state of world marine capture fisheries management: Indian Ocean. *FAO Fisheries Technical Paper*. No. 488. 458 p. Rome: FAO.

Debnath, H., (2001). Sundarbans biosphere reserve (data collection and identification of gaps). State of art report, biosphere reserve information service (BRIS), MOEF project ecology. In D. N. Guha Bakshi, P. Sanyal, & K. R. Naskar (Eds.), *Sundarbans Mangal. Naya Prakash, Calcutta* (pp. 539–547). New Delhi: Concept Publishing Company.

Debnath, H. S., & Naskar, K. R. (1999). A comparative study on the mangroves and associated flora in the ganga delta (Sunarbans) and Bay Islands (Andaman and Nicobar). In D. N. Guha Bakshi, P. Sanyal, & K. R. Naskar (Eds.), *Sundarbans Mangal* (pp. 277–292). Naya Prokash.

Deshmukh, S. V. (1991). Mangroves of India – Status report. In S. V. Deshmukh & R. Mahalingam (Eds.), *A global network of mangrove genetic resource centers, project formulation workshop* (pp. 15–25). M. S. Swaminathan Research Foundation.

DISHA. (2006). *Study on corporate abuse in Sundarban*. Society for Direct Initiative for Social & Health Action.

Doretto, A., Piano, E., & Larson, C. E. (2020). The river continuum concept: Lessons from the past and perspectives for the future. *Canadian Journal of Fisheries and Aquatic Sciences, 77*(11), 1853–1864.

Elwood, J. W., Newbold, J. D., O'Neill, R. V., & Van Winkle, W. (1983). Resource spiraling: An operational paradigm for analyzing lotic ecosystems. In T. D. Fontaine & S. M. Bartell (Eds.), *Dynamics of lotic ecosystems* (pp. 3–27). MI: Ann Arbor Science.

Ericson, J. P., et al. (2006). Effective Sea-level rise and deltas: Causes of change and human dimension implications. *Global and Planetary Change, 50*, 63–82. http://dx.doi.org/10.1016/j.gloplacha.2005.07.004

Fanchette, S. (2022). Delta vulnerability linked to climate and environmental changes. *Middle Eastern Cities in a Time of Climate Crisis, 91*.

Fergusson, J. (1863). On recent changes in the delta of the Ganges. *The Quarterly Journal of the Geological Society of London, 19*, 321–354.

Fernandes, J. A., Kay, S., Hossain, M. A., Ahmed, M., Cheung, W. W., & Lazar, A. N. (2015). Projecting marine fish production and catch potential in Bangladesh in the 21st century under long-term environmental change and management scenarios. *ICES Journal of Marine Science, 73*(5), 1357–1369.

Food and Agriculture Organization of the United Nations (2006a). Rome.

Food and Agriculture Organization of the United Nations. *Food aid for food security*. Rome (2006b).

Foufoula-Georgiou, E., Syvitski, J., Paola, C., Hoanh, C. T., Tuong, P., Vörösmarty, C., . . . Twilley, R. (2011). International year of deltas 2013: A proposal. *Eos, Transactions American Geophysical Union, 92*(40), 340–341.

Froese, R., Demirel, N., Coro, G., Kleisner, K. M., & Winker, H. (2017). Estimating fisheries reference points from catch and resilience. *Fish and Fisheries, 18*(3), 506–526.

Froese, R., &Pauly, D. (Eds.) (2025). World wide web electronic publication. www.fishbase.org (accessed 04/2025).

Gain, H., & Uddin, H. (2018). Dynamics of Sundarbans estuarine ecosystem: Eutrophication induced threat to mangroves. *Saline Systems, 68*(2), 1–16.

Ganguly, D., Mukhopadhyay, A., Pandey, R. K., & Mitra, D. (2006). Geomorphological study of Sundarban deltaic estuary. *Journal of the Indian Society of Remote Sensing, 34*(4), 431–435.

Ghosh, A. (2005). *The hungry tide*. Houghton Mifflin Harcourt.

Ghosh, P., Atreya, P., & Panigrahi, A. (2022). An overview on fish assemblages in riverine systems of southern Bengal, West Bengal, India: Diversity, distribution and potential threats. *Applied Ecology and Environmental Sciences, 10*(7), 461–471.

Gopal, B. (2005). Does inland aquatic biodiversity have a future in Asian developing countries? *Hydrobiologia, 542*, 69–75.

Gopal, B. (2006). Wetlands and biodiversity. In E. Maltby (Ed.), *The wetlands handbook*. Blackwell Science.

Gopal, B., & Chauhan, M. (2005). Biodiversity and its conservation in the Sundarban mangrove ecosystem. *Aquatic Sciences, 68*(2), 338–354.

Gopal, B., & Chauhan, M. (2020). Biodiversity and its conservation in the Sundarban mangrove ecosystem. *Aquatic Sciences, 69*(2), 338–354.

Gopal, B., & Junk, W. J. (2000). Biodiversity in wetlands: An introduction. In B. Gopal, W. J. Junk, & J. A. Davis (Eds.), *Biodiversity in wetlands: Assessment, function and conservation* (Vol. 1, pp. 1–10). Backhuys Publishers.

Gopal, B., & Junk, W. J. (2001). Assessment, determinants, function and conservation of biodiversity in wetlands: Present status and future needs. In B. Gopal, W. J. Junk, & J. A. Davis (Eds.), *Biodiversity in wetlands: Assessment, function and conservation* (Vol. 2, pp. 277–302). Backhuys Publishers.

Gopalakrishnan, V. & Rao, R. M. (1968). Observations on the distribution of juveniles of brackishwater prawns in the Hooghly-Matlah estuarine systems. In *Proceedings of the seminar on production of quality fish seed for fish culture*, sponsored by ICAR at the CIFRI, Barrackpore on Nov. 1 & 2, 1968, pp. 248–261.

Gopalakrishnan, V., Ray, P., & Ghosh, B. B. (1973). Present status of pollution in the Hooghly estuary with special reference to the adverse effects observed on the fishery resources. *Proceedings of the Symposium on Environmental Pollution, 1973*, 1–8.

Gopalkrishnan, V. (1971). The biology of the Hooghly–Matlah estuarine system (West Bengal-India) with special reference to its fisheries. *Journal of the Marine Biological Association of India ,13* (2), 182–194.

Gupta, N., Pandey, P., & Hussain, J. (2017). *Effect of physicochemical and biological parameters on the reduction by mangrove ecosystem of Bangladesh*. Bangladesh Research.

Habib, K. A., Neogi, A. K., Oh, J., Lee, Y. H., & Kim, C. G. (2019). Chelonodontops bengalensis (Tetraodontiformes: Tetraodontidae): A new species of puffer fish from the northern bay of Bengal based on morphology and DNA barcode. *Ocean Science Journal, 54*, 1–8.

Habib, K. A., Neogi, A. K., Oh, J., Lee, Y. H., & Kim, C. G. (2020). An overview of fishes of the Sundarbans, Bangladesh and their present conservation status. *Journal of Threatened Taxa, 12*(1), 15154–15172.

Habib, M. G. (1989). Wildlife management of the Sundarban – A case study. In G. M. M. E. Karim, A. W. Akonda, & P. Sewitz (Eds.), *Conservation of wildlife in Bangladesh* (pp. 161–168). German Cultural Institute, Forest Department, Dhaka University, Wildlife Society of Bangladesh, UNESCO.

Hale, R., Bain, R., Goodbred, S., Jr., & Best, J. (2019). Observations and scaling of tidal mass transport across the lower Ganges-Brahmaputra delta plain: Implications for delta management and sustainability. *Earth Surface Dynamics, 7*(1), 231–245.

Halls, A. S., Hoggarth, D. D., & Debnath, K. (1998). Impact of flood control schemes on river fish migrations and species assemblages in Bangladesh. *Journal of Fish Biology, 53*, 358–380.

Hamilton, F. (1822). *An account of the fishes found in the river Ganges and its branches*. Edinburgh & London: Archibald Constable and Company; Hurst, Robinson, and Co.

Hamilton, L., & Snedaker, S. (1984). *Handbook of mangrove area management*. East West Center.

Haque, M. E. (2003). How fishers' endeavors and information help in managing the fisheries resources of the Sundarban mangrove forest of Bangladesh. *University of British Columbia, Fisheries Center Research Report, 11*(1), 433–438.

Hazra, S., Ghosh, T., Das Gupta, R., & Sen, G. (2002). Sea level and associated changes in the Sundarbans. *Science C, 68*, 309–321.

Hazra, S., Ghosh, T., DasGupta, R., & Sen, G. (2002). Sea level and associated changes in the Sundarbans. *Science and Culture*, *68*(9–12), 309–321.

Hemprich, F.W. & Ehrenberg, C.G. (1899). *Symbolae physicae, seu, Icones et descriptiones corporum naturalium novorum aut minus cognitorum: quae ex itineribus per Libyam, Aegyptum, Nubiam, Dongolam, Syriam*, Zoologica, Berlin: *Arabiam et Habessiniam.*

Hilmi, N., Arruda, G., Broussard, D., Benitez, B. M., Sauron, L., Lamaud, T., ... Spencer, J. M. H. (2025). Blue carbon as a nature-based climate mitigation strategy for mangrove conservation in Bangladesh. *Journal for Nature Conservation*, *126885*.

Hoq, M. E. (2007). *Sundarbans mangrove: Fish & fisheries: Ecology, resources, productivity and management perspectives*. Graphic Media.

Hoq, M. E., Haroon, A. Y., & Chakraborty, S. C. (2013). *Marine fisheries of Bangladesh: Prospect and potentialities*. Support to Sustainable Management of the BOBLME Project, Bangladesh Fisheries Research Institute.

Hoq, M. E., Islam, M. N., Kamal, M., & Wahab, M. A. (2001). Abundance and seasonal distribution of Penaeus monodon postlarvae in the Sundarban mangrove, Bangladesh. *Hydrobiologia*, *457*, 97–104.

Hoq, M. E., Wahab, M. A., & Islam, M. N. (2006). Hydrographic status of Sundarbans mangrove, Bangladesh with special reference to post-larvae and juvenile fish and shrimp abundance. *Wetlands Ecology and Management*, *14*, 49–93.

Hoque, M. Z., Alam, M. N., & Nahid, K. (2018). Health consciousness and its effect on perceived knowledge, and belief in the purchase intent of liquid milk: Consumer insights from an emerging market. *Food*, *7*(9), 150.

Hoque, M. Z., Bashar, M. A., & Akhter, F. (2021). Ambiguity tolerance and confusion avoidance in the intent to purchase farmed fish. *Journal of Applied Aquaculture*, *50*(3), 45–65.

Hoque Mozumder, M. M., Shamsuzzaman, M. M., Rashed-Un-Nabi, M., & Karim, E. (2017). Social-ecological dynamics of the small scale fisheries in Sundarban mangrove Forest, Bangladesh. *Aquaculture and Fisheries*, *3*(1), 38–49. https://doi.org/10.1016/j.aaf.2017.12.002

Hora, S. L. (1934). Brackish water animals of the Gangetic delta. *Current Science*, *2*, 426–427. http://www.wetlands.org/reports/ris/2BD001en.pdf

Hossain, M. S., Chowdhury, S. R., Sharifuzzaman, S. M., & Sarker, S. (2015). *Vulnerability of the bay of Bengal to ocean acidification*. IUCN, International Union for Conservation of Nature, Bangladesh Country Office.

Hossain, N. I., Rahman, S. M., Hossain, M. A., Hafiz, K. B., Sowrav, S. F. F., Masud-Ul-Alam, M., & Choudhury, T. R. (2025). Decoding phytoplankton quality in the waterways of the Sundarbans mangrove forest of Bangladesh. *Marine Pollution Bulletin*, *213*, 117669.

Hussain, Z., & Acharya, G. (1994). *Mangrove of the Sundarbans, volume two: Bangladesh*. IUCN Southeast Asia Regional Office.

Hutchison, J., Philipp, D. P., Claussen, J. E., Aburto-Oropeza, O., Carrasquilla-Henao, M., Castellanos-Galindo, G. A., ... Spalding, M. (2015). *Building an expert judgment-based model of mangrove fisheries*. American Fisheries Society.

Iftekhar, M. S., & Islam, M. R. (2004). Managing mangroves in Bangladesh: A strategy analysis. *Journal of Coastal Conservation*, *10*(1), 139–146.

Islam, M. M., & Shamsuddoha, M. (2018). Coastal and marine conservation strategy for Bangladesh in the context of achieving blue growth and sustainable development goals (SDGs). *Environmental Science & Policy*, *87*, 45–54.

Islam, M. S. (2003). Perspectives of the coastal and marine fisheries of the bay of Bengal. *Bangladesh. Ocean & Coastal Management*, *46*(8), 763–796.

IUCN. (1997). *Sundarban wildlife sanctuaries (Bangladesh): World Heritage nomination – IUCN technical evaluation.* http://hc.unesco.org/archive/advisory_body_evaluation/798.pdf

IUCN-BD. (2002). *Molluskan biodiversity of Sundarban – Final report.* IUCN-Bangladesh.

Jana, Zaman, S., Pramanick, P., Mukhopadhyay, N., Bose, R., Mitra, A., Ray Chaudhuri, T., & Raha, A. K. (2014). Signal of climate change through decadal variation of aquatic pH in Indian Sundarbans. *Journal of Marine Science Research and Development, S11*, 003.

Jennerjahn, T. C. (2012). Biogeochemical response of tropical coastal systems to present and past environmental change. *Earth-Science Reviews, 114*(1-2), 19–41.

Jhingran, V. G. (1982). *Fish and fisheries of India.*

Junk, W. J., Bayley, P. B., & Sparks, R. E. (1989). The flood pulse concept in river-floodplain systems. *Canadian Special Publication of Fisheries and Aquatic Sciences, 106*(1), 110–127.

Kemp, S. W. (1917). Notes on the fauna of the Matlah River in the Gangetic delta. *Records of the Indian Museum, 13*(4), 233–241.

Khan, E. (2011). *The Bangladesh Sundarbans.* Wildlife Trust of Bangladesh (WTB).

Khan, M. J. U., Durand, F., Emanuel, K., Krien, Y., Testut, L., & Islam, A. S. (2022). Storm surge hazard over Bengal delta: A probabilistic–deterministic modelling approach. *Natural Hazards and Earth System Sciences, 22*(7), 2359–2379.

Krien, Y., Karpytchev, M., Ballu, V., Becker, M., Grall, C., Goodbred, S., ... Khan, Z. (2019). Present-day subsidence in the Ganges-Brahmaputra-Meghna Delta: Eastern amplification of the Holocene sediment loading contribution. *Geophysical Research Letters, 46*(19), 10764–10772.

Krien, Y., Mayet, C., Testut, L., Durand, F., Tazkia, A. R., Islam, A. K. M. S., ... Ballu, V. (2016). Improved bathymetric dataset and tidal model for the northern bay of Bengal. *Marine Geodesy, 39*(6), 422–438.

Kristensen, E., Bouillon, S., Dittmar, T., & Marchand, C. (2008). Organic carbon dynamics in mangrove ecosystems: A review. *Aquatic Botany, 89*(2), 201–219.

Lauria, V., Das, I., Hazra, S., Cazcarro, I., Arto, I., Kay, S., ... Fernandes, J. A. (2018). Importance of fisheries for food security across three climate change vulnerable deltas. *Science of the Total Environment, 640*, 1566–1577.

MacFarlane, G. R., & Burchett, M. D. (2000). Cellular distribution of copper, lead and zinc in the grey mangrove, Avicennia marina (Forsk.) Vierh. *Aquatic Botany, 68*(1), 45–59.

Mandal, A. K. (2003). *The Sundarbans of India – A development analysis.* Indus Publishing Company.

Mandal, A. K., & Ghosh, R. K. (1989). *Sundarban: A socio bio-ecological study.*

Mandal, A. K., & Nandi, N. C. (1989). *Fauna of Sundarban mangrove ecosystem, West Bengal, India. Fauna of conservation areas: 3* (116 pp). Kolkata: Zoological Survey of India.

Mandal, B., Mukherjee, A., & Banerjee, S. (2013). A review on the ichthyofaunal diversity in mangrove based estuary of Sundarbans. *Reviews in Fish Biology and Fisheries, 23*(3), 365–374.

Mandal, R. N., & Naskar, K. R. (1994). Studies on periphytic algae on the aerial roots of the mangroves of Sindarbans in West Bengal. In M. Roy (Ed.), *Environmental pollution and impact of technology on life* (pp. 91–104). Viswa Bharati.

Marchand, C., Allenbach, M., & Lallier-Vergès, E. (2016). Relationships between heavy metals and organic matter in mangrove sediments: The case of Rhizophora stylosa forests in New Caledonia. *Chemosphere, 154*, 239–249.

Mitra, A. (2013). Climate change and plankton spectrum of mangrove ecosystem. In *Sensitivity of mangrove ecosystem to changing climate* (pp. 161–190). Springer India: New Delhi.

Mitra, A. (2019). Climate change and pollution. In A. Mitra (Ed.), *Estuarine pollution in the lower Gangetic Delta* (pp. 207–227). Springer Nature.

Mitra, A., Banerjee, K., & Sengupta, K. (2011). Impact of AILA, a tropical cyclone on salinity, pH and dissolved oxygen of an aquatic sub-system of Indian Sundarbans. *National Academy of Science Letters India, 81*(Part II), 198–205.

Mitra, A., Gangopadhyay, A., Dube, A., Schmidt, A. C., & Banerjee, K. (2009). Observed changes in water mass properties in the Indian Sundarbans (northwestern bay of Bengal) during 1980–2007. *Current Science*, 1445–1452.

Mitra, A., & Homchaudhuri, S. (2013). Trophic niche and nutritional profile of two mudskippers Periophthalmus novemradiatus and Boleophthalmus boddarti in their native estuarine habitat. *Science and Culture*, *79*, 248–251.

Mozumder, M., Uddin, M., Schneider, P., Islam, M., & Shamsuzzaman, M. (2018). Fisheries-based ecotourism in Bangladesh: Potentials and challenges. *Resources*, *7*(4), 61–75.

Mukherjee, P., Mitra, A., Zaman, S., & Mitra, A. (2021). Impact of climate-change-induced salinity alteration on Ichthyoplankton diversity of Indian Sundarbans. In *Handbook of climate change management: Research, leadership, transformation* (pp. 1–27). Cham: Springer International Publishing.

Mumby, P. J., Edwards, A. J., Ernesto Arias-González, J., Lindeman, K. C., Blackwell, P. G., Gall, A., . . . Llewellyn, G. (2004). Mangroves enhance the biomass of coral reef fish communities in the Caribbean. *Nature*, *427*(6974), 533–536.

Mustafa, A., Zulfiqar, U., Mumtaz, M.Z., Radziemska, M., Haider, F.U., Holatko, J., Brtnicky, M., et al. (2023) Nickel (Ni) Phytotoxicity and detoxification mechanisms: A review. *Chemosphere*, *328*, 138574. https://doi.org/10.1016/j.chemosphere.2023.138574

Naiman, R. J., & Decamps, H. (1997). The ecology of interfaces: Riparian zones. *Annual Review of Ecology and Systematics*, *28*(1), 621–658.

Naskar, K. R. (1999). Status of mangroves in the Indian Sundarbans in the perspective of Indian and world mangals. In D. N. Guha Bakshi, P. Sanyal, & K. R. Naskar (Eds.), *Sundarbans Mangal* (pp. 20–89). Naya Prokash.

Naskar, K. R. (2004). *Manual of Indian Sundarbans*. New Delhi: Daya Publishing House.

Naskar, K. R. (2004). *Manual of Indian Sundarbans*. New Delhi: DayaPublishing House.

Naskar, K. R., & Guha Bakshi, D. N. (1987). *Mangrove swamps of the Sundarbans*. Calcutta: Naya Prokash.

Naskar, K. R., & Guhabakshi, D. N. (1987). *Mangrove swamps of the Sundarbans – An ecological perspective*. Naya Prokash.

Naskar, K. R., & Mandal, R. N. (1999). *Ecology and biodiversity of Indian mangroves*. Day Publishing House.

Nishat, B. (2019). *Landscape narrative of the Sundarban: Towards collaborative management by Bangladesh and India*.[World Bank Report No. 133378].

Panja, S., Podder, A., & Homechaudhury, S. (2020). Evaluation of aquatic ecological systems through dynamics of Ichthyofaunal diversity in a Himalayan torrential river, Murti. *Limnologica*, *82*(2), 125–135.

Pantulu, V. R. (1966). *Contribution to the study of the biology and fishery of some estuarine fishes*. PhD thesis,. University of Calcutta.

Pethick, J., & Orford, J. D. (2013). Rapid rise in effective sea-level in Southwest Bangladesh: Its causes and contemporary rates. *Global and Planetary Change*, *111*, 237–245.

Pillay, T. V. R. (Ed.). (1967). *FAO fisheries report* (Vol. 1, No. 44). FAO.

Poff, N. L., Allan, J. D., Bain, M. B., Karr, J. R., Prestegaard, K. L., Richter, B. D., . . . Stromberg, J. C. (1997). The natural flow regime. *Bioscience*, *47*(11), 769–784.

Prasad, M. B. K., Kumar, A., Ramanathan, A. L., & Datta, D. K. (2017). Sources and dynamics of sedimentary organic matter in Sundarban mangrove estuary from indo-Gangetic delta. *Ecological Processes*, *6*(1), 8.

Pusey, B. J., & Arthington, A. H. (2003). Importance of the riparian zone to the conservation and management of freshwater fish: A review. *Marine and Freshwater Research*, *54*(1), 1–16.

Rahaman, S. M. B., & Asaduzzamn, B. (2020). Seasonal nutrient distribution in the Rupsha-Passur tidal river system of the Sundarbans mangrove forest, Bangladesh. *Ecological Processes*, *3*, 18.

Rahman, A. K. A. (1989). *Freshwater fishes of Bangladesh*. Zoological Society of Bangladesh.

Rahman, M. F., Al, M. A., Akhtar, A., Hoque, M. N., Shafiqul Islam, M., Sikder, M. N. A., & Barman, P. P. (2019). Structural patterns of fisheries communities with relation to ecological condition in subtropical coastal waters of the northern bay of Bengal, Bangladesh. *Indian Journal of Geo-Marine Sciences, 48*, 1720–1732.

Reaks, H. G. (1919). Report on the physical and hydraulic characteristics of the delta. In C. J. Stevenson-Moore, C. H. D. Ryder, M. C. Nandi, R. C. Law, H. H. Hayden, J. Campbell, A. R. Murray, C. Addams-Williums, & E. A. Constable (Eds.), *Report on the Hooghly River and its headwaters* (Vol. 1, pp. 29–132). Bengal Secretariat Book Depot.

Sahana, M., Hong, H., Ahmed, R., Patel, P. P., Bhakat, P., & Sajjad, H. (2019). Assessing coastal island vulnerability in the Sundarban biosphere reserve, India, using geospatial technology. *Environmental Earth Sciences, 78*(10), 304.

Salagrama, V. (2012). *Climate change and fisheries: Perspectives from small-scale fishing communities in India on measures to protect life and livelihood*. International Collective in Support of Fishworkers.

Sanyal, P. (1999). Sundarbans-the largest mangrove diversity on globe. In D. N. Guha Bakshi, P. Sanyal, & K. R. Naskar (Eds.), *Sundarbans Mangal* (pp. 428–448). Calcutta: Naya Prakash.

Sarkar, D., Chattopadhyay, G. N., & Naskar, K. R. (1999). Nature and properties of coastal saline soils of Sundarbans with relation to mangrove vegetation. In D. N. Guha Bakshi, P. Sanyal, & K. R. Naskar (Eds.), *Sundarbans Mangal* (pp. 199–204). Naya Prokash.

Sarkar, S. K., Singh, B. N., & Choudhury, A. (1986). Composition and variations in the abundance of zooplankton in the Hooghly estuary, West Bengal, India. *Proceedings: Animal Sciences, 95*(2), 125–134.

Sarker, S. U. (1985a). Ecological observation on the endangered whitebellied sea eagle Haliaeetus leucogaster (Gmelin) in the Sundarban, Bangladesh. In *Symposium on endangered marine animals and marine parks* (Vol. 4, Paper No. 58). Marine Biological Association of India.

Sarker, S. U. (1985b). Density, productivity and biomass of raptoral birds of the Sundarbans, Bangladesh. In *Proceedings of SAARC seminar on biomass production* (pp. 84–92).

Sarker, S. U. (1989). Fish eating wildlife and some fishes of the Sundarbans, Bangladesh. *The Journal of Noami, 6*(1-2), 17–29.

Sarker, S. U., & Sarker, N. J. (1986). Status and distribution of birds of the Sundarbans, Bangladesh. *The Journal of Noami, 3*, 19–33.

Seidensticker, J., & Hai, M. A. (1983). *The Sundarbans wildlife management plan: Conservation in the Bangladesh coastal zone*. IUCN.

Sen, N., Thakur, G. C., Sarkar, D., & Naskar, K. R. (1999). Identification and impact assessment studies of the algal flora of the Indian Sundarbans. In D. N. Guha Bakshi, P. Sanyal, & K. R. Naskar (Eds.), *Sundarbans Mangal* (pp. 350–354). Naya Prokash.

Sen, T. K. (1975). Fish and fisheries aspects of the Sundarbans, West Bengal. *Science and Culture, 41*(10), 500–502.

Shah, M. S., & Hossain, M. H. (2006). A checklist of fish and crustaceans of the south-west coast of Bangladesh. In *Khulna university studies, special issue, 1st research cell conference* (pp. 129–140).

Shaha, B., Alam, M. M., Islam, H. M., Alam, L., Ghosh, A., Ahmed, K., & Mokhtar, M. (2014). Assessment of gear efficiency for harvesting artisanal giant freshwater prawn (macrobrachium Rosenbergii De man) fisheries from the Sundarbans mangrove ecosystem in Bangladesh.

Shamsuzzaman, M. M., Islam, M. M., Tania, N. J., Al-Mamun, M. A., Barman, P. P., & Xu, X. (2017). Fisheries resources of Bangladesh: Present status and future direction. *Aquaculture and Fisheries, 2*(4), 145–156.

Shamsuzzaman, M. M., Rashid, A. H. A., Mamun, M. A. A., Mazumder, S. K., & Haque, M. A. (2005). Present status of marine puffer fishes in Bangladesh. *Journal of Aquaculture Research and Development, 6*(10), 1–5. https://doi.org/10.4172/2155-9546.1000370

Shetty, H.P.C., Saha, S.B. and Ghosh, B.B. (1961). Observations on the distribution and fluctuations of plankton in the Hooghly–Matlah estuarine system with notes on their relation to commercial fish landings. *Indian Journal of Fisheries*, 8 (8), 326–363.

Sinha, M., Mukhopadhyay, M. K., Mitra, P. M., Bagchi, M. M., & Karmakar, H. C. (2020). Impact of Farakka barrage on the hydrology and fishery of Hooghly estuary. *Estuaries, 19*(3), 710–722.

Snedaker, S. (1982). Mangrove species zonation: Why? In D. N. Sen & K. S. Rajpurohit (Eds.), *Contributions to the ecology of halophytes* (pp. 111–126). W. Junk Publishers.

Stanford, J. W. J., & Ward, J. V. (1983). *The serial discontinuity concept of lotic ecosystems*. Ann Arbor: Science Publishers.

Strahler, A. N. (1952). Dynamic basis of geomorphology. *Geological Society of America Bulletin, 63*(9), 923–938.

Strahler, A. N. (1957). Quantitative analysis of watershed geomorphology. *Eos, Transactions American Geophysical Union, 38*(6), 913–920.

Talwar, P. K., & Joglekar, A. (1970). On a new species of 'Wak' Lin 1938 (Pisces: Sciaenidae) from Hooghly estuary. *Journal of the Inland Fisheries Society of India, 1*, 32–36.

Tam, N. F. Y., & Wong, Y. S. (1995). Mangrove soils as sinks for wastewater-borne pollutants. *Hydrobiologia, 295*(1–3), 231–241.

Thom, B. G. (1984). Transgressive and regressive stratigraphies of coastal sand barriers in Southeast Australia. *Marine Geology, 56*(1-4), 137–158.

Thorpe, J. H., & Delong, M. D. (1994). The riverine productivity model: An heuristic view of carbon sources and organic processing in large river ecosystems. *Oikos, 70*, 305–308.

Ullah, H., Gibson, D., Knip, D., Zylich, K., & Zeller, D. (2014). Reconstruction of total marine fisheries catches for Bangladesh: 1950–2010. In *Fisheries Centre Working Paper #2014-15*, Vancouver: University of British Columbia.

Vannote, R. L., Minshall, G. W., Cummins, K. W., Sedell, J. R., & Cushing, C. E. (1980). The river continuum concept. *Canadian Journal of Fisheries and Aquatic Sciences, 37*(1), 130–137.

Vass, K., Das, M., Tyagi, R., Katiha, P., Samanta, S., Srivastava, N., . . . Gopal, B. (2011). Strategies for sustainable fisheries in the Indian part of the ganga-Brahmaputra river basins. *International Journal of Ecology and Environmental Sciences, 37*(4), 157–218.

Wahl, T., Brown, S., Haigh, I. D., & Nilsen, J. E. Ø. (2018). Coastal Sea levels, impacts, and adaptation. *Journal of Marine Science and Engineering, 6*(1), 19. https://doi.org/10.3390/jmse6010019

Walker, B. (1995). Conserving biological diversity through ecosystem resilience. *Conservation Biology, 9*(4), 747–752.

Welcomme, R. L. (1985). *River fisheries*.

Whitley, G.P. (1934). Supplement to the check-list of the fishes of New South Wales. *Australian Zoologist, 7*(4), 345–364.

Wilson, C., Goodbred, S., Small, C., Gilligan, J., Sams, S., Mallick, B., & Hale, R. (2017). Widespread infilling of tidal channels and navigable waterways in the human-modified tidal delta plain of Southwest Bangladesh. *Elemental Science of the Anthropocene, 5*(78), 12.

Winemiller, K. O. (2004). Floodplain river food webs: generalizations and implications for fisheries management. In *Proceedings of the second international symposium on the management of large rivers for fisheries* (Vol. 2, pp. 285–309).

Wu, H., Liu, J., Bi, X., Lin, G., Feng, C. C., Li, Z., Qi, F., Zheng, T., & Xie, L. (2019). Trace elements in sediments and benthic animals from aquaculture ponds near a mangrove wetland in southern China. *Marine Pollution Bulletin, 117*(1–2), 486–491.

Yadav, S. K., & Majumdar, K. (2020). Mangrove associated seaweeds in Sundarban biosphere reserve, West Bengal, India. *International Journal of Advanced Research in Biological Sciences, 7*(12), 53–62.

Zaky, M. M. M., Salem, M. A. M., Persson, K. M. M., & Eslamian, S. S. (2011). Incidence of Aeromonas species isolated from water and fish sources of Lake Manzala in Egypt. *International Journal of Hydrology Science and Technology, 1*(1/2), 47–62.

3 Hooked in the Net of Nature and Necessity: *Coping with Ecological Risks and Socioeconomic Traps*

3.1 INTRODUCTION

The interpretative value of the term "vulnerability" is contingent upon the contexts in which it is used (Miller et al., 2010). In recent decades, assessing vulnerability to natural catastrophes and climate change has become significantly important (Birkmann et al., 2013). Viability, as defined by the International Strategy for Disaster Reduction (2014), refers to the conditions shaped by physical, social, economic, and environmental elements or processes that intensify a community's vulnerability to the effects of hazards (Chatterjee et al., 2024a). This technique classifies vulnerability into many components or parts (physical or social), which are further associated with other features (ISDR, 2014). The definitions and explanations of vulnerability provide a restricted array of the several existing meanings. However, it demonstrates that the interpretation of the term varies, even among followers of a certain scientific discipline. This complexity is aptly reflected in Vincent van Gogh's observation: "The fishermen know that the sea is dangerous and the storm terrible, but they have never found these dangers sufficient reason for remaining ashore." This underscores that vulnerability does not imply passivity or helplessness; rather, it coexists with resilience, agency, and livelihood necessity. Communities may remain exposed to hazards not due to ignorance, but because social, economic, and cultural imperatives compel continued engagement with risk—an insight central to contemporary vulnerability and disaster-risk scholarship.

Birkmann et al. (2013) contend that the assessment of vulnerability presents a paradoxical situation, as our objective is to quantify vulnerability but lacks a specific definition. Nevertheless, the capacity to quantify vulnerability is progressively recognized as a crucial metric for successful risk reduction and the advancement of a culture of disaster resilience (Eslamian, 2009). The significance of social vulnerability lies in its correlation with socioeconomic factors and individual traits that impact the ability of a community to effectively prepare for, respond to, and recover from disasters (Kienberger et al., 2014; Lee, 2014). Consequently, it elucidates the reasons behind the varying experiences of the same hazardous event among different communities. Yoon (2012) states that understanding the variability in the impacts of hazard events is essential for mitigating the detrimental effects of natural disasters. There are several methodologies for comprehending vulnerability and risk across

DOI: 10.1201/9781003729990-3

diverse scientific disciplines, namely (1) The pressure and release (PAR) model in political economics analyses the structural determinants of vulnerability (Cai et al., 2025; Wisner et al., 2014). (2) Social ecology frameworks amalgamate human–environmental interactions (Carreño et al., 2012; Villagran, 2006). (3) IPCC climate frameworks concentrate on exposure, sensitivity, and adaptive capacity (Chen et al., 2014; Cutter 2020). (4) The dual structure context differentiates external shocks from internal coping strategies (Bohle, 2001; Chambers, 1989). (5) The Sendai Framework (UNDRR, 2023), the HAZUS model (FEMA, 2024), and the MOVE (Methodical, Operational, Visionary, Empathic) framework are all operational tools. The MOVE framework looks at exposure, susceptibility, and resilience holistically (Birkmann et al., 2013).

3.2 LIVING ON THE EDGE

The Sundarbans is acknowledged as a critical climate change hotspot for mangrove ecosystems, as recognized by a worldwide scientific consensus (Chowdhury et al., 2016; Dubey et al., 2017; Gupta, 2021). Recent decades have seen substantisal threats to the Indian Sundarbans area due to climatic circumstances such as increasing temperatures, increased salinity (Aziz & Paul, 2015; Chamberlain et al., 2021), sea level rise (SLR), and an increased occurrence of tropical cyclones (Neogi et al., 2016). The transformation of agricultural land or river floodplains into aquaculture systems, driven by anthropogenic factors such as population development, presents escalating issues for climate change adaptation (Giri et al., 2022; Jayanthi et al., 2022) (Table 3.1). Therefore, it is essential to comprehend the regional implications of these ecosystems to enhance global knowledge of climate change's effects on mangrove ecosystems. The local factors affecting the ecological and biological variety of an ecosystem will be given more attention as a result. According to Jayanthi et al. (2022), to enhance adaptation and planning, the Sundarbans, an integrated human–natural ecosystem, necessitates a regional understanding of numerous physical and socioeconomic sensitivity elements. Despite comprehensive studies evaluating the vulnerability of the coastal Sundarbans region (Bera & Maiti, 2021; Ghosh & Mistri, 2021; Majumdar et al., 2021; Mondal et al., 2022), the ecological sensitivity of the mangrove ecosystem remains little explored. Focused research (Majumdar et al., 2021; Mondal et al., 2022) has aimed to examine the vulnerability of mangrove forests; nevertheless, the selected variables overlooked the extended biological reactions of the ecosystem.

The focus is drawn to social systems, and vulnerability is conceived as having two sides: an external side encompassing the perturbations and risks the system is subjected to, and an internal side that includes the system's own capacity to cope (Aly et al., 2022) and respond to hazardous events (Sahana & Sajjad, 2019). Consequently, factors such as resilience, sensitivity, resistance, and coping ability are prevalent components in these investigations. Complex systems, like the human–environment system, are characterized by elements like multitude of scales, irreducible uncertainty, nonlinearity, and a range of legitimate perspectives (Sahana et al., 2019; Sahoo & Bhaskaran, 2018). The comprehensive assessment of the ecological susceptibility of the Indian Sundarbans Region (ISR) indicated that the western region of the Sundarbans protected area faces significant risk. In line with earlier

TABLE 3.1
Stressors of Climate Change and Its Effect on Fisheries

Stressors of Climate Change	Effect On	Potential Impacts
Warmer water	Biological productivity	- Disruption of marine ecosystems - Reduced growth of fish stocks - Altered species distribution
Water level rise	Biological productivity	- Fish kills due to low oxygen levels - Habitat loss for certain species - Increased salinity
Stronger storms	Capture fisheries	- Damage to critical coastal habitats - Spread of waterborne diseases - Destruction of fishing gear
Unpredictable sea conditions	Capture fisheries	- Lower catch yield - Altered species composition in catches - Greater risk to safety at sea
Change to ocean currents	Capture fisheries	- Altered migratory patterns of marine species - Damaged infrastructure and gear - Increased costs
Extreme weather events	Occupation/livelihood	- Loss of income for fishing communities - Increased occupational hazards - Infrastructure damage
Ocean acidification	Occupation/livelihood	- Reduced shellfish population due to weaker shells - Threats to food security - Decreased well-being

studies on precipitation uncertainty in the Indian subcontinent, the current study found a declining trend in future precipitation (Katzenberger et al., 2021). Prolonged alterations in precipitation patterns may affect the biotic composition of the mangrove ecosystem and modify the community structure of mangroves under forthcoming climate conditions (Simard et al., 2019). The projected climatic scenario for temperature indicates that the disparity between maximum and lowest temperatures in the future will be minimal, consequently placing the mangrove community near the threshold of its thermal limit. The research conducted by Lovelock et al. (2017a) indicates that warm tropical mangroves are operating near their thermal thresholds. Thus, any additional rise in temperature may profoundly impact their physiological functions. The anticipated net SLR by the end of the twenty-first century is to exceed 1 m, especially in the ISR, attributed to the heightened subsidence of its western segment (Brown & Nicholls, 2015).

Globally, the primary policy response to rising sea levels has been the construction of embankments; however, increase in sea level is projected to persist beyond 2100, rendering embankment construction a temporary adaptation strategy (Dasgupta et al., 2020). An alternate approach involves the zonation of especially vulnerable shorelines, as described by Mondal et al. (2022), alongside the expedited establishment of mangrove plantations in these regions, which may function as a nature-based remedy and enhance the overall ecological condition of the ISR. It is essential to acknowledge that nature-based solutions (NBS) offer numerous advantages; yet, significant obstacles hinder their incorporation into mangrove ecosystems for adaptation techniques. Conflicting land use objectives, inadequate monitoring and assessment, and a lack of focus on sociocultural factors may hinder the integration of mangroves as NBS. A community-oriented strategy for mangrove planting is crucial for effective execution and oversight.

The rise in sea level and storm surge will considerably jeopardize islands like Sagar, G-Plot, Lothian, and Mousuni, located at the southernmost boundary of the ISR. The nearshore bathymetric depth significantly influences the magnitude of the storm surge. A scatterplot depicting historical storm surge heights and bathymetric depths indicates that reduced nearshore bathymetric depth exacerbates storm surge height, corroborating the findings of Sahana and Sajjad (2019). The elevation of these coastal regions was determined to be significantly low, rendering them susceptible to rainwater incursions. Additionally, to safeguard the densely populated areas of the southern Indian Sundarbans (Sardar & Samadder, 2023), national and local authorities have constructed elevated coastal embankments in these places. These embankments obstruct the influx of tidal water and hinder the natural sediment deposition process in river floodplains, leading to silt accumulation on the riverbed and a reduction in river depth. The absence of natural silt deposition from tidal floods results in a lowland topography that significantly increases the risk of flooding during severe storm surges.

The influence of human activities on climate forcing is extensively documented in climate research (Ghil, 2016). In ecology, anthropogenic pressure correlates with the proximity of protected areas to human-made structures (Dutta et al., 2017). The demographic map of the Indian Sundarbans indicates that the westernmost and northernmost islands are significantly threatened by human encroachment. Moreover, forest regions adjacent to the Basanti, Gosaba, and Canning blocks demonstrate intensified vulnerability to anthropogenic disturbances. Humans currently reside on specific regions of South-Katuajpuri, as these islands are situated outside the defined protected zone, resulting in insufficient protection of these sites.

Kanan et al. (2023a) found that the expected flooding of the Sundarbans from SLR by the end of the twenty-first century may be significantly less than earlier estimates from the past few decades. Jabir et al. (2021) indicated that 75% of the Sundarbans will be submerged due to a 45 cm rise in sea level. Colette (2017) projected that a 45 cm increase in global sea level by 2100 might eliminate 75% of the Sundarbans. Additionally, an SLR of 100 cm (1.0 m) will obliterate the entire Sundarbans (World Bank, 2020). Kanan et al. (2023a) illustrate that an SLR of 35 cm, 52 cm, and 70 cm, accounting for the current subsidence rate of −2.4 mm/year, will submerge 1%, 1.8%, and 3.4% of the Sundarbans landmass, respectively. The findings suggest that a 147 cm increase in sea level, along with a net subsidence rate of −2.4 mm/year, might inundate as much as 23% of the terrestrial area. Except in extreme scenarios, the low, mid, and high SLRs with a subsidence rate of −0.4 mm/year will submerge riverbanks and beaches.

The extreme scenario is deemed realistic yet improbable (Jevrejeva et al., 2014). Also, the elevation distribution indicates that 4% of the Sundarbans is situated at elevations exceeding 0–100 cm, whereas 32% is within the 101–200 cm height range (Colette, 2017; Jabir et al., 2021; World Bank, 2020) due to differing assumptions, thereby challenging the validity of these assumptions and, consequently, the conclusions of past studies. Brammer (2014) asserts that a prevalent fallacy prevails regarding the degradation of coastal areas in Bangladesh, particularly the Sundarbans, by the conclusion of the twenty-first century because of SLR. Loucks et al. (2010) projected that 96% of the Bangladesh Sundarbans will be eradicated if SLR reaches 28 cm by 2070; this assessment is overly optimistic. The potential cause may be that the Digital Elevation Model[1] utilized in their analysis is outdated or that there are errors in the inundation calculations. Approximately 96% of the Sundarbans region will be submerged if sea levels increase by 3.5 m; however, this scenario is projected to occur in approximately 600 years, given the present SLR rate of 3.6 mm per year and a net subsidence rate of −2.4 mm yearly. The projected flood zones in the Sundarbans will differ from the prior predictions established by Payo et al. (2016) and Ghosh et al. (2019) at the conclusion of the twenty-first century. Ghosh et al. (2019) forecasted that the mangrove region will be submerged by 26 km^2, 96 km^2, and 747 km^2 by the end of the twenty-first century under low, medium, and high SLR scenarios, respectively, with a net subsidence rate of −2.4 mm/year. However, Kanan et al. (2023a) forecast that projected inundation will attain 40 km^2, 72 km^2, and 918 km^2 by the conclusion of the twenty-first century, contingent upon stable scenarios and subsidence rates of −2.4 mm/year by 2100. Results vary according to the application of Representative Concentration Pathway (RCP) 4.5 and 8.5 scenarios in assessing inundation in the Sundarbans of Bangladesh and West Bengal, India. The IPCC predicts a more significant rise in sea level for West Bengal than for the Bangladesh Sundarbans in both scenarios. Payo et al. (2016) and Akbar Hossain et al. (2024) reported that the sea level in the Indian Sundarbans is increasing at an annual rate of +0.36 cm. By the end of the twenty-first century, SLR will attain 29 cm. The Sundarbans landmass, encompassing offshore islands and coastal areas, is markedly diminishing due to tidal surges, with soil erosion occurring at a rate surpassing that of accretion (Akbar Hossain et al., 2022). So, the rise in sea level may not be the primary cause contributing to the degradation of the Sundarbans landmass. Barua et al. (2022a) indicated that the lack of clean water is an escalating concern in the Sundarbans of Bangladesh and India due to the increasing depletion of freshwater sources. As freshwater supplies in coastal areas progressively become brackish and saline, the scarcity of clean water is emerging as a more pressing concern. While most coastal families depend on pond water as their principal water supply, some coastal communities have access to enhanced water sources, such as tubewells. Due to the susceptibility of surface water to pollution, coastal families may have elevated incidence of waterborne diseases compared to inland homes, as well as challenges in accessing enough clean water. Researchers recognized many reasons for increased salinity, some principally attributed to human activities and others directly linked to climate extremes (climatic variables) (non-climatic factors). To meet and/or reduce the total demand for drinking water, several technologies have been developed and adapted to the evolving environment.

Sardar and Samadder (2023) indicated that mangrove ecosystems have considerable adaptation and reactivity to environmental changes via several natural resilience mechanisms at the ecosystem level. They revealed that SLR is projected to range from 0.7 m to 0.9 m under baseline climatic circumstances. The western Sundarbans is especially vulnerable to climate sensitivity.

Mitra et al. (2024) analyzed 20 years of daily precipitation data of this research region, revealing that monthly precipitation varied between 125 mm and 141 mm. They used Shared Socioeconomic Pathways (SSPs)[2] that suggest a reduction in precipitation throughout the latter two decades of the twenty-first century, ranging from 89 mm to 110 mm. This scenario forecasts a 4°C increase in maximum temperatures and a 7°C rise in winter low temperatures by the century's conclusion, resulting in a more consistently warm environment in the future (Table 3.2).

Within the fisheries sector, the fisherfolk being the main actor (Béné, 2009), the resilience of the fisherfolk is influenced not only by the characteristics of the fishery resources but also by the socioeconomic and institutional environment in which these systems function. This encompasses risks related to climate change, health, market dynamics, sociopolitical elements, and security (Smederevac-Lalić et al., 2023; Visser et al., 2021). Fishermen are more vulnerable to these hazards than individuals in other socioeconomic categories (Béné & Friend, 2011). Also, their ability to adapt to these vulnerabilities is sometimes regarded as inadequate.

In numerous fields and traditions, including economics, anthropology, psychology, and engineering, the notion of vulnerability has been employed (Adger, 2006). Consequently, numerous definitions, paradigms, and frameworks/methodologies regarding vulnerability have been developed from diverse perspectives (Räsänen et al., 2016). Bohle et al. (2014) argue that vulnerability is a dynamic and complex social concept shaped by individuals' specific political, economic, and institutional capacities within spatial and temporal circumstances. The complex and diverse nature and features of vulnerability provide significant challenges. The discourse on risk assessment methods can be attributed to five primary intellectual traditions: (1) risk/hazard or biophysical methods, which relate to vulnerability resulting from biophysical threats; (2) political–ecological/political–economic frameworks, interpreting vulnerability as a condition arising from varying degrees of exposure, impact, and capacity; (3) the notion of resilience as a systemic phenomenon, analyzing vulnerability and capacity while taking into account their dynamic characteristics; (4) integrated hazards of connected human–environmental systems (Islam & Chuenpagdee, 2022). In climate change literature, vulnerability is generally classified into two categories: social vulnerability, also referred to as socioeconomic vulnerability, and biophysical vulnerability, known as physical or natural vulnerability (Barua, 2021; Simić et al., 2020; Radenković et al., 2023).

Thus, contextual information (regarding environmental risks exposure) and PAR model (concentrating on causal elements, sources of pressure, and socioeconomic circumstances) described by Adger (2006), Füssel (2012), Eakin and Luers (2006) are crucial to clarify the impacts of climate change and guide decision-making regarding the expenses of adaptation in comparison to mitigation expenditures (Jatoi et al., 2022; O'Brien et al., 2007). Thus, the central emphasis is on the origin of risk or hazard, which dictates the degree of vulnerability. Vulnerability study is frequently characterized by the scale, duration, and consequences of the climatic event.

TABLE 3.2
Risk Level Rating of Different Parameters for Vulnerability Assessment for Sundarbans

Indicators	Very High (5)	High (4)	Moderate (3)	Low (2)	Very Low (1)
Monthly precipitation (mm)					
IMD	>141	141–136	136–130	130–125	<125
Mean maximum temperature (°C)					
IMD	>31.18	31.18–31.12	31.12–31.05	31.05–30.96	<30.96
Mean minimum temperature (°C)					
IMD	>23.75	23.75–22.64	22.64–22.53	22.53–22.43	<22.43
Wind speed (m/s)	>4.72	4.72–4.55	4.55–4.44	4.44–4.3	<4.3
Coastal inundation due to SLR					
7 m SLR	Inundated areas	–	–	–	Non-inundated areas
9 m SLR	Inundated areas	–	–	–	Non-inundated areas
Cyclone density (sq. km)	>6.21	6.21–4.66	4.66–3.11	3.11–1.56	<1.56
Storm surge inundation height (m)	6.5–5	5–3	3–1.5	1.5–0.5	<0.5
Population growth rate	>21	21–17	17–13	13–10	<10
Elevation	<1	1–2.5	2.5–3.5	3.5–5.5	>5.5
Slope	<0.5	1–1.25	1.25–1.75	>1.75	
Bathymetry (values in negative)	>6	14–6	27–14	50–27	<50
Shoreline change (LRR)	<−37.5	−16.5	−12.5	0–91.3	No change
Distance from shoreline (km)	0–1	3–6	6–12	>12	
Distance from human habitat (km)	0–5	10–25	25–35	>35	
Mangrove forest density	0–20	30–45	45–65	65–85	>85
Mangrove forest stability	−0.8 to −0.6	−0.6 to −0.4	−0.4 to 0	0 to 0.4	>0.2

Modified after Jana and Bhattacharyaa (2013); Bhaskaran et al. 2013; Bhadra et al. 2023; Census of India 2001; Das et al.2020; Deb and Ferreira 2017; Ghosh et al. 2019; 2020; IMD; Kanan et al. 2023; Karim et al. 2008; Saha et al. 2025; Sahoo and Bhaskaran 2018; Samanta et al. 2021; Sardar and Samadder 2023.

These solutions are referred to as Alternative Risk-Hazard strategies (Eakin & Luers, 2006; Turner et al., 2003) or impact-oriented techniques. International institutions are progressively embracing and advocating for them as an effective policy framework to mitigate poverty and vulnerability, hence promote growth and development (Mukhopadhyay et al., 2024). Moreover, within national social development and environmental protection agencies, they provide support in formulating community-specific strategies that effectively address the environmental, socioeconomic, and sociocultural requirements and rights of rural communities whose prosperity relies on natural resources (Eriksson et al., 2017; Fabinyi, 2010; Kolding et al., 2014). The concepts of vulnerability theories and the livelihood approach, as presented by Bennett et al. (2015) and Eriksen et al. (2021), have been extensively analyzed by the five capitals of vulnerability, namely: physical, natural, social/political, financial, and human capitals.

The sustainability of a livelihood is characterized by its ability to withstand and recuperate from unexpected events while preserving or augmenting its capabilities and resources without compromising the natural resource foundation. In traditional fishing villages, social and cultural artifacts, encompassing social networks and informal institutions, are frequently distinctive and may not be easily replaced by the acquisition of alternative forms of "wealth." This method imposes certain constraints on the compensability of various forms of "capital" (natural, social, etc.) (Etingoff, 2019). Ecosystem pressures may disproportionately impact the security and social cohesion of vulnerable and marginalized groups, exacerbating underlying tensions that manifest as political instability outside their borders (Jentoft et al., 2017). The detrimental effects of poverty will ultimately influence the entire global system (Pasini et al., 2018). Such scenarios are likely to surpass mild damage and lead to catastrophic collapse (Cernexv & Fenner, 2020). A comprehensive evaluation of the social dimensions of sustainability identifies six relevant data categories concerning fishing: (1) Historical context of fishing; (2) Sociocultural characteristics; (3) Quality of life; (4) Social capital; (5) Values and beliefs of fishing communities and the broader society; (6) Geospatial analysis of communities concerning the fisheries resources they utilize (Kolding et al., 2014; Madhanagopal & Pattanaik, 2020).

The Sundarbans SSFs, as a coastal system, exhibit considerable climatic and oceanographic variability and are particularly vulnerable to disruptions from severe natural events and climate change trends (Defeo et al., 2013; Ortega et al., 2018). The artisanal fishers of the Indian part of Sundarbans are least adaptive to climate change, and this region is least "developed" among the state-wise SSFs of India (Mandal et al., 2017; Salagrama, 2012). Integrating climate change and its consequences is crucial for attaining sustainable development in fishing communities. The tropical SSFs are increasingly threatened by climate dynamics where fish productivity, abundance, and distribution are expected to decrease across the latitudes (Sumaila et al., 2011) due to ocean warming, SLR, precipitation, and oceanographic variables (such as wind velocity, wave action, and increases in the frequency and severity of extreme weather events) (Chatterjee, 2017; Saha et al., 2019; Mitra et al., 2020a) and can impact the fisher's livelihood through multiple pathways (Ali et al., 2020; Sadrekarimi et al., 2022). The fishers being the main actor in SSFs, "the fishers' perceptions about climate change rather than about climate change per se" is more important to address the consequences of climate change to define sustainability (Figure 3.1).

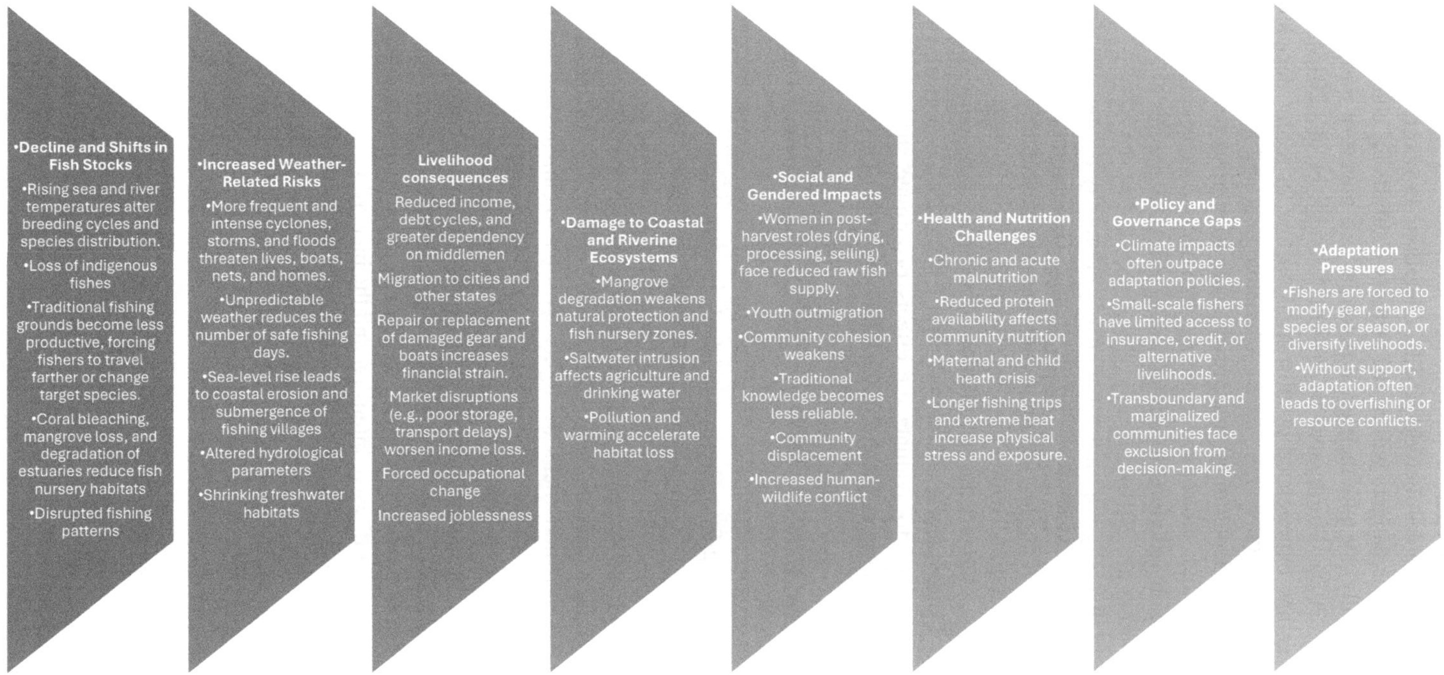

FIGURE 3.1 Impact of climate change on small-scale fishing communities of Sundarbans.

Climate change will exert a multifaceted influence on mangrove ecosystems globally. Climatic variables including SLR, alterations in ocean currents, temperature rises, and cyclonic dangers will be the primary catalysts for change in mangrove communities globally. Despite their great potential as a means of mitigating climate change (Alongi, 2020; Murdiyarso et al., 2015; Sidik et al., 2018), the global mangrove cover has experienced a substantial decline in recent decades (Sahana et al., 2021; Sahoo & Bhaskaran, 2018). Climatological and anthropogenic causes both contribute to the depletion of mangrove cover. Mangrove communities, mostly located in tropical and subtropical regions, demonstrate considerable geographic heterogeneity in their reactions to climatic stressors (Devaney et al., 2021; Gupta, 2016, 2021; Ward et al., 2016).

The degree of diversity within the mangrove community is crucial in determining the response of these ecosystems to climate change. Geographical regions characterized by monospecific mangrove communities are more susceptible to climate fluctuations compared to regions with a high level of mangrove diversity. The diversity of ecological responses hinders the formulation of a universal framework for assessing mangrove risk. Consequently, a regional framework for ecological vulnerability evaluation may be used to evaluate the effects of climate stressors on mangrove ecosystems alone (Biswas et al., 2020; Chand et al., 2021; Das et al., 2022).

3.2.1 Addressing Climatic Hazards and Oceanographic Changes

Ecosystem stresses are anticipated to disproportionately damage the security and social cohesion of fragile and impoverished communities, intensifying existing tensions that might result in political instability, perhaps extending beyond the impacted areas. Pasini (2018) asserts that "the misfortune of the poor will ultimately impact the entire global system." Such consequences may surpass gradual damage, potentially triggering a rapid collapse (Cernexv & Fenner, 2020). Coastal small-scale fisheries (SSFs) in the Sundarbans are highly sensitive to climatic and oceanographic changes (Chavez et al., 2003; Defeo et al., 2013). Addressing climate change and its effects is thus crucial for promoting sustainable development in fishing communities. Tropical SSFs face increasing threats from shifting climate dynamics, where fish productivity, abundance, and distribution are projected to decline across latitudes due to ocean warming, rising sea levels, changes in precipitation, and oceanographic factors such as wind speed, wave intensity, and the increasing frequency and severity of extreme weather events (Perry et al., 2005; Pinsky et al., 2013; Pratchett et al., 2011; Sumaila et al., 2011; Woodruff et al., 2013). These factors can affect fishers' livelihoods through various pathways (Allison et al., 2012; Bedjeck et al., 2010; Kittinger et al., 2013). In the following section we will discuss the oceanographic fluctuations and climatic variability along with other ecosystem stresses to SSFs of Sundarbans.

Sea Level Rise: The Sundarbans ecoregion located in the tidally active lower deltaic plain of the Ganges–Brahmaputra–Meghna basin (GBD) is undergoing one of the most pronounced effects of climate change resulting in the form of SLR at an average rate of 3.14 mm per year (maximum centennial-scale relative SLR is estimated to be 0.9 ± 3.3 cm/yr). The estuaries are particularly sensitive to changes in the mean sea level because, apart from enhancing vulnerabilities to erosion and storm inundation, it alters tidal forcing and influences sedimentation by setting up flood-dominated tidal asymmetry (Chowdhury et al., 2021; Kumar et al., 2022). The configuration of a deltaic shoreline illustrates the interplay between sediments carried by rivers and their modification by tidal and wave motions, which are markedly amplified during storms. The waves generated by the winds produce diverse currents that function obliquely, transversely, or longitudinally along the shoreline. Tidal variations modify the positioning of wave activity and generate significant volumes of water referred to as tidal prisms that flow into and out the coastal plains twice daily. The Sundarbans' tidal cycle is semi-daily, with a slight diurnal variation (Guha & Mitra, 2020). The influx of elevated water levels into the estuaries is intensified by resonance and other factors that obscure any discernible pattern in the magnitude of tidal variations. The minimum tidal range along the GBD coastline occurs near the mouth of the Pussur estuary (Hiron Point, 2.95 m). The range extends to its boundaries at the mouths of the Hugli (Sagar Island, 4.32 m) and Meghna estuaries (Sandwip, 6.01 m). The increasing landward morphological constriction of hyper-synchronous estuarine waterways elevates the tidal range by as much as 2.5 m in certain areas (Mitra, 2018; Mitra, 2020). The tides of the Sundarbans display asymmetry, marked by a pronounced predominance of flooding. This suggests that the ascending tide requires less duration within a cycle, resulting in a more rapid landward movement of the bidirectional tidal current, thus transforming the estuaries into sediment repositories. Analogous to the tidal range, the asymmetry amplifies in a northward direction, signifying a more heightened sedimentation rate in the upper sections of the estuaries. Tidal pumping generates a substantial net influx of sediments, significantly influencing vertical accretion in the Sundarbans region, as it receives minimal or no sediment discharge from upstream rivers, except for the Hugli and Baleswar (Rahman et al., 2018; Barua & Rahman, 2019; Barua & Chakraborty, 2011; Barua et al., 2024; Danda, 2019; Khan, 2013).

The SLR in the Sundarbans areas is higher than the global average rate of sea rise, and many studies assume that most of the dry land of the Sundarbans will be inundated by the end of the twenty-first century. Kanan et al. (2023a) found that low (35 cm), mid (52 cm), high (70 cm), and extreme (147 cm) SLR scenarios indicate that the Sundarbans landmass area will be flooded up to 40 km^2 (1%), 72 km^2 (1.8%), 136 km^2 (3.4%), and 918 km^2 (23%), respectively, under the current net subsidence rate of −2.4 mm/year by 2100. Except for the extreme scenarios, the low, mid, and high SLR will result in riverbank and beach areas to be covered by water. They concluded that inundation aspect of SLR will not directly affect the Bangladesh Sundarbans; however, indirectly related threats and anthropogenic disturbances can be major drivers of the Sundarbans' degradation by the end of the twenty-first century. Moreover, Ghosh et al. (2019) and Payo et al. (2016) stated that sea level is increasing in Indian Sundarbans according to a +0.36 cm/year annual rate. By the end of the twenty-first century, the SLR would be 29 cm.

Kanan et al. (2023b) found that predicted flooding of the current dense, moderate, and sparse forest regions will not be substantial by the end of the twenty-first century for low, medium, and high SLR scenarios. The projected inundation of barren land areas is expected to exceed that of the thick, moderate, and sparse forest classes because of the predominant coverage of unproductive land along riverbanks and beaches and its location in low-elevation regions. The continuous flooding of riverbanks and beaches augments the salt of the surrounding soil and modifies the sequential trajectory of mangroves. Ghosh et al. (2019) assert that SLR linked to higher salinity leads to the substitution of low-tolerant species with halo-tolerant coastal ones, hence altering the composition of species. Regions characterized by rapid SLR and limited sediment availability have heightened flooding, leading to tree death at the edges of seaward mangroves and migration toward land (Xie et al., 2022). The extent of inundation in thick, moderate, and sparse forest types is generally greater in extreme scenarios. However, the Sundarbans zones do not consistently exhibit extreme conditions. The station-based SLR increase data (1977–2020) of Sundarbans showed the low (35 cm) SLR scenarios at a subsidence rate of −2.4 mm/year. Hence, SLR may not play a crucial role in the deterioration of Sundarbans. Yet, the degradation of Sundarbans biodiversity and its ecosystem is primarily threatened by other climate change-related hazards like cyclones, storm surges, salinity, soil erosion, pests, and illnesses (Mandal & Hosaka, 2020; Kibria et al., 2018).

Salinity: The Sundarbans is experiencing a significant increase in salinity, driven by climate change and SLR. Neogi et al. (2017) documented a 26% rise in cyclonic storms in the Sundarbans coastal region from 1881 to 2001. This escalation led to prolonged waterlogging, increased soil salinity, and reduced biodiversity. Cyclone disruptions can severely damage forest structural frameworks (Moon et al., 2019; Ortolano et al., 2016) and tree ecosystems. Predictions indicate that the effects of tropical cyclones will be significant in the next years (Alam & Dominey-Howes, 2015; Walsh et al., 2016). Das et al. (2016) highlighted the human activities, including shrimp farming, dam construction, and excessive exploitation of forest resources, contributing to the ecological degradation of the Sundarbans. The construction of the Farakka Barrage in 1975 resulted in a decline of upstream freshwater flow from the Ganges River, measured at 3700 m^3/sec in 1962, 364 m^3/sec in 2006, and 250 m^3/sec in 2022. The reduction in flow results in an increase in the salt concentration of the Sundarbans (Islam & Gnauck, 2023). The population reliant on the Sundarbans is rising, resulting in increased demand for forest resources (Ghosh et al., 2022), and consequently, there is an elevated risk of overexploitation (Uddin et al., 2013), jeopardizing the ecological balance of the Sundarbans (Ahmed et al., 2021a). Akbar Hossain et al. (2022) asserted that SLR will significantly reduce the mangrove vegetation regions of the Sundarbans owing to flooding by the end of the twenty-first century. The resulting salinity stratification inhibits vertical mixing of heat and contributes to maintaining warm sea surface temperature (SST) and high climatological rainfall over the Bay of Bengal (Jabir et al., 2021;). The variance in salinity in the aquatic subsystem has triggered a compositional variation of aquatic fauna. The commercially important fish species (*Tenualosa ilisha*, *Polynemus paradiseus*, *Sillaginopsis panijus*, *Pama pama*, *Arius jella*, *Osteogeneiosus militaris*, etc.) that prefer freshwater for breeding have changed their course from central to western sector of Indian Sundarbans due to increased dilution factor in the western sector because of barrage discharge. Hyposaline environment of western Indian Sundarbans may be attributed to Farakka barrage discharge situated in the upstream region of

Ganga–Bhagirathi–Hooghly River system. There are 23-year surveys (1999–2022) on water discharge from Farakka dam, which revealed an average discharge of $(3.7 \pm 1.15) \times 10^3$ m^3/sec. Throughout the monsoon, elevated discharge levels were documented, averaging $(3.81 \pm 1.23) \times 10^3$ m^3/sec, with a maximum of 4524 m^3/sec seen during September. Discharge values were markedly diminished during the premonsoon period, averaging $(1.18 \pm 0.08) \times 10^3$ m^3/sec, with a minimum of 846 m^3/sec documented in May. Postmonsoon discharge demonstrated moderate amounts, averaging $(1.98 \pm 0.97) \times 10^3$ m^3/sec, as recorded by prior studies (Mitra, 2013). However, the trash fishes (*Stolephorus* sp., *Thryssa* sp., *Harpodon nehereus*, *Trichiurus* sp., etc.), adept at surviving and reproducing in challenging saline conditions, were noted to predominate the catch composition in the central sector relative to the western sector of Sundarbans (Mitra et al., 2014; Mitra & Gati, 2015). Research revealed the extinction of Chandana Hilsa (last observed 25 years ago), sea-conch as well as the endangerment of pangas and sharks, among others, in Namkhana, Kakdwip, Diamond Harbour, Roydighi, and Canning within the coastal area of the Indian Sundarbans (Das et al., 2022). The rise in sea level negatively impacts overall fishing production, encompassing fish landing, processing, and marketing facilities. As a result, the seaports, fish jetties, and storage facilities located along the coastal margins are vulnerable to recurrent tidal and storm floods in Sundarbans (Mitra, 2018; Mitra et al., 2019).

Sea surface temperatures (SSTs): The surface air temperature over the Bay of Bengal is rising at 0.019°C/year (Dhar et al., 2021). The rise in surface air temperature causes the thermal expansion of ocean water by increasing the temperature of the ocean's surface (Chowdhury et al., 2021). Studies by Chakraborty et al. (2016), Banerjee et al. (2018), and Fahad et al. (2016) indicate that changes in SST affect ocean dynamics, including salinity, pH, dissolved oxygen, biological production, and nutrient upwelling. These alterations result in modifications in fish distribution, either by permanent relocation or shifts in migration patterns, fish abundance, phenology, or the timing of life-cycle events such as growth, reproduction, spawning, and recruitment, ultimately affecting the quantity of fish landings in coastal regions. Sengupta and Mitra (2020) noted that the water in the Sundarbans has not undergone rigorous evaluation, and there is scanty information regarding its quality. Agarwal et al. reported that in the Sundarbans region, the SST has risen by 0.5°C per decade over the past 30 years, a rate eightfold greater than the global warming rate of 0.06°C per decade (Daly & Winter, 2012), thus identifying the Sundarbans as one of the most critical climate change hotspots globally (Agarwal et al., 2019). Changes in wind intensity and patterns modify ocean currents and upwelling processes, affecting fish migration and navigation, ultimately resulting in a 40-year drop in catch per unit effort (CPUE)[3] (Jana et al., 2020; Ni et al., 2019).

The Sundarbans' Outstanding Universal Value (OUV)[4] includes its rich estuarine and mangrove-associated fish fauna. The decrease in OUV in the context of ecological and environmental significance results from a decline in fish diversity. Declining fish diversity disrupts food webs, degrades habitats, and weakens ecological processes that justify OUV status. Loss of species also affects cultural practices and livelihoods, threatening a site's recognition and increasing the risk of being listed as "in danger" (Dasgupta et al., 2021; Gopal & Chauhan, 2006).

In contrast, no reliable catch data are available in that area. Consequently, climate change leads to diminished fishing effort and overcapacity, which may result in increased resource rent. As most of the people in this region rely on fish and fisheries as a source of livelihood and protein, climate change would intensify their socio-economic and food-security problems. Gain and Uddin examined the effects of salinity in Sundarbans coastal regions. The Bangladesh Water Development Board provided data on the rise in salinity in the Sibsa and Passur rivers affecting fish diversity in the Sundarbans between 1975 and 2014. The report indicates that increased salinity prevents fish species commonly found in freshwater environments within the Sundarbans—such as *Pangasius pangasius* and *Lates calcarifer*—as well as those from moderately saline environments, including *T. ilisha*, from reproducing. Only fish adapted to higher salinity levels, such as *Trichiurus savala* and *H. nehereus*, can survive under these conditions (Mandal et al., 2017).

Natural disasters: Natural disasters are irregular events in society that lead to societal disruption and human distress. Tierney et al. (2010) contend that catastrophes function as catalysts for collective action in physical and social contexts before, during, and after their manifestation. Natural phenomena (or hazards) alone do not define catastrophes. They become catastrophes when they impact communities and regions vulnerable to their effects (Aguirre, 2015). Disasters are not intrinsically natural; they arise from decisions made by collectives, often disconnected from their eventual consequences, and originate from their inability or unwillingness to adopt sustainable living patterns. Unfortunately, so-called natural disasters are often misinterpreted as mere natural occurrences, when they represent an interaction between the natural environment and sociocultural systems, particularly regarding societal vulnerability. It is imperative to develop conceptual frameworks that tackle this type of vulnerability. According to Cutter et al. (2003), using the model of disaster (DROP model—Disaster Resilience of Place) suggests that social vulnerability is a multidimensional concept that helps to identify those characteristics and experiences of communities (and individuals) that allow them to respond and recover from natural disasters, and it is associated with the concept of resilience.

The rhythm of the seasons, reflected in the yearly cycles of wind, temperature, and precipitation regimes, has an important bearing on the tidal and supratidal landforms of the Sundarbans. In general, wind speeds decrease eastward. During April and May, peak southerly and southwesterly wind speeds of around 25 km h^{-1} are observed in the unshielded western half of the area. This time aligns with the hottest months of the year (30°C, also arid), leading to significant changes in the morphology of landforms on the southern sea-facing islands. The southern breezes endure during the monsoon season (June–September). During periods of regular precipitation, sand movement decreases due to increased moisture, while higher local water levels from freshets, along with wind-driven waves and tropical cyclones, contribute to greater coastal erosion and the reworking of tidal sediments. From the month of October, the winds change direction and velocity (northerly, approximately 5 km/hr), leading to significant alterations in the wave environment, and with the onset of winter, depositional processes prevail on the beaches. The area can be described to have three seasons in a year: premonsoon (February–May), monsoon

(June–September), and postmonsoon (October–January). The temporal units, characterized by specific meteorological and tidal conditions, provide a framework for examining the medium- and large-scale landforms of the Sundarbans' annual patterns. The gently inclined shelf of the GBD encounters winds from cyclones that produce waves, leading to water accumulation along the coast and causing storm surges that can completely inundate small, low-lying islands of the Sundarbans. Sea surface elevation rises by about 1 cm for every millibar decreases in pressure associated with cyclones, which are low-pressure systems. As a storm surge progresses inland, its levels are exacerbated by the narrowing estuaries to the north (Pal et al., 2017; Pramanick et al., 2016; Yadav & Majumdar, 2020).

The timely arrival of the southwest monsoon is essential for the sustainability of fisheries in the Bay of Bengal. The inconsistent and reduced patterns of the northeast monsoon from October to December, along with the irregular monsoon that continues into November, adversely impact fish availability through the decrease in the catches of key seasonal fish species in the Sundarbans. The modification of precipitation patterns and sudden, intense rainstorms release numerous land pollutants from industrial areas into the ocean, leading to extensive mortality among nearshore fish and significantly affecting fish-drying activities. Nine of the fourteen deadliest tropical cyclones have struck the Bay of Bengal (World Bank Group, 2014). Bangladesh ranks sixth in cyclone fatalities, while India is fourteenth; notably, India ranked fourth in 2014 and tenth in 2015 due to significant weather events (Lauria et al., 2018). The incidence of severe, recurrent cyclonic storms in the Bay of Bengal has increased by 26% over the past two decades, as evidenced by recent cyclones Aila, Amphan, Yaas, and Rimel. Cyclonic storm surges in coastal areas extensively disturb estuarine systems by infusing seawater. Such changes lead to modify fish assemblage composition through increased mortality, displacement, and alterations in recruitment or habitat structure, influencing fishing revenue (Mukherjee et al., 2022). A study by Mukherjee et al. (2012) demonstrated that the composition of fish assemblages in the Matla River significantly changed following Cyclone Aila. The study revealed that the presence of different species had decreased in the catchment areas, which led to a significant reduction of species diversity. The species that became more abundant after Aila were *Glossogobius giuris*, *Cynoglossus lingua* (which prefers rocky, gravelly, or sandy substrates), *T. ilisha*, *H. nehereus*, *Setipinna taty*, *Panna microdon*, *Pampus argenteus*, *Toxotes chatareus*, *Callionymus sagitta*, *Coilia ramcarati*, and *Liza parsia*, all of which reside in habitats marked by substantial organic detritus and low dissolved oxygen concentrations. However, there was a substantial decline in the population of freshwater species (e.g., *Gudusia chapra*). Specific areas of Sundarbans, such as Sagar Island, Jambu Deep, Namkhana, Patharpratima, and Bkangaduni, demonstrate increased susceptibility to variable seasonal tidal amplitudes and cyclones (Mukhopadhyay, 2022). A common natural hazard faced by fishermen in Sundarbans is frequent floods. Embankments are essential in the Sundarbans for protection from cyclonic storms, SLR, and annual floods; nevertheless, the structural integrity of several embankments, built in the nineteenth century, requires enhancement. In the rehabilitated Sundarbans, adequately maintained peripheral dykes offer sufficient protection against the maximum spring tides. The probability of overtopping significantly escalates if a storm's landfall coincides with the region's high water level (Pal et al., 2017; Pramanick et al., 2016; Yadav & Majumdar, 2020).

Coastal erosion in the Indian Sundarbans transpired at a rate of approximately 6 km^2 annually during the past decade. The total land area, originally 6,402 km^2 in 2001, diminished to 6,358 km^2 by 2009 and thereafter to 4,890 km^2 by 2022, culminating in a net loss of 44 km^2 (Hazra et al., 2002). The total length of the embankments in the Indian Sundarbans is 3,638 km, of which 470 km were designated as vulnerable before Aila. The embankments built in the Sundarbans to protect human settlements from flooding have been counterproductive, as flooding promotes siltation at river mouths and creeks, disrupting fish breeding and nursery habitats, diminishing crab populations, and ultimately affecting fishing activities. Due to siltation and the fact that their estuarine character is only maintained by monsoonal runoff and tidal actions, the majority of the river distributaries that helped form this deltaic region—from west to east: Muriganga, Saptamukhi, Thakuran, Matla, Gosaba, and Bidya—have lost their original connections with the Ganga as a result of the ongoing tectonic activity. Attrition and accretion transpire concurrently in several areas of the Indian Sundarbans. Reclamation activities started in 1770 significantly enhanced sediment reworking in an environment dominated by flood tides, according to studies. Phased embankment removal may thereby restore height in the restored Sundarbans. The western Indian Sundarbans' islands are steadily deteriorating, according to satellite imagery. Conversely, the Sundarbans islands in central India are exhibiting accretion (Raha et al., 2012). The dynamic process of char formation is regarded as the most fertile shrimping area of Bay of Bengal. However, recently, climate change has caused major issues for the prawn and prawn fisheries. Community members observed that salty water intrusion on agricultural land has enhanced the potential for prawn and prawn farming. The alternative livelihood of coastal landless individuals (about 86%) has intensified fishing pressure on postlarvae. The distribution of postlarvae along riverbanks was affected by SLR. The average fishing time for seed collectors decreased from 6.25 hours in 2006 to 6.07 hours in 2011 because of rising temperatures and heat waves. Cyclones often diminish fishing days and, therefore, the catch rate. The availability of postlarvae after storms was significantly reduced for a duration ranging from 1 week to several weeks. During the wet season, the marketing of postlarvae has challenges owing to limited accessibility and inadequate transportation infrastructure. Consequently, the average revenue from postlarvae fishing has diminished by 22% over the last 5 years due to a decrease in catch rate and market price. In 2022, the average price of postlarvae was projected to be US$50 per 1,000, compared to US$38 in 2006. The capture of prawn postlarvae has progressively decreased by around 15% in recent years (Ahmed et al., 2024). The newly introduced species are claimed to alter water courses, modify biological conditions, and disrupt fishing patterns. The emergence of a new char off Kakdwip poses a major problem to fishing migration, which extended the longer journey durations, causing increased fuel and ice expenses contingent upon catch volumes and patterns, as well as the prevailing management framework (Sumaila et al., 2011). Along with the increased turbidity in these areas fish shoals are no longer discernible to fishermen based on water color, unlike in the past. Construction operations next to streams and estuaries, siltation, oil spills from boats and vessels, and extensive seabed drilling contribute to water turbidity and obstruct nets and fish gills, resulting in significant mortality rates among tiny fish (Salagrama, 2012). Aside from reductions, the main target species remain unchanged across the border. Consequently, the fisheries of *T. ilisha*, *P. pangasius*, *Plotosus* spp., and *Scylla serrata* were subjected to excessive exploitation. The utilization of *L. calcarifer* is expected to be at its maximum, producing around 150 MT. The fishery for *Johnius argentatus* is almost entirely exploited. The population of riverine catfish *P. pangasius and Plotosus canius* in the SRF is critically low and nearing extinction. The numbers, particularly per vessel, have diminished, with a significant percentage (50% of the catch during postmonsoon, exceeding 70%) including juveniles and smaller fish, indicating biological overfishing. The seasonal diversity of fish is becoming unpredictable. The presence of fish out of season is becoming problematic due to insufficient postharvest and trading procedures (Hoq et al., 2024). Consequently, climate change may influence the economics of fishing due to the amount and quality of marine fish catches and their distribution inside and among countries' exclusive economic zones. Variations in fishing costs, fishermen's incomes, fishing companies' earnings, discount rates, economic rent (the surplus remaining after all expenses, including standard profits, are accounted for), and capital expenditures such as those for boats, gear, processing facilities, and other assets may indicate the economic impacts of climate change on fisheries. Adapting to the effects of climate change on the amount, composition, and distribution of fisheries resources will need more funding for fishing and processing activities. Fishing is regarded as a high-investment endeavor for artisans due to rising expenditures on ice, transportation of catches via motorized boats especially after these climatic disasters, and monofilament nets that necessitate more frequent replacement than the previously utilized nylon or cotton nets, with uncertain revenue prospects in the Bay of Bengal. Occasionally, concurrent expeditions to the coast and shoreline incite disputes among fishermen from adjacent states and nations. The conventional relationships of reciprocity are being rejected. The total fishing days for small-scale vessels have decreased, although the fleet size has expanded, hence exacerbating the physical and psychological stress on the fishers (Thoya et al., 2022; Sumaila et al., 2020).

3.2.2 Industrial Pollution and Corporate Environmental Abuse

The Hooghly-Matla estuary outlet, including the Sundarbans, is contaminated by the lower Bhagirathi-Hooghly basin's rapid industrialization and urbanization. According to Sinha (1998), the Hooghly estuary discharges 1,125 million liters of wastewater per day, while lower portions get 396×10^6 m^3 of sewage and waste load each hour, with an annual flow of 493 km^3 of urban domestic trash. The extensive development of the deep-sea port located off the Sundarbans coast at Sandhead, in conjunction with the port's operational activities, has jeopardized the livelihoods of local fishermen (DISHA, 2006) by restricting access to key fishing zones. The state-owned Oil and Natural Gas Corporation (ONGC) has discovered a significant gas reservoir in the Bay of Bengal, located 55 km off the Sundarbans coast, with preliminary estimates indicating around 21 trillion cubic feet of reserves. Fish habitat is destroyed, and mass mortality is caused by the negative effects of seismic testing for petroleum deposits and the waste from toxic drilling discharges (lead, mercury and cadmium, benzene, toluene, and arsenic), which affects the distribution, abundance, and catch rates of commercial fish over a wide geographic area. Large-scale developments in the Sundarbans region, together with forthcoming projects at Kulpi (located north of Sagar Island in Sundarbans) and SAHARA Mega tourism might jeopardize the livelihoods of fishing communities by deteriorating the ecosystem, reducing onshore working areas, and hindering fishermen access to the sea (Chacraverti, 2015a; IWA, 2008).

Upstream of the Sundarbans, at least 20 pesticides, 18 fungicides, 2 rodenticides, and several fertilizers are reportedly being applied. Another major pollutant in the Hooghly-Matla estuarine mouth is agricultural waste that contains pesticide and insecticide residues, such as chlorinated hydrocarbon and organophosphates (Bhattacharya et al., 2018; Chakraborty et al., 2019; Das et al., 2021b; Nath et al., 2021; Mondal & Pal, 2023; Mukherjee et al., 2021a; Sarkar et al., 2022). These agrochemicals are transported downstream in the Sundarbans and integrated into the food chain, resulting in biological amplification at elevated trophic levels. Heavy metal contamination from industrial waste over 84 townships in the lower Hooghly basin has the highest concentrations of arsenic, cadmium, and mercury, in addition to manganese (732 ppm), nickel (49 ppm), copper (44 ppm), zinc (151 ppm), chromium (98 ppm), and lead (32 ppm). Industrial trash is indiscriminately discharged into the river upstream of the Sundarbans from textile mills, chemical plants, pharmaceutical facilities, plastic manufacturers, detergent producers, jute mills, and tire factories within the Hooghly Industrial Belt, significantly contaminating the estuary and coastal waters of the Sundarbans (Taher et al., 2023). Oil spills from ships result in immediate death of fish, prawns, and several other creatures. In 1994 and 2015, two significant oil spills transpired in the Bangladesh Sundarbans, resulting in immediate harm to the environment and wildlife, while the long-term effects remain mostly undetermined. The recurrent capsizing of vessels at the Hooghly's mouth (with up to 30 incidents recorded from 2005 to 2022) has been observed to release substantial amounts of liquid and solid waste into the sea, adversely impacting both the navigation of fishers and the migratory breeding patterns of fish (Mitra, 2020). The increasing prevalence of microplastics in this ecosystem poses significant risks to both the ecology and the health of the fishing communities dependent on these resources.

The microplastic pollution in the Sundarbans has been extensively documented over the last few years. The presence of these pollutants in several environmental matrices, including water, sediment, and animals, has become a major concern. Microplastics are found in the aquatic ecosystem and are also being consumed by local fish and crab populations, which are the primary source of economy and diet for the local people (Chatterjee et al., 2024a; Das et al., 2024; Fatema et al., 2023; Kumar et al., 2022). Research indicates that the consumption of microplastics by fish species in the Sundarbans serves as a potential indicator of health risks to aquatic organisms and human consumers of these contaminated species (Fatema et al., 2023; Sarker et al., 2022). The trophic transfer of microplastics within the food web exacerbates the issue, as they accumulate and magnify across various trophic levels, leading to significant ecological consequences (Sarker et al., 2022). Islam (2023) conducted a socioeconomic study of the small-scale fishermen in the Sundarbans, revealing that diminished fish populations due to pollution and habitat degradation led to reduced income and increased livelihood instability. The fish from the Sundarbans estuary contain significant levels of plastic particles, which might pose long-term health risks. These results align with the works of Rahman et al. (2020) and Mishu et al. (2025), who confirmed that microplastic contamination in the Pasur River—one of the Sundarbans' main water bodies—has rendered fisheries a possible source of human exposure to microplastics. Das et al. (2024) investigated the ecological impact of microplastics in shell water systems and sediment of the Sundarbans, revealing that plastic pollution is not only harming fish but also other marine life vital to the food web. The socioeconomic impacts of microplastic pollution are especially catastrophic for small-scale fishers in the Sundarbans. These communities depend mostly on the well-being of their marine ecosystem for their livelihood. The deterioration of fish stocks from pollution can cause reduced catches, compromising food security and economic stability (Roy et al., 2023c, 2023a, 2024). Besides, microplastic presence may influence the fish quality, discourage consumers, and decrease market price, thus contributing to the further exacerbation of the economic strain in these communities (Das et al., 2024; Nawar et al., 2022). Apart from direct effects on fisheries, microplastic pollution also threatens the environment. Microplastic interaction with the mangrove ecosystem can disrupt nutrient processes and damage biodiversity that is fundamental to ecological stability (Ahmed, 2025; Fatema et al., 2023).

Barua and Barua (2024) ventured into studying sustainable management of fisheries in Bangladesh, calling for the minimization of plastic use in fishing practices and encouraging the use of biodegradable substitutes (Barua & Barua, 2024). In addition, Ahmed et al. (2025) emphasized the significance of environmental policy in addressing microplastic pollution in coastal wetlands, recommending more stringent plastic waste disposal regulations in the Sundarbans area. Even with these initiatives, there remains limited cooperation among government agencies, conservation groups, and local fishing communities. Research studies indicate that, despite the existence of policy regulations, the enforcement of these regulations is weak due to limited resources and that the local communities are not aware of the situation (Islam et al., 2024; Kumar et al., 2022). Most of the studies are short-term, but long-term monitoring of microplastic pollution and its impact on marine species and fisheries is necessary. Despite the regulations for better waste management, very few are the indications of the effectiveness of the current policies and the potential barriers to their implementation. The part played by fishing nets and gears in the generation of microplastic pollution is well documented but alternative, environmentally friendly materials have not been extensively investigated. Control of microplastic pollution in the Sundarbans needs a holistic approach of policy interventions, sustainable fishing techniques, and waste management practices by the community.

3.3 SOCIOECONOMIC TRAP

Fishing is much more than fish. It is the great occasion when we may return to the fine simplicity of our forefathers

(Herbert Hoover).

Fishing in the Sundarbans is a communal activity involving men, women, and children. However, the collection of prawn fry is often conducted independently. Societal and religious limitations prohibit women from fishing collectively. A significant number of fishers within the artisanal fishing community, however, are unable to be fully engaged in the occupation, hindered by insufficient equipment. Consequently, the existing economic challenges faced by the ultra-poor have escalated significantly disproportionately.

The fisheries in Bangladesh's mangrove forests are very diverse and rich in endemism. Fishing in mangrove areas is mostly defined by artisanal fisheries, which use diverse fishing gear and boats (Islam & Wahab, 2020). The condition of mangrove fisheries in the SMF over several past decades is marked by two different phases. The first phase, which was the original one, showed a rising trajectory until 2015–2016, while the second phase that followed is figured by a continuous deterioration thereafter. This pattern implies that the mangrove fisheries have suffered from overfishing, degradation of their habitats, or a combination that has led to their decline (Islam & Haque, 2021).

Heavy fishing pressure and resource exploitation in the Sundarbans have resulted in the large-scale removal of juvenile prawns and fish from their natural habitats before they reach reproductive maturity. This practice has significantly reduced the population of these species, as juveniles are harvested before they can reproduce and replenish their numbers. The ongoing overexploitation has led to a continuous decline in the annual availability of prawn fry, affecting both artisanal and commercial fishing operations. As a result, fishing activities have been adversely affected, with both small-scale and larger fisheries experiencing decreased catches and diminished yields. This trend not only threatens the sustainability of the coastal ecosystem but also impacts on the livelihoods of the fishers (Deb, 2018). One of the major factors that has been the most significant contributor to the overexploitation of nearshore fisheries in the Sundarbans environment is the extensive use of set bag nets for the collection of prawn fry. This poses a substantial threat to coastal ecosystems, jeopardizing the nursery habitats of several aquatic species and negatively impacting both newly established and old protected mangrove forests (Hoq, 2020). Artisanal fisheries are characterized by labor intensity, little capital investment, traditional practices, and objectives of subsistence, commercial profit, or both. The inshore, estuarine, and coastal fisheries of the Sundarbans support around 200,000 fishermen who engage in daily operations inside its waterways (Islam & Wahab, 2020). Siddique et al. (2024) reported that about two million inhabitants depend on the Sundarbans for direct or indirect subsistence. Many disadvantaged and resource-dependent communities have become very vulnerable to the scarcity of other jobs and income opportunities, and thus they are highly dependent on fishing as their source of livelihood.

Various international agreements impact national policy, obliging state forest managers to conserve the Sundarbans to protect the iconic faunal species, the Royal Bengal tiger, and the pristine ecology. A variety of people engage in fishing activities, since capture fisheries substantially enhance the state's economic growth. These are presently being reduced at an alarming pace due to anthropogenic influences (deforestation for aquaculture and agriculture, extraction for construction materials, paper pulp, and fuelwood) and natural factors (recurring storms, erosion, and SLR) (Chacraverti, 2015b; Chakraborty et al., 2019; Ortolano et al., 2019; Partelow et al., 2018). These factors have intensified the problems of SSF communities, leading to increased migration, overfishing, occupational displacement, human–wildlife conflict, and lack of livelihood options (Abdullah-Al-Mamun et al., 2017; Giri et al., 2017; Guha & Roy, 2016; Vivekananda et al., 2018). Fishers have increasingly embraced intensive aquaculture to make foreign currency, mitigate poverty, promote economic growth and development, and improve food security (Singh et al., 2017). And as industrial aquaculture has been the main contributor to meet the worldwide demand for marine products (Thomas et al., 2017), SSF communities have been forced into a condition of vulnerability. The primary drivers for change are outlined in Table 3.3.

The intensifying pressure on land, swift physical alterations at the peripheries of the islands, land loss due to erosion, governmental land reforms, and distribution exacerbate the economic vulnerabilities of the Sundarbans fishing community by progressively undermining their occupation (Béné & Friend, 2020), ultimately resulting in a comprehensive lack of sustainability within the sector (Zacharia & Najmudeen, 2012). The research said that the region's overall population is projected to exceed 5 million by 2020. From 2001 to 2005, numbers of fisherfolk increased at a faster rate and after that it has been declining rapidly. Over the last 15 years, fishing efforts have doubled, resulting in a decrease in CPUE from 200–250 kg per haul to 65–70 kg per haul. Between 1995 and 2020, prawn harvesting became a principal income source for the inhabitants of Sundarbans. The fishing community experiences unemployment and underemployment, resulting in inadequate food for their families, distress migration, exploitative labor practices, and trafficking. Migration is a complex phenomenon influenced by "macro" factors (political, demographic, socioeconomic, and environmental contexts), "meso" factors (social networks, diasporic connections, technology), and "micro" factors (individual characteristics), collectively shaping the ultimate decision to migrate, in accordance with

TABLE 3.3
Natural and Anthropogenic Drivers of Change in Sundarbans

Drivers	Hazards	References
Natural	Cyclones, flooding, salinization, erosion, sea level rise, geohydrological changes	Blythe et al. (2014); Hossain et al. (2018); Lara et al. (2009); Malakar et al. (2018); Mendelsohn et al. (2012); Moniruzzaman et al. (2018); Thomas et al. (2017)
Anthropogenic	Conversion of lands for aquaculture, agriculture, coastal development, unsustainable development practices, deforestation, pollution, tourism	Aburto-Oropeza et al. (2008); Banerjee et al. (2012); S. K. Chakraborty (2011); Knowler et al. (2009); Kumar (2012); Osland et al. (2017); Salunke et al. (2020); Shaffril et al. (2017)

the push–pull theory (Mitra, 2017; Rahman et al., 2020). In Sundarbans' artisanal fishers, the pattern of migration is individually oriented.

Most artisans in Sundarbans are concurrently engaged in operating small enterprises in the neighboring towns. However, there is a notable influx into the trawl fisheries, characterized by a lower number of in-migrants relative to the out-migrants pursuing other occupations. The privatization of fisheries resources in Sundarbans, frequently coinciding with the rapid degradation and privatization of other primary common-pool resources (forests, grasslands), results in subsequent increase in vulnerability for fisher communities, as individuals transition to alternative livelihoods (Rahman et al., 2020). The prohibition on wood collecting has prompted several former wood collectors to transition into fisheries, consequently intensifying the strain on fishery resources. The rising salinity and scarcity of surface freshwater are significantly impacting traditional agriculture practices in the areas around the Sundarbans. The region's agricultural activities are entirely reliant on monsoon rains due to the lack of artificial irrigation infrastructure. The rising water levels in the estuary pose a risk of flooding to settlements encircled by embankments (Agarwal & Mitra, 2018; Chakraborty et al., 2017). Upon the breach of the barrier, floodwaters inundate the settlement, which stays waterlogged for many days until evacuated by artificial pumping. Consequently, primary crop yields are compromised due to the current farmed types lacking tolerance to deep water and salinity. Consequently, farmer families, regardless of caste or religion, are being coerced into participating in the fishery, particularly in the collecting of tiger shrimp larvae. Fishermen work as laborers or under profit-sharing arrangements without ownership of fishing equipment, vessels, or land (Banerjee et al., 2017; Paszkowski et al., 2021). Consequently, caste fishermen transitioned into marginal farmers, while caste farmers shifted to fishing. Hale et al. (2019) indicated that in the Indian section of the Sundarbans, namely in Jharkhali and Gosaba, Kultali, the quantity of traditional boats has decreased substantially by 50% over the last decade.

Numerous fishing families exist in the state characterized by Krishna (2010) as "merely one illness away from poverty." The village communities are densely populated, and sanitary facilities are inadequately developed. A significant number of homes remain without hygienic toilets, exposing them to the danger of waterborne diseases. These communities often exhibit a deficiency in access to medicines and health services. The manual or gear-based fishing technique for extended periods, often in severe weather conditions, requires considerable physical exertion. The physical debilitation linked to disease significantly disrupts the fishing enterprise and adversely influences family economic stability (Gain et al., 2017). The collecting times for meendhara fluctuate according to the tides, totaling around 13–14 hours daily, with 6–8 hours spent in the water. Lead, nickel, and cadmium are the primary anthropogenic contaminants in the Indian Sundarbans, and seed collectors often experience heavy metal contamination. Health professionals stated that dermatological, ocular, and respiratory ailments, as well as urinary and gynecological issues, sunstroke, and hepatitis are prevalent. Fishers in locations impacted by protracted flooding often encounter several health issues, resulting in diminished fishing efforts owing to sickness (Auerbach et al., 2015a; Bain et al., 2019). The availability of essential services and infrastructure for the fishers remains inadequate. The family income of a fisher, including the earnings of all family members from various sources, showed significant variability. Ninety-five percent of fisher households earn between Rs. 4000 and Rs. 8000. The remaining 5% of households possess an income ranging from Rs. 8000 to Rs. 10000 (Bomer et al., 2020).

Research by Dasgupta et al. (2021) indicated that the average monthly income varies between 5000 and 7000 BDT and 6000 and 9000 Taka, which is below Bangladesh's general per capita income. Consequently, most artisans rely on fish merchants, traders, middlemen, local private creditors, and moneylenders (Dadondars) for access to equipment, vessels, and financial resources (including daily expenditures) at exorbitant rates. The Dadondars and dealers (Arotdars) dominate the fish markets by manipulating market prices, and reports indicate that fishermen endure verbal and sometimes physical harassment when they do not meet supply demands. The market-based assessments indicated that artisanal fishermen seldom obtained fair compensation for their catches. If the catch includes larger specimen they are sold to the "middlemen." The influence of export agents also causes fluctuations on the profit level. The reliance is particularly pronounced for the relocated fishermen who are housed by local merchants or dealers. Recent reports indicate that, due to rising overhead costs in fishing and a recent prohibition on fry collection, fishers have resorted to illegal methods to offset their losses, such as augmenting their catch size illegally (Black trips) or bribing Forest Department officials to obtain permits in the Reserve forest (Salehin et al., 2018; Sen & Homechaudhuri, 2017a). Fishing in the tide-ruled waters of Sundarbans is a more hazardous vocation than agriculture or any other land-based employment. The Sundarbans area has significant morbidity and death due to its ecological sensitivity. The waterways of the Sundarbans harbor many species of sharks, colloquially referred to as kamote, and crocodiles that often attack fishermen. Fishers are the largest occupational category among tiger victims in forested regions, aiming for the most profitable fish harvests. The Forest Department compensates the family of fisherman killed by a tiger with Rs. 100,000, only to licensed fishers, whereas no compensation is awarded to illegal fishers. Annually, several fishermen are killed by the tigers, incidents that often go unreported (Banerjee et al., 2018; Biswas et al., 2019; Chakraborty et al., 2018), resulting in a significant population of tiger widows in each community. They live in an isolated enclave known as "widow-hamlet," enduring profound bodily and psychological distress. Numerous survivors of tiger and crocodile assaults, among physical disabilities and disfigurements, also have posttraumatic stress disorder.

River piracy is prevalent in the confluence of the Matla, Bidya, and Thakuran rivers next to the Bay of Bengal (Sarkar et al., 2021). While river piracy in this area is mostly perpetrated by Indian pirates in the Bay region, international pirate boats from Bangladesh, Myanmar, and Thailand are also operational. The 70 km trans-border river between India and Bangladesh has almost no Border Security Force installation. Reports indicate that around 1,000 fishing vessels were plundered by pirates in the Bay of Bengal and the Sundarbans from June to October 2022; over 600 fishermen were kidnapped, with their families paying Tk 150,000 (US$2,000) as ransom for everyone's release. Reports indicate several abductions of fishing vessels accompanied by substantial ransom demands, particularly during the Hilsa season. Most of the piracy is perpetrated by local gangs; a network of gangs exists.

The isolated Sundarbans islands are vulnerable to criminal syndicates from across the border, who have perpetrated thefts and kidnappings for ransom inside the community (Agarwal et al., 2020; Ahmed et al., 2021a).

3.3.1 Womenfolk Perspective: Through the Gender Lens

The lack of acknowledgment of women's fishing participation or of the significant contribution to the livelihoods of coastal people is due, in part, to the non-remuneration of their fishing activities. The lack of data and appropriate economic valuation of subsistence fisheries result in women's fishing activities not being included in most official statistics (Datta & Mete, 2022)

Gender discrepancies are apparent in the SSFs of the Sundarbans, like other impoverished countries, owing to the patriarchal and patrilineal sociocultural framework (Barua, 2021). Despite both genders engaging in analogous activities such as fishing, net mending, fish processing, and trading (Siason, 2001), males are often perceived as principal providers due to their emphasis on "cash crops," whereas females are seen as "hunters and gatherers," enhancing dietary stability. Men mostly participate in offshore and long-distance fishing, which are more hazardous but linked to higher-value fish species. The women of Sundarbans have restricted control over profitable fishing activities. Women's fishing activities are restricted to coastal regions or minor streams, mostly occurring during the rainy season, using rudimentary fishing equipment. This activity enhances men's fishing endeavors, catering to both personal consumption and commercial selling, akin to worldwide patterns (Barua et al., 2022a; Mitra & Barua, 2023) (Figure 3.2).

Women engaged in fry collection and fish drying/processing have intensified difficulties stemming from extreme poverty, substandard housing, health and sanitation concerns, restricted access to potable water, low literacy rates, limited support services, scarce resource opportunities, and social marginalization (Figure 3.3) (Roy et al., 2023b). In the Indian region of the Sundarbans, namely in North and

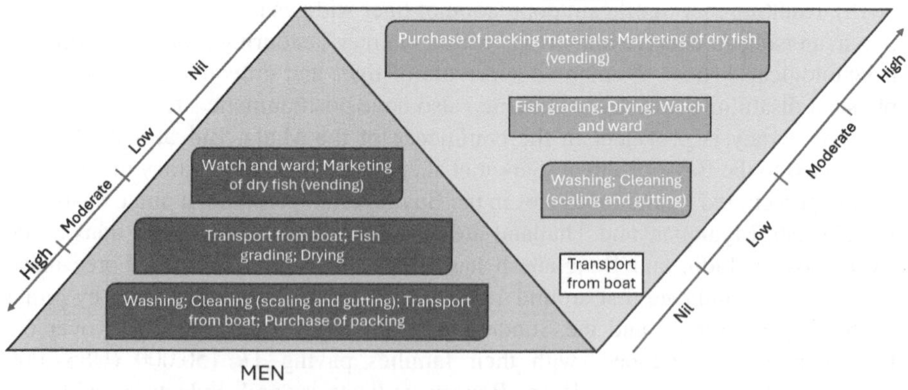

FIGURE 3.2 Gender participation matrix of Sundarbans SSFs men and women fisher. (Modified after Mitra and Barua (2023).)

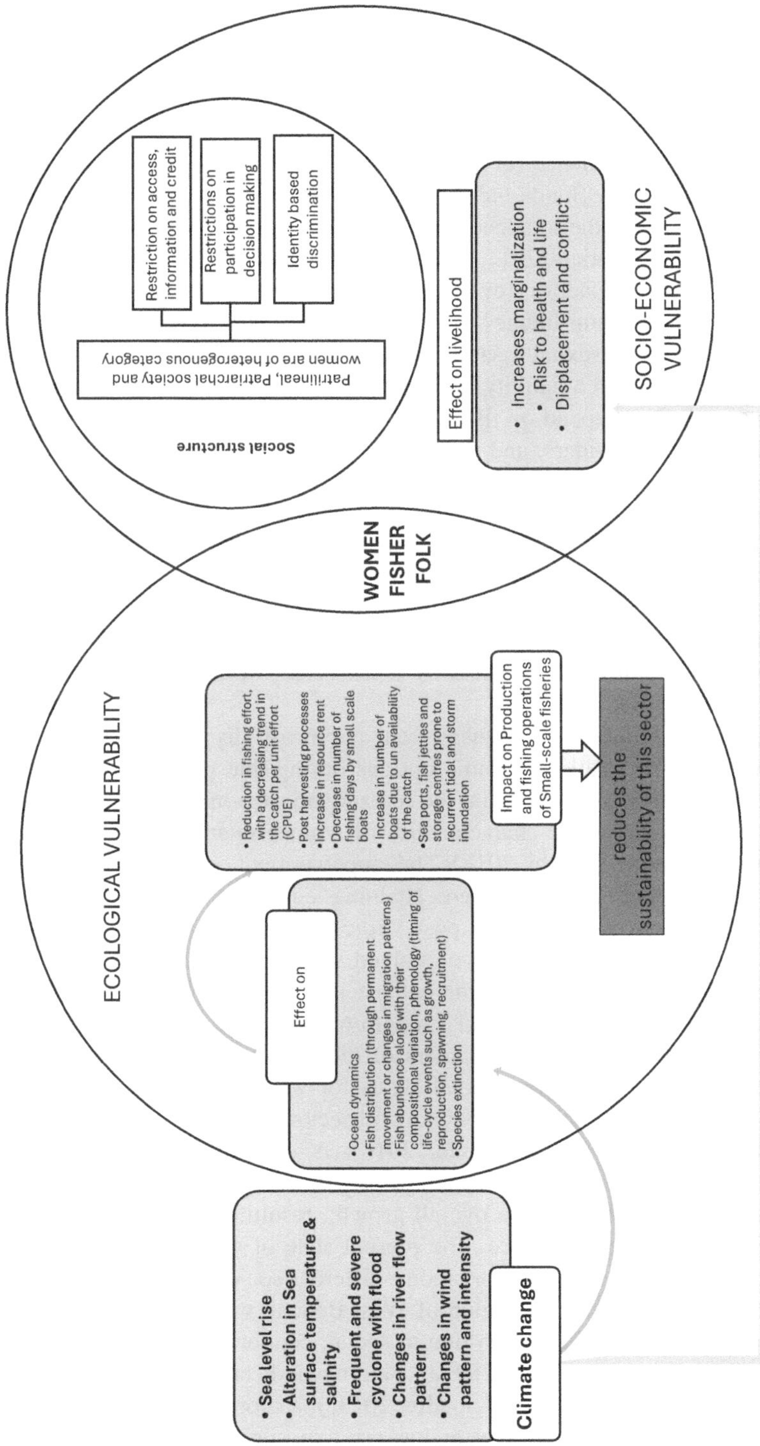

FIGURE 3.3 Climate change-associated vulnerability of Sundarbans small-scale women fisherfolk.

South 24-Parganas districts, over 50,000 individuals are harvesting fish seeds. Around 75% of this group are women (Roy et al., 2016). The investigation of the age distribution among fisherwomen indicated that the majority, accounting for about 30% of the entire sample examined, engage in fish seed collecting during the age range of 10–20 years (Chatterjee et al., 2009). The fishing industry serves as a crucial means of sustenance for the populations, particularly those residing in the peripheral regions of the Sundarbans Reserve Forests. However, abundant resources exist in this region, and the socioeconomic status of rural women could be improved. Regional research by Sarkar (2011) in Satkhira district, next to the Sundarbans West Reserve Forest in Bangladesh, reveals that women fisherfolk have a higher prevalence of illiteracy and semi-literacy (47%) than their male counterparts (7%). Female fishermen often endure sexual assault, and their prospects for marriage face societal disfavor, rendering them a liability to their family due to the custom of dowry (EJF, 2014). Artisans often depend on fish merchants and dealers (Arotdars), middlemen, local private credit providers, and money lenders (Dadondars) at excessive rates to acquire equipment, boats, and finances for daily necessities. These people have power over the fish markets by determining the price. Market-based studies indicate that fishermen face verbal and physical abuse when their catches are insufficient and seldom get equitable prices for their harvests (Chacraverti, 2014; Hoq, 2007). It has been recorded that fishermen are compelled to retire early from fishing if climate change adversely impacts their fisheries (Barua et al., 2022a). Climate change-associated vulnerability of Sundarbans small-scale women fisherfolk has been depicted in Figure 3.3.

Women in the Sundarbans frequently engage in fish drying for 4 months due to the absence of other viable economic options during that period. Considering this background, a study was done in the Frezarganj region of the Sundarbans to elucidate the socioeconomic characteristics of women engaged in fish drying as a supplementary income (Karabi, 2018). The empowerment of women is a contentious issue. In ancient India, women were attaining equal status with men, yet they encountered challenges during the post-Vedic and epic periods. Since the early twentieth century, their statuses have evolved slowly and progressively. The term "women's empowerment" denotes the enhancement and augmentation of women's social, economic, political, and legal strength to guarantee equal rights for women (Finkl & Makowski, 2017). The framers of the Constitution and national leaders emphatically advocate for equal social status for women and men. Contemporary women hold esteemed positions throughout all sectors. Women are vital contributors to the expanding economy (Samanta, 2018). A significant land mass of India comprises coastal area, around 7500 km in length. Their socioeconomic constraints constitute significant obstacles to overall growth, resulting in women encountering numerous issues in the workforce. The present state of women's empowerment in coastal regions is markedly uneven among different societal sectors (Abedin et al., 2014). The Sundarbans coastal region of West Bengal, with a female population of 2.14 million with adverse situation due to social, economic, and educational backwardness. (Mondal and Tripathy, 2023a). Findings indicate that women's empowerment remains limited (Mondal & Tripathy, 2023a). According to Roy et al. (2023b), limited access to financial resources and assets, exposure to abuse and intimidation, lower social standing, and lack of representation in policy development and decision-making are factors associated with the situation of women's communities. A total of 37% of employees said their engagement in their present positions is attributable to the lack of viable alternatives. Therefore, it is imperative to empower women fishers within the business to improve their socioeconomic status (Mukherjee, 2018).

The examination of age demographics of fisherwomen in Sundarbans indicated that the age range of 10–20 years is responsible for fish seed collecting, including over 30% of the total respondents (Majumder, 2019). Simultaneously, individuals aged 22–54 years mostly participate in fish drying operations, comprising 65% of the population. The demographic under 21 years is considered a sporadic workforce for fish drying operations. The predominant proportion (40%) of mud crab collectors was within the age class of 31–40 years, followed by a significant (24%) age group of 54–65 years (Mondal & Tripathy, 2022; Roy et al., 2017). The creation of the Sundarbans tiger reserve in 1973 resulted in several women losing their livelihoods, since they were barred from fishing in the creeks next to critical tiger habitats and reserve forests, as well as from gathering prawn seeds along the coasts of populated islands. Women, accompanied by their children, relocate to different regions to engage in domestic labor during the off-peak season (Das et al., 2016). The restricted potential for income production and salary disparity relegates women-headed families to the lowest economic tier within the community. Women fishers often encounter sexual assault, and for unmarried girls, marriage possibilities are stigmatized, rendering them a familial obligation owing to the dowry system (Mitra & Barua, 2023; Salagrama, 2012). Women and children engaged in spawning collection are at increased risk of encountering crocodiles and sharks and may develop dermatological conditions due to prolonged exposure to saltwater. Additionally, the physical demands of bending can lead to spinal discomfort, and there is heightened susceptibility to various reproductive tract illnesses (Pitchaikani, 2020; Rabbitt et al., 2019). While women are indispensable to the sustenance and resilience of Sundarbans fishing communities, persistent gender-based barriers limit their economic advancement and social well-being. Addressing these challenges requires a holistic, gender-sensitive approach that integrates socioeconomic, cultural, and institutional reforms. There is a marked lack of recognition for women's fishing-related activities, largely due to the non-remunerative nature of their work and the absence of systematic data or economic valuation of subsistence fisheries. As a result, women's contributions are typically omitted from official statistics and policy considerations, perpetuating their invisibility within the sector.

3.3.2 Child Labor and Educational Challenges in the Small-Scale Fishing Communities of Sundarbans: A Struggle for Survival

Child labor is prevalent in the Sundarbans. Children aid their parents and relatives on fishing trips, engage in activities at fish landing centers, and actively participate in the capture and sorting of fry to enhance family income. Barua et al. (2024) also identified a bigger family size (5–6 individuals) that mostly relies on fishing, namely the collection of prawn fry. This practice demonstrates a systemic dependence on child labor due to a lack of other income-generating options. Along the southeastern coast of Bangladesh, children do not go fishing in the sea; rather, they perform individual fishing along the coast or work in fish landing centers or collect discarded fish to earn around 50 cents to $1 per day (Mukherjee et al., 2022; Neogi et al., 2016). Children in these areas are especially susceptible; drowning is a predominant cause of fatality, with rates reaching 243.8 per 100,000 for those aged 1–4 years (Gupta, 2021). Moreover, persistent undernutrition and health complications are widespread, exacerbated by climate disturbances and insufficient healthcare facilities (Mukherjee et al., 2012).

Illiteracy is widespread among the artisanal fishing populations of Sundarbans. To fulfill familial obligations, the children often forgo their education, with over 74% having discontinued their study. Fifty-six percent of minors and 80% of adult fry collectors employed in this industry are functionally illiterate (Mukherjee, 2018). A survey in the Sundarbans impact zone revealed that around 50% are illiterate, 35% have completed elementary education, and 5% are enrolled in secondary school (Mozumder, 2018). Molla et al. (2020) indicated that in the Satkhira district near the Sundarbans West Reserve Forest, women exhibit greater rates of illiteracy and semi-literacy (47%) compared to males (7%). A year-long study conducted in the Mongla Upazila and the vicinity of Mongla port within the Sundarbans region revealed that 77.21% of the surveyed fishermen were illiterate, while 13.97% had attained primary education, 5.15% had completed secondary education (grades 6–8), and 3.68% had reached the secondary level (grades 9–10). The current studies indicate that no individuals attained higher secondary or advanced education (Mishra et al., 2021). Khan et al. (2018) examined two artisanal fishing communities in Bamna Nil Dumor village (Sathkira district) and Joymunir Gul (Bagerhat) in Bangladesh, revealing that most fishing villages lacked schools, and the nearest educational institutions were often unreachable due to considerable distances or inadequate road infrastructure.

3.3.3 The Lure of White Gold

The prawn collectors had 'de-sacralised' the forest, transforming it into 'a big marketplace' where prawn collectors unscrupulously disturbed animals by walking in and out at all times of the day and night. Kais and Islam (2019)

Shrimps and prawns (White Gold) are referred to as "living dollars" (Nandi & Misra, 1987) in the Sundarbans. The surge in brackish (Bagda) and freshwater (Golda) prawn hatchery production over the past two decades has coincided with the decline of wild stocks (Kundu & Santhanam, 2025), attributed to their superior quality and elevated survival rates, particularly in the Bangladeshi region of the Sundarbans (Hasan et al., 2020). Annually, 1,500–2,000 million postlarvae (bagda and golda), valued at US$30 million, are harvested from the wild because of their superior hardiness compared to hatchery-produced fry. The initiation of fry collection (meendhara) took place in the early 1970s in the southwestern area of Bangladesh (Gupta et al., 2020) using conventional cotton nets. The traditional expertise in capturing prawn larvae was conveyed to other residents by the inhabitants of the Satkhira region, located on the border of Bangladesh and India (Bay of Bengal

Program) (Ahmed et al., 2010; Dutta et al., 2017; Roy & Sen, 2024). The collection of prawn seeds gained momentum after the disastrous cyclone of 1981 and 1988, when numerous islands were inundated by saline water, damaging crops worth Rs. 830 million and leaving the land infertile. Prawn seed collection peaked during 1993–1994 to 1996–1997 as (1) prawn seeds were easily available with high market value; (2) collection of the seeds was considered legal and practiced along the banks of one's island and without the stigma of other forest venture such as wood collection and poaching; (3) it could be undertaken during one's leisure time and became a major source of income for thousands of landless poor and unemployed people. The prawn collection resembles a "business" or a lottery (which can either "make them" or "break them"), relying on the variable prices set by shrimp dealers and entrepreneurs (Dasgupta et al., 2022). Over the past 30 years, rising demand has driven widespread commercial shrimp harvesting in coastal estuaries and creeks (Islam & Wahab, 2020). Modern PVC gill nets and mosquito nets have replaced traditional cotton nets due to their adaptable mesh sizes, durability, non-biodegradability, resilience, and low visibility in marine environments. The optimal time frame for prawn fry harvesting is from mid-February to mid-March. The fry collectors are regarded as "the lowest of the low" among the most destitute and marginalized individuals in the coastal settlements of Sundarbans. The collection of prawn fry in estuarine and riverbank environments involves around 423,000 individuals, including 40% women and children. Annual income from prawn fry collecting is projected at US$375, representing a 16% increase over the annual salary of day laborers in this location. The capture of 20–40 fry or postlarvae help them to earn between US$0.45 and US$1.10 per day. However, female collectors earn half that of male (around US $60 compared to around US$110 per year) and sometimes younger than their male counterparts (10.6 and 12.3 years, respectively) (Danda, 2019). The harvesting of tiger prawn seed generates immediate income, which mostly facilitates the spread of excessive alcohol use in the area. Chantarasri (1994) documented that in the peak season throughout the three coastal districts of Bangladesh (Satkhira, Khulna, and Cox's Bazar), the total collection of prawn larvae was estimated at 75,000 in 1990, escalating to 1,453 million in 1993, involving over 100,000 fry collectors. Currently, over 400,000 individuals, mostly women and children, are engaged in prawn and prawn postlarvae fishing along the coast of Bangladesh. Annually, around 1,500–2,000 million postlarvae, with a value of US$30 million, are harvested from the wild. The fishing of postlarvae substantially enhances both the local and national economy, as well as foreign currency revenues. During the 2010–2011 fiscal year, Bangladesh exported 54,891 shellfish worth US$446 million, with prawns accounting for 30% of this value (Chowdhury et al., 2012). The rising demand has a possible adverse impact on biodiversity due to elevated bycatch levels. An estimation indicates that 3 billion *Penaeus monodon* are captured each year, accompanied by 300 billion bycatches; for every Golda or Bagda fry captured, an average of 1,341 individuals of other species are also caught, significantly affecting biodiversity and capture fisheries production (Chatterjee et al., 2018). The capture of *Macrobrachium rosenbergii* larvae leads to the incidental catch of economically significant prawn species such as *M. malcolmsonii*, *M. villosimanus*, *M. mirabilis*, *M. birmanicum*, *M. impolitesse*, and *M. dayanum*; shrimp—*P. monodon*, *P. indicus*, *P. merguiensis*, and

Metapenaeus monoceros; fish—*Tenualosa ilisha*, *G. chapra*, and *L. calcarifer*; as well as crabs, snails, oysters, mollusks, turtle (*Lepidochelys olivacea*), and many juvenile species. Approximately 240 operative days result in the extinction of 181.4 million seeds of noneconomic kinds of shell finfish (Berkes, 2015; Brugere et al., 2019). Research done along a 40 km segment of the Hooghly estuary, from Kulpi to Namkhana, revealed that 5,000 seed collectors from 21 communities used 7,000 nets to harvest 2,500 seeds of *P. monodon* each day (over 8 hours), constituting 28% of their entire mixed fry catch. Between 1994 and 1998, the average yearly capture of seeds in six principal estuaries of Sundarbans ranged from 8,174 to 10,062 individuals per net per day. Following the segregation of tiger shrimps (2.9–4.0%), the remaining seeds were eradicated in the estuaries of Hooghly (930.0 million), Matla (568.8 million), Saptamukhi (775.0 million), Thakuran (233.4 million), Bidya (196.2 million), and Jhilla (160.0 million). An annual decline of 456 million shrimp was documented in the estuaries of the Matla, Thakuran, and Saptamukhi rivers (Bhaumik et al., 1992). The harvesting of prawn larvae impacts the recruitment of migratory species, which are seen as significant contributors to the progressive constriction of the wide base of the mangrove food pyramid, ultimately resulting in the collapse of the whole food chain over time (Das, 2009). The prawn collectors honor the forest as they approach the forest "in peace," consider tigers as "brothers," and thereby distribute the forest's resources equitably between themselves and wildlife, viewing the forest as a "common food storehouse" and as a sort of economic "agreement" concerning the equitable allocation of food and resources, with "proper management of the forest commons." Despite the prohibition enacted by the Indian Government in 1999 and the Department of Fisheries (DOF) in Bangladesh since 2000 (Nordhagen et al., 2020), the Government has inadequately enforced this injunction in the Sundarbans due to a lack of alternative livelihoods, a preference for wild varieties, reluctance, and insufficient resources (Ahmed & Troell, 2010; Hoq et al., 2024).

3.3.4 The Collectors of Crabs and Shellfish

The Sundarbans is home to a diverse array of crab species: 405 samples of brachyuran crabs comprising 6 families, 7 genera, and 14 species has been documented by Gokul (2017), with the mud crab (*S. serrata*), locally known as "nona kankra," being the most economically significant. Other notable species include *Scylla tranquebarica*, *Scylla olivacea*, the mangrove crab (*Metopograpsus* spp.), *Ocypode* spp., as well as smaller varieties such as *Uca* (fiddler crabs). These crabs play a vital ecological role in the mangrove ecosystem by aiding nutrient cycling, sediment aeration, and serving as prey for a range of wildlife, including birds and fish. Crab harvesting occurs in Sundarbans, with a minimum of five species consistently captured. The mud crab (*S. serrata*) has significant commercial value. The projected crab catch from the Sundarbans is around 1,200–1,500 tons/year. The crab fishery has seen significant fishing pressure, resulting in diminished harvests since 1998, and is now overexploited (Rahman et al., 2018). From a socioeconomic perspective, crab collection and fattening have emerged as important livelihood activities for many coastal communities in the Sundarbans. The mud crab is highly valued for export due to its large size and high meat content, fetching premium prices in both domestic and international markets, especially in Southeast Asian countries. Crab farming and wild collection provide crucial supplementary income, especially for women and marginalized groups, who are often engaged in the collection, sorting, and fattening processes. This sector not only enhances household income but also contributes to the region's economic resilience and food security. However, unsustainable harvesting practices and habitat degradation pose significant threats to crab populations, highlighting the need for responsible management to ensure the continued ecological and economic benefits of these species in the Sundarbans.

The predominant technique in artisanal crab fishing relies on conventional or low-tech equipment, including hand nets, seine nets, gill nets, baited traps, and indigenous hooks used in the stick method, employed by over 60% of crab collectors. Fishing is carried out on the open mudflats during the low tides by hook and stick method and baited line method (by using shrimps, small fishes, bycatch, and even juvenile fiddler crabs); however, the use of a small, traditional sailing vessel (dingi) is also commonly used by professionals. Catches are landed at small landing sites located in or close to local communities, usually without any significant infrastructure for crab storage. The mud crab market in Sundarbans is inadequately controlled and recorded; the underreporting of landings mostly results from an absence of rules governing exploitation and fisheries management. Despite its artisanal characteristics, the crab fishery has a complex market chain that incorporates intermediaries at various levels, resulting in diminished income for the fishermen (Sen & Homechaudhuri, 2017a). There exists a notable price disparity for this species in the market, resulting in selective capture: berried females command more value than non-berried females, while males are the least valued. Females devoid of eggs are sold for fattening in ponds, thereafter, exported as berried females. Mud crab fishing occurs year-round inside the SRF, with 30% consisting of berried females (Hoq, 2007). Crab harvesting occurs in Sundarbans, where a minimum of five species is routinely captured. The mud crab (*S. serrata*) has significant commercial value. The projected crab yield from the Sundarbans is around 1,200–1,500 tons annually. The crab fishery has seen significant fishing pressure, resulting in diminished harvests since 1998, and is deemed overexploited (Rahman et al., 2018). The Thopa method employs a hook and stick approach, utilizing bamboo sticks measuring 5–6 feet in length. One end of each stick is fitted with a 10–12 inch fine nylon rope, while the opposite end is prepared with crab bait. The Thopa is engineered to accommodate brick weights ranging from 100 to 120 g. The activity mostly occurs at high tide, known as Bhora Jowar (between Troyodashi and Saptami of the Bengali calendar), but crab harvesting during low tide, referred to as Mora Kotal, is predominantly undertaken by women. During Marani (between Panchami and Dasami of the Bengali calendar), local inhabitants selectively harvest crabs referred to as Kankra kurano. Among the crab collectors, 80% are male and 20% are female. The crab catchers gather crabs in teams of four to six individuals or, more often, in groups of 10 persons. Occasionally, adolescents aged 13–14 engage actively with adults. The crab-catching expedition last for 6–10 days. Crab catchers get 25–50 rupees for collecting around 140–160 crabs. Catches are brought to small landing locations situated inside or near local settlements, often without substantial infrastructure for crab storage. Canning has been the predominant landing point in the Indian Sundarbans since 1990, exceeding 12 carats. The mud crab market in Sundarbans is inadequately controlled and recorded; the underreporting of landings mostly results from an absence of rules governing exploitation and fisheries management. Crab chamber aquaculture is mostly situated at Sandeshkhali, Gosaba, Basanti, Minakha, Hingalgunj, and Hasnabad inside the Indian segment of the Sundarbans. Crabs are cultivated for 15– 20 days for export. The current unemployment rate in several industries is compelling numerous jobless individuals to engage in traditional crab harvesting. The "middlemen" (crab mongers) engaged in small-scale artisanal crab fishing maintain direct communication with crab collectors and often serve as brokers for larger collectors (Gibbon, 1997). A ban on crab collection often exists during certain months; however, compliance with this restriction is frequently lacking. During the off-season, poor crab collectors have few employment prospects; but some transition to fishing, prawn fry collection, gastropod and oyster harvesting, or agricultural wage labor (Barua et al., 2022a). Hence, the Sundarbans' crab fishery, vital for local livelihoods—especially women and marginalized groups—faces over-exploitation and market inequities. Traditional, low-tech harvesting persists amid weak regulation and habitat threats. Income opportunities are limited and seasonal, requiring urgent, sustainable management to ensure ecological balance and economic resilience for coastal communities.

3.4 CONCLUSION

Deltas are important social, political, and environmental systems that significantly impact national economies (Szabo et al., 2016). Climate change has increased the vulnerability of the Sundarbans, where large populations rely on natural resources but struggle with inadequate coping strategies (Faisal & Parveen, 2004; Wang et al., 2017). Evaluating small-scale fishers' vulnerability hinges on socioeconomic factors and realistic assessments of adaptation measures. Ecosystem-based adaptation is increasingly relevant globally, yet merely projecting multi-hazards is insufficient to combat climate change effects. Reliable predictions require databases of multifactorial experiments on local communities, considering biological acclimation. Forecasting future shorelines is challenging despite advanced technologies; factors like sediment supply and offshore geology must be included. Collaborative efforts among governments, media, the private sector, and civil society are vital for effective disaster preparedness. Identifying vulnerable hotspot regions in the Sundarbans is crucial for developing targeted emergency strategies. A study reported that vulnerable communities are concentrated along the delta's seaward edge (Mandal et al., 2025). Economic considerations for ecosystem-based adaptation and coastal resource conservation are essential, alongside site-specific evaluations of ecosystem services. Assessing the costs and benefits of adaptation strategies for SSFs against climate risks is necessary to reduce vulnerability. Local initiatives can enhance resilience, making locally driven adaptation pathways (Micheli et al., 2024).

NOTES

1 A 3D digital representation of a planet's surface, showing "bare earth" elevation data without objects like trees or buildings.
2 Are a set of scenarios that describe potential future developments for global society, economics, and demographics up to 2100.
3 CPUE, or Catch Per Unit Effort, is a fisheries metric that measures the amount of fish caught in relation to the amount of fishing effort used, such as hours, gear, or trips. It serves as an indirect index of a fish species' abundance or population size. By analyzing changes in CPUE over time, researchers and fisheries managers can infer shifts in stock health, identify trends in resource availability, and make informed decisions for effective fisheries management.
4 Is a term used by UNESCO to recognize sites with exceptional cultural or natural significance to humanity.

BIBLIOGRAPHY

Abdullah-Al-Mamun, Masum, K. M., Raihan Sarker, A. H. M., & Mansor, A. (2017). Ecosystem services assessment using a valuation framework for the Bangladesh Sundarbans: Livelihood contribution and degradation analysis. *Journal of Forestry Research*, *28*(1), 50–75. https://doi.org/10.1007/s11676-016-0275-5

Abedin, M. A., Habiba, U., & Shaw, R. (2014). Community perception and adaptation to safe drinking water scarcity: Salinity, arsenic, and drought risks in coastal Bangladesh. *International Journal of Disaster Risk Science*, *5*(2), 110–124. https://doi.org/10.1007/s13753-014-0021-6

Aburto-Oropeza, O., Ezcurra, E., Danemann, G., Valdez, V., Murray, J., & Sala, E. (2008). Mangroves in the Gulf of California increase fishery yields. *Proceedings of the National Academy of Sciences of the United States of America*, *105*(30), 10456–10459. https://doi.org/10.1073/pnas.0804601105

Adger, W. N. (2006). Vulnerability. *Global Environmental Change*, *16*(3), 268–281. https://doi.org/10.1016/j.gloenvcha.2006.02.006

Agarwal, S., Fazli, P., Zaman, S., Pramanick, P., & Mitra, A. (2019). Seasonal variability of acidification in major estuaries of Indian Sundarbans. *Global Journal of Engineering Research*, *6*(4), 493–498.

Agarwal, S., Islam, S., Zaman, S., & Mitra, A. (2020). Carbon stocks in the mangrove vegetation of Lothian wild life sanctuary of Indian Sundarbans: Implication for climate change mitigation. *Global Journal of Engineering Science Research*, *4*(10), 1–9.

Agarwal, S., & Mitra, A. (2018). Bioaccumulation pattern of heavy metals in water, sediment and barnacle (Balanus balanoides) in coastal West Bengal. *International Journal of Scientific Basic and Applied Research*, *8*(11), 683–693.

Agarwal, S., Pramanick, P., Zaman, S., Roy, A., & Mitra, A. (2016). Status of stored carbon in mangroves of lower Gangetic delta. *Global Journal of Engineering Research*, 7–16.

Aguilera, S., Cole, J., Finkbeiner, E., Cornu, E., Ban, N., Carr, M., Cinner, J., Gelcich, S., Lopes, P. F., Pomeroy, R., Starr, R., & Broad, K. (2015). Managing small-scale commercial fisheries for adaptive capacity: Insights from dynamic social-ecological drivers of change in Monterey Bay. *PLoS One*, *10*(3), Article e0118992. https://doi.org/10.1371/journal.pone.0118992

Aguirre, B. E. (Ed.). (2015). *Disaster studies*. Palgrave Macmillan, Springer.

Ahmad, O. A., Jamal, M. T., Almalki, H. S., Alzahrani, A. H., Alatawi, A. S., & Haque, M. F. (2025). Microplastic pollution in the marine environment: Sources, impacts, and degradation. *Journal of Advanced Veterinary and Animal Research*, *12*(1), 260.

Ahmed, A. (2025). *Microplastic characterization and factors influencing its abundance in coastal wetlands: Insights from the world's largest mangrove ecosystem, Sundarbans*. Environmental Science and Pollution Research. https://doi.org/10.1007/s11356-025-36044-9

Ahmed, A., Mahmud, H., & Sohel, M. S. I. (2021a). DPSIR framework to analyze anthropogenic factors influence on provisioning and cultural ecosystem services of Sundarbans east reserve Forest. *Bangladesh. Regional Studies in Marine Science*, *48*, 102042. https://doi.org/10.1016/j.rsma.2021.102042

Ahmed, A., Mahmud, H., & Sohel, M. S. I. (2021c). DPSIR framework to analyze anthropogenic factors influence on provisioning and cultural ecosystem services of Sundarbans east reserve Forest, Bangladesh. *Regional Studies in Marine Science*, *48*, Article 102042. https://doi.org/10.1016/j.rsma.2021.102042

Ahmed, N., & Troell, M. (2010). Fishing for prawn larvae in Bangladesh: An important coastal livelihood causing negative effects on the environment. *Ambio*, *39*(1), 20–29.

Ahmed, N., Troell, M., Allison, E. H., & Muir, J. F. (2010). Prawn postlarvae fishing in coastal Bangladesh: Challenges for sustainable livelihoods. *Marine Policy*, *34*(2), 218–227.

Ahmed, S., Pramanick, P., Zaman, S., & Mitra, A. (2024). Spatial variability in carbon storage among dominant mangrove species in the Indian Sundarbans. In *Forests and climate change: Biological perspectives on impact, adaptation, and mitigation strategies* (pp. 691–707). Singapore: Springer Nature Singapore.

Ahmed, S., Zaman, S., & Mitra, A. (2021b). Carbon sequestration in a common true mangrove flora, Excoecariaagallocha of Indian Sundarbans. In A. Mitra, M. M. Calma, & S. P. Chakrabarty (Eds.), *Natural resources and their ecosystem services* (Vol. II, pp. 138–143). HSRA Publication.

Akbar Hossain, K., Masiero, M., & Pirotti, F. (2022). Land cover change across 45 years in the world's largest mangrove forest (Sundarbans): The contribution of remote sensing in forest monitoring. *European Journal of Remote Sensing*, *57*(1), 1–17. https://doi.org/10.1080/22797254.2022.2097450

Akbar Hossain, K., Masiero, M., & Pirotti, F. (2024). Land cover change across 45 years in the world's largest mangrove forest (Sundarbans): the contribution of remote sensing in forest monitoring. *European Journal of Remote Sensing*, *57*(1), 2097450.

Akter, S. (2020). Social cohesion and willingness to pay for cyclone risk reduction: The case for the coastal embankment improvement project in Bangladesh. *International Journal of Disaster Risk Reduction*, *48*(2), 101–120. https://doi.org/10.1016/j.ijdrr.2020.101580

Alam, E., & Dominey-Howes, D. (2015). A new catalogue of tropical cyclones of the northern bay of Bengal and the distribution and effects of selected landfalling events in Bangladesh. *International Journal of Climatology*, *35*(6), 801–835. https://doi.org/10.1002/joc.4035

Ali, M. Z., Carlile, G., & Giasuddin, M. (2020). Impact of global climate change on livestock health: Bangladesh perspective. *Open Veterinary Journal*, *10*(2), 178–188. https://doi.org/10.4314/ovj.v10i2.9

Allison, E. H., Ratner, B. D., Åsgård, B., Willmann, R., Pomeroy, R., & Kurien, J. (2012). Rights-based fisheries governance: from fishing rights to human rights. *Fish and Fisheries*, *13*(1), 14–29.

Alongi, D. M. (2020). Global significance of mangrove blue carbon in climate change mitigation. *Science*, *2*(3), 67. https://doi.org/10.3390/sci2030067

Aly, E., Elsawah, S., & Ryan, M. J. (2022). Aligning the achievement of SDGs with long-term sustainability and resilience: An OOBN modelling approach. *Environmental Modelling & Software, 150*, 105360. https://doi.org/10.1016/j.envsoft.2022.105360

Amin, G., Biswas, S., Zaman, S., Pramanick, P., Trivedi, S., & Mitra, A. (2015). Prediction of dissolved oxygen in Indian Sundarbans: Vision 2050. *International Advanced Research Journal of Science, Engineering and Technology, 2*(12), 31–33.

Auerbach, L. W., Goodbred Jr, S. L., Mondal, D. R., Wilson, C. A., Ahmed, K. R., Roy, K., Steckler, M. S., Small, C., Gilligan, J. M., & Ackerly, B. A. (2015b). Flood risk of natural and embanked landscapes on the Ganges–Brahmaputra tidal delta plain. *Nature Climate Change, 5*(2), 153–157. https://doi.org/10.1038/nclimate2472

Auerbach, L. W., Goodbred, S. L., Jr., Mondal, D. R., Wilson, C. A., Ahmed, K. R., Roy, K., Steckler, M. S., Small, C., Gilligan, J. M., & Ackerly, B. A. (2015a). Flood risk of natural and embanked landscapes on the Ganges-Brahmaputra tidal delta plain. *Nature Climate Change, 5*(2), 153–157. https://doi.org/10.1038/nclimate2472

Aziz, A., & Paul, A. R. (2015). Bangladesh Sundarbans: Present status of the environment and biota. *Diversity, 7*(3), 242–269. https://doi.org/10.3390/d7030242

Bain, R. L., Hale, R. P., & Goodbred, S. L. (2019). Flow reorganization in an anthropogenically modified tidal channel network: An example from the southwestern Ganges-Brahmaputra-Meghna Delta. *Journal of Geophysical Research: Earth Surface, 124*, 2141–2159. https://doi.org/10.1029/2018JF004996

Banerjee, K., Amin, G., Chakraborty, S., Pramanick, P., Ghosh, R., Fazli, P., Zaman, S., Biswas, P., Pal, N., & Mitra, A. (2014). Bioaccumulation pattern of heavy metals in saltmarsh grass (Porteresiacoarctata) of Indian Sundarbans. *Journal of Energy, Environment and Carbon Credits, 4*(3), 1–10.

Banerjee, K., Gatti, R. C., & Mitra, A. (2017). Climate change-induced salinity variation impacts on a stenoecious mangrove species in the Indian Sundarbans. *Ambio, 46*, 492–499. https://doi.org/10.1007/s13280-016-0839-9

Banerjee, K., Gatti, R. C., & Mitra, A. (2018). Climate change-induced salinity variation impacts on a stenoecious mangrove species in the Indian Sundarbans. *Ambio, 46*(2), 492–499. https://doi.org/10.1007/s13280-016-0839-9

Banerjee, K., Roy Chowdhury, M., Sengupta, K., Sett, S., & Mitra, A. (2012). Influence of anthropogenic and natural factors on the mangrove soil of Indian Sundarbans wetland. *Archives of Environmental Science, 6*, 80–91.

Barratt, C., & Allison, E. H. (2014). Vulnerable people, vulnerable resources? Exploring the relationship between people's vulnerability and the sustainability of community-managed natural resources. *Development Studies Research, 1*(1), 16–27. https://doi.org/10.1080/21665095.2014.904079

Barua, P. (2021). Coping practices of coastal fishermen in response to climate change for southern coastal belt of Bangladesh. *Social Values and Society, 3*(2), 74–80.

Barua, P., & Barua, C. (2024). Sustainable fisheries management approach of Bangladesh: A potential blue economy path. *Asian Journal of Research and Review in Agriculture, 6*(1), 109–125.

Barua, P., & Chakraborty, S. (2011). Assessment of aquatic health index for coastal aquaculture activity in and around South-east coast of Bangladesh. *Current Biotica, 5*(2), 180–195.

Barua, P., Chowdhury, S. S., & Mitra, A. (2024). Review on conservation approach study of largest mangrove ecosystem as the mitigation of climate change. *Journal of Climate Change, 10*(3), 31–42.

Barua, P., Islam, M., & Mitra, A. (2022a). Developing a sustainable coastal aquaculture value chain for the rural communities of Bangladesh. *IUP Journal of Supply Chain Management, 19*(1), 7–44.

Barua, P., & Rahman, S. H. (2019). Impact of river erosion on livelihood and coping strategies of displaced people in South-Eastern Bangladesh. *International Journal of Migration and Residential Mobility*, *2*(1), 34–55. https://doi.org/10.1504/IJMRM.2019.103275

Barua, P., Rahman, S. H., & Barua, M. (2022b). Building resilience towards climate change vulnerability: A case study of fishing communities in southern Bangladesh. In M. M. Islam (Ed.), *Small in scale big in contributions: Advancing knowledge of small-scale fisheries in Bangladesh*. TBTI Global Publication.

Barua, P., Rahman, S. H., Barua, S., & Rahman, I. M. M. (2020). Climate change vulnerability and responses of fisherfolk communities in the south-eastern coast of Bangladesh. *Journal of Water Conservation and Management*, *4*(1), 45–65.

Bedjeck, M.C., et al. (2010). 'Impacts of climate variability and change on fishery-based livelihood'. *Journal of Marine Policy*, 34, 375–383.

Béné, C. (2009). Are fishers poor or vulnerable? Assessing economic vulnerability in small-scale fishing communities. *The Journal of Development Studies*, *45*(6), 911–933.

Béné, C., & Friend, R. M. (2011). Poverty in small-scale fisheries: Progress in development studies. *Progress in Development Studies*, *11*, 119–144. https://doi.org/10.1177/146499341001100201

Béné, C., & Friend, R. M. (2020). Poverty in small-scale fisheries: Old issue, new analysis. *Progress in Development Studies*, *11*(2), 119–144. https://doi.org/10.1177/146499341001100203

Bennett, N. J., Dearden, P., & Peredo, A. (2015). Vulnerability to multiple stressors in coastal communities: A study of the Andaman coast of Thailand. *Climate and Development*, *7*(2), 124–141. https://doi.org/10.1080/17565529.2014.886993

Bera, A., Meraj, G., Kanga, S., Farooq, M., Singh, S. K., Sahu, N., & Kumar, P. (2022). Vulnerability and risk assessment to climate change in Sagar Island, India. *Water*, *14*(5), Article 823. https://doi.org/10.3390/w14050823

Bera, R., & Maiti, R. (2021). Multi hazards risk assessment of Indian Sundarbans using GIS based analytic hierarchy process (AHP). *Regional Studies in Marine Science*, *44*, 101766. https://doi.org/10.1016/j.rsma.2021.101766

Berenji, S., Nayak, P., & Shukla, A. (2021). Exploring values and beliefs in a complex coastal social-ecological system: A case of small-scale fishery and dried fish production in Sagar Island, Indian Sundarbans. *Frontiers in Marine Science*, *8*. https://doi.org/10.3389/fmars.2021.795973

Berkes, F. (2015). Coasts for people: Interdisciplinary approaches to coastal and marine resource management. *Marine Policy*, *39*(4), 920–935. https://doi.org/10.1016/j.marpol.2015.01.013

Bhadra, T., Saikia, P., & Das, A. P. (2023). Monitoring the mangroves of Indian Sundarbans using geospatial techniques. *ISPRS Archives*, *XLVIII-1/W2-2023*, 405–411. https://doi.org/10.5194/isprs-archives-XLVIII-1-W2-2023-405-2023

Bhaskaran, P. K., Nayak, S., Bonthu, S. R., Murty, P. L. N., & Kumar, P. (2013). Performance and validation of a coupled parallel ADCIRC–SWAN model for THANE cyclone in the Bay of Bengal. *Environmental Fluid Mechanics*, *13*(6), 601–623.

Bhattacharya, B., Sarkar, S. K., Bhattacharya, A., Chatterjee, M., Alam, A., Satpathy, K. K., & Jonathan, M. P. (2018). Effects of persistent organic pollutants on biodiversity in the Sundarbans biosphere reserve. *Environmental Monitoring and Assessment*, *190*(2). https://doi.org/10.1007/s10661-018-6469-x

Bhaumik, U., S.K.Saha & J.G.Chatterjee. 1992. Need for conservation to protect the Brackish-water Finfish & Shellfish Seed Resources in Sundarbans. Environment & Ecology 10(4): 919–922.

Birkmann, J., Cardona, O. D., Carreño, M. L., Barbat, A. H., Pelling, M., Schneiderbauer, S., Kienberger, S., Keiler, M., Alexander, D., Zeil, P., & Welle, T. (2013). Framing vulnerability, risk and societal responses: The MOVE framework. *Natural Hazards*, *67*, 193–211. https://doi.org/10.1007/s11069-013-0558-5

Biswas, C., Channarayapatna, S., & Pandey, C. N. (2024). Livelihood and beyond relations of local communities with Sundarban mangroves. *Journal of Environmental Management*, *340*(1), 118480. https://doi.org/10.1016/j.jenvman.2023.118480

Biswas, S., Pal, N., Zaman, S., & Mitra, A. (2019). Samosa and kachuri from mangrove associate species: An innovative utilization of coastal flora. *International Journal of Research in Applied Science and Engineering Technology*, *7*(1), 73–79. https://doi.org/10.22214/ijraset.2019.1014

Biswas, S., Zaman, S., & Mitra, A. (2020). Development of food products from mangrove associate species: A step towards health security. *Our Heritage*, *68*(8), 69–77.

Bitew, A. B., & Minale, A. S. (2025). Smallholder farmers' perceptions of climate variability and its risks across agroecological zones in the Ayehu watershed, upper Blue Nile Basin, Ethiopia. *Environmental and Sustainability Indicators*, *25*. https://doi.org/10.1016/j.indic.2024.100546

Blythe, J. L., Murray, G., & Flaherty, M. (2014). Strengthening threatened communities through adaptation: Insights from coastal Mozambique. *Ecology and Society*, *19*(2), 50–70. https://doi.org/10.5751/ES-06408-190206

Bohle, H. (2001). *Vulnerability and criticality: Perspectives from social geography*. IHDP Update.

Bohle, H. G., Downing, T. E., & Watts, M. (2014). Climate change and social vulnerability: Toward a sand geography of food. *Global Environmental Change*, *25*(1), 37–48. https://doi.org/10.1016/j.gloenvcha.2014.01.005

Bomer, E. J., Wilson, C. A., Hale, R. P., Hossain, A. N. M., & Rahman, F. M. A. (2020). Surface elevation and sedimentation dynamics in the Ganges-Brahmaputra tidal delta plain, Bangladesh: Evidence for mangrove adaptation to human-induced tidal amplification. *Catena*, *187*(104312), 1–12. https://doi.org/10.1016/j.catena.2019.104312

Brammer, H. (2014). Bangladesh's dynamic coastal regions and sea-level rise. *Climate Risk Management*, *1*, 51–62.

Brown, S., & Nicholls, R. J. (2015). Subsidence and human influences in mega deltas: The case of the Ganges–Brahmaputra–Meghna. *Science of the Total Environment*, *527*, 362–374. https://doi.org/10.1016/j.scitotenv.2015.04.124

Brugère, C., Aguilar-Manjarrez, J., Beveridge, M. C., & Soto, D. (2019). The ecosystem approach to aquaculture 10 years on–a critical review and consideration of its future role in blue growth. *Reviews in Aquaculture*, *11*(3), 493–514.

Brugère, C., Holvoet, K., & Allison, E. H. (2008). Livelihood diversification in coastal and inland fishing communities: Misconceptions, evidence and implications for fisheries management. *Coastal Management*, *40*(4), 55–75.

Cai, Q., Fang, D., & Chen, B. (2025). Social vulnerability assessment for air pollution in China. *Environmental Impact Assessment Review*, *114*, 107925. https://doi.org/10.1016/j.eiar.2025.107925

Carreño, M. L., Cardona, O. D., & Barbat, A. H. (2012). New methodology for urban seismic risk assessment from a holistic perspective. *Bulletin of Earthquake Engineering*, *10*, 547–565. https://doi.org/10.1007/s10518-011-9302-2

Census of India. (2001). *District census handbook — South 24 Parganas*. Office of the Registrar General & Census Commissioner, Government of India.

Cernexv, T., & Fenner, R. (2020). The importance of achieving foundational sustainable development goals in reducing global risk. *Futures*, *115*(9), Article 120–140. https://doi.org/10.1016/j.futures.2019.102492

Chacraverti, S. (2014). *The Sundarbans fishers coping in an overly stressed mangrove estuary*. SIDALC.

Chacraverti, S. (2015a). *The Sundarbans fishers: Coping in an overly stressed mangrove estuary.*

Chacraverti, S. (2015b). *The Sundarbans fishers: Coping in an overly stressed mangrove estuary.*

Chakraborty, A., Datta, U., Zaman, S., Pal, N., Pramanick, P., & Mitra, A. (2018). Spatio-temporal variation of stored carbon in Sonneratia apetala: A case study from Indian Sundarbans. *International Journal of Innovative Research in Science and Engineering Technology*, *5*(3), 2975–2980.

Chakraborty, P., Zhang, G., Eckhardt, S., Li, J., Breivik, K., Lam, P. K., Tanabe, S., & Jones, K. C. (2019). Occurrence and distribution of pesticides in sediments and water from the Hooghly estuary and Sundarbans mangrove ecosystem. *Science of the Total Environment*, *682*, 646–656. https://doi.org/10.1016/j.scitotenv.2019.05.154

Chakraborty, S., Rudra, T., Guha, A., Ray, A., Pal, N., & Mitra, A. (2016). Spatial variation of heavy metals in Tenualosailisha muscle: A case study from the lower Gangetic delta and coastal West Bengal. *International Journal of Innovative Science, Engineering & Technology*, *3*(4), 1–14.

Chakraborty, S., Zaman, S., & Mitra, A. (2017). Bioaccumulation of heavy metals by dominant oyster in Indian Sundarbans. *Journal of Science, Engineering and Health Management*, *1*(2), 34–39.

Chakraborty, S., Zaman, S., Pramanick, P., Raha, A. K., Mukhopadhyay, N., Chakravartty, D., & Mitra, A. (2013). Acidification of Sundarbans mangrove estuarine system. *Discovery Nature*, *6*(14), 14–20.

Chakraborty, S. K. (2011). Mangrove ecosystem of Sundarbans, India: Biodiversity, ecology, threats and conservation. In *Mangroves: Ecology, biology and taxonomy* (pp. 87–116). Nova Science Publishers.

Chakravartty, D., Mondal, A., Raychowdhury, P., Bhattacharya, S. B., & Mitra, A. (2017). Role of aquaponics in the sustenance of coastal India–aquaponics is a solution for modern agriculture in ecologically sensitive Indian mangrove Sundarbans: A review. *International Journal of Fisheries and Aquatic Studies*, *5*(2), 441–448.

Chamberlain, D. A., Phinn, S. R., & Possingham, H. P. (2021). Mangrove forest cover and phenology with landsat dense time series in Central Queensland, Australia. *Remote Sensing*, *13*(15), 3032. https://doi.org/10.3390/rs13153032

Chambers, R. (1989) Editorial introduction: Vulnerability, coping and policy. *IDS Bulletin*, *20*, 1–7. http://dx.doi.org/10.1111/j.1759-5436.1989.mp20002001.x

Chand, B. K., Trivedi, R. K., Biswas, A., Dubey, S. K., & Beg, M. M. (2021). Study on impact of saline water inundation on freshwater aquaculture in Sundarban using risk analysis tools. *Exploratory Animal and Medical Research*, *2*(II), 170–178.

Chantarasri, S. (1994). Fisheries Resources Management for Sundarban Reserved Forest. Integrated resource Development of the Sundarbans Reserved Forest, Bangladesh. BGD/84/056, UNDP and FAO.

Chatterjee, N., Manna, S., Ray, A., Das, S., Rana, N., Banerjee, A., Basu, M., Das, M., Sarkar, J., Khan, P., Sen, S., Mukherjee, S., & Ray, S. (2024a). Microplastics contamination in two species of gobies and their estuarine habitat of Indian Sundarbans. *Marine Pollution Bulletin*, *198*, Article 115857. https://doi.org/10.1016/j.marpolbul.2023.115857

Chatterjee, P., Bhuinya, N., & Mondal, S. (2009). Traditional fishers in the Sundarban tiger reserve: A study on livelihood practice under protected area. Society for Direct Initiative for social and health action (DISHA). https://www.dishaearth.org/Traditional%20Fishers%20in%20The%20Sundarban%20Tiger%20Reserve.pdf

Chatterjee, P., Bhuinya, N., & Mondal, S. (2018). Traditional fishers in the Sundarban Tiger Reserve: A study on livelihood practice under protected area. *Wetland Management*, *60*(3), 80–95.

Chatterjee, T. (2017). Climate change and its collision on fisheries resource in Sundarbans region of south 24 pgns, West Bengal. *Suresh Gyan Vihar University International Journal of Environment, Science and Technology*, *3*(2), 10–13.
Chatterjee, U., Shaw, R., Kumar, S., Raj, A. D., & Das, S. (2024b). *Climate crisis: Adaptive approaches and sustainability*. Springer Nature.
Chavez, F. P., Ryan, J., Lluch-Cota, S. E., & Ñiquen, C. M. (2003). From anchovies to sardines and back: multidecadal change in the Pacific Ocean. *Science*, *299*(5604), 217–221.
Chen, W., Cutter, S. L., Emrich, C. T., & Shi, P. (2014). Measuring social vulnerability to natural hazards in the Yangtze River Delta region, China. *International Journal of Disaster Risk Science*, *4*, 169–181. https://doi.org/10.1007/s13753-013-0018-6
Chowdhury, A. (2017). Environmental impact of coal based power plant of Rampal on the Sundarbans (world largest mangrove forest) and surrounding areas. *MOJ Ecology & Environmental Sciences*, *2*(3), Article 00022. https://doi.org/10.15406/mojes.2017.02.00022
Chowdhury, A., Naz, A., Bhattacharyya, S., & Sanyal, P. (2021). Dynamics of salinity intrusion in the surface and ground water of Sundarban biosphere reserve, India. *IOP Conference Series: Earth and Environmental Science*, *944*, Article 012061. https://doi.org/10.1088/1755-1315/944/1/012061
Chowdhury, A., Sanyal, P., & Maiti, S. K. (2016). Dynamics of mangrove diversity influenced by climate change and consequent accelerated sea level rise at Indian Sundarbans. *International Journal of Global Warming*, *9*(4), 486–506. https://doi.org/10.1504/IJGW.2016.076333
Chowdhury, M. S. N., Hossain, M. S., Das, N. G., & Barua, P. (2011). Small-scale fishermen along the Naaf River, Bangladesh in crisis: A framework for management. *Mesopotamian Journal of Marine Science*, *26*(2), 146–169.
Chowdhury, S. R., Hossain, M. S., Shamsuddoha, M., & Khan, S. M. M. H. (2012). *Coastal fishers' livelihood in peril: sea surface temperature and tropical cyclones in Bangladesh*. Dhaka, Bangladesh: Center for Participatory Research and Development.
Coldren, G. A., Langley, J. A., Feller, I. C., & Chapman, S. K. (2019). Warming accelerates mangrove expansion and surface elevation gain in a subtropical wetland. *Journal of Ecology*, *107*(1), 79–90. https://doi.org/10.1111/1365-2745.13049
Colette, A. (2017). *Case studies on climate change and world heritage*. UNESCO World Heritage Centre.
Cutter, S. L. (2020). Community resilience, natural hazards, and climate change: Is the present a prologue to the future? *Norsk Geografisk Tidsskrift-Norwegian Journal of Geography*, *74*(3), 200–208.
Cutter, S. L., Boruff, B. J., & Shirley, W. L. (2003). Social vulnerability to environmental hazards. *Social Science Quarterly*, *84*, 242–261. https://doi.org/10.1111/1540-6237. 8402002
Daly, P. T., & Winter, T. (Eds.). (2012). *Routledge handbook of heritage in Asia*. Routledge.
Danda, A. A. (2019). Environmental security in the Sundarban in the current climate change era: Strengthening India-Bangladesh cooperation. *Fisheries Management*, *35*(3), 80–95.
Das, P., Das, A., & Roy, S. (2016). Shrimp fry (meen) farmers of Sundarban mangrove Forest (India): A tale of ecological damage and economic hardship. *International Journal of Agriculture and Food Research*, *5*(2), 45–65.
Das, S., Chanda, A., & Ghosh, T. (2021a). Socio-ecological system. In *Pond ecosystems of the Indian Sundarbans: An overview*, 1.
Das, S., Chatterjee, N., Choudhury, A., Ray, A., Rana, N., Banerjee, A., Khan, P., Basu, M., Sarkar, J., Das, M., Sen, S., Mukherjee, S., & Ray, S. (2024). Characterization and ecological risk assessment of microplastics accumulated in sea water, sand, sediment, shell water and selected tissues of hermit crab of Sundarban biosphere reserve. *Environmental Pollution*, *357*. https://doi.org/10.1016/j.envpol.2024.124484

Das, S., Ghosh, A., Chakraborty, P., & Banerjee, K. (2021b). Impact of agricultural pesticide residues on the estuarine ecosystem of the Sundarbans. *Environmental Science and Pollution Research, 28*(15), 18945–18960. https://doi.org/10.1007/s11356-021-12389-9

Das, S. C., Pullaiah, T., & Ashton, E. C. (Eds.). (2022). *Mangroves: Biodiversity, livelihoods and conservation*. Springer.

Das, T., Mukherjee, R. N., & Chaudhuri, K. S. (2009). Bioeconomic harvesting of a prey–predator fishery. *Journal of Biological Dynamics, 3*(5), 447–462.

Das, K., Bhattacharya, S., & Alam, N. M. (2020). Precipitation extremes and anomalies of the Indian Sundarban. *MAUSAM, 71*(4), 849–858. https://mausamjournal.imd.gov.in

Dasgupta, S., Wheeler, D., Bandyopadhyay, S., Ghosh, S., & Roy, U. (2021). Coastal dilemma: Climate change, public assistance and population displacement. *World Development, 150*(1), 200–220. https://doi.org/10.1016/j.worlddev.2021.105707

Dasgupta, S., Wheeler, D., & Santadas, G. (2022). Fishing in salty waters: Poverty, occupational saline exposure, and women's health in the Indian Sundarban. *Journal of Management and Sustainability, 12*(1), 50–65.

Dasgupta, S., Wheeler, D., Sobhan, M. I., Bandyopadhyay, S., Nishat, A., & Paul, T. (2020). *Coping with climate change vulnerability in the Sundarbans: Lessons from multidisciplinary studies*. International Bank for Reconstruction and Development.

Datta, R., & Mete, R. (2022). The socio-economic status of females of Sundarban regions in West Bengal. *Journal of Applied Development Economics, 1*(1), 33–41.

Deb, A. K. (2018). Fake blue revolution: Environmental and socio-economic impacts of shrimp culture in the coastal areas of Bangladesh. *Ocean & Coastal Management, 65*(1), 63–88. https://doi.org/10.1016/j.ocecoaman.2012.04.017

Deb, M., & Ferreira, C. M. (2017). Potential impacts of the Sunderban mangrove degradation on future coastal flooding in Bangladesh. *Journal of Hydro-Environment Research, 17*, 30–46. https://doi.org/10.1016/j.jher.2016.11.005

Defeo, O., Hyguiie, H., & Castilla, J. C. (2013). Governance and governability of coastal shellfisheries in Latin America and the Caribbean: Multi-scale emerging models and effects of globalization and climate change. *Current Opinion in Environmental Sustainability, 4*, 344–350. https://doi.org/10.1016/j.cosust.2013.07.003

Devaney, J. L., Pullen, J., Feller, I. C., & Parker, J. D. (2021). Low humidity and hypersalinity reduce cold tolerance in mangroves. *Estuarine, Coastal and Shelf Science, 248*, Article 107015. https://doi.org/10.1016/j.ecss.2020.107015

Dhar, I., Sengupta, G., Biswas, S., Sinha, M., & Mitra, A. (2021). Salinity: A major environmental factor in sustainability of the blue carbon. *Journal of Mechanics Continua and Mathematical Sciences, 16*(11), 34–42. https://doi.org/10.26782/jmcms.2021.11.00004

DISHA. (2006). *A study on – Corporate abuse in Sundarban.*

Douglas, I., Anderson, P. M. L., Goode, D., Houck, M. C., Maddox, D., Nagendra, H., & Yok, T. P. (2020). *The Routledge handbook of urban ecology*. Routledge. https://doi.org/10.4324/9780429506758

Dubey, S. K., Trivedi, R. K., Chand, B. K., Mandal, B., & Rout, S. K. (2017). Farmers' perceptions of climate change, impacts on freshwater aquaculture and adaptation strategies in climatic change hotspots: A case of the Indian Sundarban delta. *Environmental Development, 21*, 38–51. https://doi.org/10.1016/j.envdev.2016.12.002

Dutta, S., Chakraborty, K., & Hazra, S. (2017). Ecosystem structure and trophic dynamics of an exploited ecosystem of bay of Bengal, Sundarban estuary, India. *Fisheries Science, 83*(2), 145–159. https://doi.org/10.1007/s12562-016-1049-x

Eakin, H., Benessaiah, K., Barrera, J. F., Cruz-Bello, G. M., & Morales, H. (2012). Livelihoods and landscapes at the threshold of change: Disaster and resilience in a Chiapas coffee community. *Regional Environmental Change, 12*, 475–488. https://doi.org/10.1007/s10113-011-0263-4

Eakin, H., & Luers, A. L. (2006). Assessing the vulnerability of social-environmental systems. *Annual Review of Environment and Resources*, *31*, 365–394. https://doi.org/10.1146/annurev.energy.30.050504.144352

Egbueri, J. C., Ighalo, J. O., & Pande, C. B. (2024). *Microplastics in African and Asian environments*. Springer.

Environement Justice Foundation 2014 report

Eriksen, S., Schipper, E. L. F., Scoville-Simonds, M., Vincent, K., Adam, H. N., Brooks, N., Harding, B., Lenaerts, L., Liverman, D., Mills-Novoa, M., & Mosberg, M. (2021). Adaptation interventions and their effect on vulnerability in developing countries: Help, hindrance or irrelevance? *World Development*, *141*, Article 105383. https://doi.org/10.1016/j.worlddev.2020.105383

Eriksson, H., Albert, J., Albert, S., Warren, R., Pakoa, K., & Andrew, N. (2017). The role of fish and fisheries in recovering from natural hazards: Lessons learned from Vanuatu. *Environmental Science & Policy*, *76*, 50–58. https://doi.org/10.1016/j.envsci.2017.06.012

Eslamian, S. S. (2009). Basin ecology and environment (BEE). *International Journal of Ecological Economic and Statistics*Ed., Special Issue Volume, CESER, *13*(W09), 1–99.

Etingoff, K. (2019). *Ecological resilience: Response to climate change and natural disasters*. Apple Academic Press.

Fabinyi, M. (2010). The intensification of fishing and the rise of tourism: Competing coastal livelihoods in the Calamianes Islands, Philippines. *Human Ecology*, *38*(3), 415–427. https://doi.org/10.1007/s10745-010-9329-z

Fahad, S., Hussain, S., Saud, S., Hassan, S., Ihsan, Z., Shah, A. N., Wu, C., Yousaf, M., Nasim, W., Alharby, H., Alghabari, F., & Huang, J. (2016). Exogenously applied plant growth regulators enhance the morpho-physiological growth and yield of rice under high temperature. *Frontiers in Plant Science*, *7*, Article 1250. https://doi.org/10.3389/fpls.2016.01250

Faisal, I. M., & Parveen, S. (2004). Food security in the face of climate change, population growth, and resource constraints: Implications for Bangladesh. *Environmental Management*, *34*(4), 487–498.

Fatema, K., Hawa, M., Masnoon, S., Alam, M., Islam, M., Hasan, M., Sujan, K. M., & Rashid, H. (2023). Microplastic pollution in surface waters and sediments matrices of the Sundarbans – The largest single block of tidal halophytic mangrove forest in the world. *Regional Studies in Marine Science*, *67*, Article 103226. https://doi.org/10.1016/j.rsma.2023.103226

FEMA. *National disaster recovery framework* (3rd ed. – Amended December 10, 2024

Finkl, C. W., & Makowski, C. (Eds.). (2017). *Coastal wetlands: Alteration and remediation* (Vol. 21). Springer. https://doi.org/10.1007/978-3-319-56179-0

Füssel, H. M. (2012). Vulnerability to climate change and poverty. In *Climate Change, Justice and sustainability: Linking climate and development policy* (pp. 9–17). Dordrecht: Springer Netherlands.

Gabler, C. A., Osland, M. J., Grace, J. B., Stagg, C. L., Day, R. H., Hartley, S. B., & McLeod, J. L. (2017). Macroclimatic change expected to transform coastal wetland ecosystems this century. *Nature Climate Change*, *7*(2), 142–147. https://doi.org/10.1038/nclimate3203

Gain, A. K., Benson, D., Rahman, R., Datta, D. K., & Rouillard, J. J. (2017). Tidal river management in the south West Ganges-Brahmaputra delta in Bangladesh: Moving towards a transdisciplinary approach? *Environmental Science & Policy*, *75*, 111–120. https://doi.org/10.1016/j.envsci.2017.05.020

Ghasemi, M., Badsar, M., Falahati, L., & Karamidehkordi, E. (2021). The mediation effect of rural women empowerment between social factors and environment conservation (combination of empowerment and ecofeminist theories). *Environment, Development and Sustainability*, *23*(6), 8473–8493. https://doi.org/10.1007/s10668-020-00977-7

Ghil, M. (2016). A mathematical theory of climate sensitivity or, how to deal with both anthropogenic forcing and natural variability? In *Climate change: Multidecadal and beyond* (pp. 31–51). World Scientific. https://doi.org/10.1142/9789814579933_0002

Ghosh, A., & Bhowmik, D. (2021). Contaminant pathways of organochlorine pesticides in the Sundarbans delta. *Water Research, 203*, Article 117450. https://doi.org/10.1016/j.watres.2021.117450

Ghosh, M. K., Kumar, L., & Langat, P. K. (2019). Geospatial modelling of the inundation levels in the Sundarbans mangrove forests due to the impact of sea level rise and identification of affected species and regions. *Geomatics, Natural Hazards and Risk, 10*(1), 1028–1046. https://doi.org/10.1080/19475705.2018.1564373

Ghosh, M. K., Kumar, L., & Roy, C. (2022). Mapping long-term changes in mangrove species composition and distribution in the Sundarbans. *Forests, 7*(12). https://doi.org/10.3390/f7120305

Ghosh, T., Bhandari, G., & Hazra, S. (2019). Geospatial modelling of inundation levels in the Sundarbans mangrove forests due to the impact of SLR. *Geomatics, Natural Hazards and Risk, 10*(1), 1802–1819. https://doi.org/10.1080/19475705.2018.1564373

Ghosh, S., Das, A., & Hembram, T. K. (2020). Analyzing climate variability and its effects in Sundarban Biosphere Reserve, India. *Environment, Development and Sustainability, 23*, 5157–5185. https://doi.org/10.1007/s10668-020-00682-5

Ghosh, S., & Mistri, B. (2021). Assessing coastal vulnerability to environmental hazards of Indian Sundarban delta using multi-criteria decision-making approaches. *Ocean & Coastal Management, 209*, 105641. https://doi.org/10.1016/j.ocecoaman.2021.105641

Gibbon, P. (1997). Prawns and piranhas: the political economy of a Tanzanian private sector marketing chain. *The Journal of Peasant Studies, 25*(1), 1–86.

Giri, C., Long, J., Abbas, S., Murali, R., Qamer, F. M., Pengra, B., & Thau, D. (2017). Distribution and dynamics of mangrove forests of South Asia. *Journal of Environmental Management, 148*, 101–111. https://doi.org/10.1016/j.jenvman.2014.01.020

Giri, S., Daw, T. M., Hazra, S., Troell, M., Samanta, S., Basu, O., & Chanda, A. (2022). Economic incentives drive the conversion of agriculture to aquaculture in the Indian Sundarbans: Livelihood and environmental implications of different aquaculture types. *Ambio, 51*, 1835–1849. https://doi.org/10.1007/s13280-022-01709-z

Gokul, A. (2017). Taxonomy and ecology of brachyuran crabs in Sunderbans. *Records of the Zoological Survey of India, 117*(2), 131–139. https://doi.org/10.26515/rzsi/v117/i2/2017/118011

Gopal, B., & Chauhan, M. (2006). Biodiversity and its conservation in the Sundarban Mangrove Ecosystem. *Aquatic Science* 68, 338–354. https://doi.org/10.1007/s00027-006-0868-8

Guha, I., & Roy, C. (2016). Climate change, migration and food security: Evidence from Indian Sundarbans. *Coastal Management, 50*(3), 50–65. https://doi.org/10.1080/08920753.2016.1116678

Guha, T., & Mitra, A. (2020). Salinity—a crucial factor in ecological sustainability for Sundarbans Mangrove ecosystem. In *Proceedings: Natural resources and their ecosystem services. HSRA Publication*, 27–35.

Gupta, N., Kanagavel, A., Dandekar, P., Dahanukar, N., Sivakumar, K., Mathur, V. B., & Raghavan, R. (2016). God's fishes: Religion, culture and freshwater fish conservation in India. *Coastal Management, 50*(2), 244–248. https://doi.org/10.1080/08920753.2016.1208882

Gupta, S. (2021). Disturbance of mangrove forest due to climate change: The prospects of Sundarban. *International Journal of Multidisciplinary: Applied Business and Education Research, 2*(12), 1306–1313.

Gupta, T., Booth, H., Arlidge, W., Rao, C., Manoharakrishnan, M., Namboothri, N., ... & Milner-Gulland, E. J. (2020). Mitigation of elasmobranch bycatch in trawlers: a case study in Indian fisheries. *Frontiers in Marine Science*, *7*, 571.

Hale, R., Bain, R., Goodbred, S., Jr., & Best, J. (2019). Observations and scaling of tidal mass transport across the lower Ganges-Brahmaputra delta plain: Implications for delta management and sustainability. *Earth Surface Dynamics*, *7*(1), 231–245. https://doi.org/10.5194/esurf-7-231-2019

Hasan, M., Monjur, B. M., Sajjad, H. M., Jobaer, A., Azam, K. C., Karim, A. A., & Nuruddin, K. H. (2018). The prospects of blue economy to promote Bangladesh into a middle-income country. *Open Journal of Marine Science*, *8*(3), 355–369. https://doi.org/10.4236/ojms.2018.83019

Hasan, M. M., Tulloch, R. L., Thomson, P. C., Raadsma, H. W., & Khatkar, M. S. (2020). Meta-analysis of genetic parameters of production traits in cultured shrimp species. *Fish and Fisheries*, *21*(6), 1150–1174.

Hasan, R., & Naser, M. N. (2018). Fishermen livelihood and fishery resources of the Sundarbans reserved forest along the Mongla port area Bangladesh. *International Journal of Fish and Aquatic Studies*, *4*(3), 468–475.

Hazra, S., Ghosh, T., DasGupta, R., and Sen, G. (2002). Sea level and associated changes in the Sundarbans. *Science and Culture*, 68(9–12), 309–321.

Hoq, M. E. (2007). An analysis of fisheries exploitation and management practices in Sundarbans mangrove ecosystem, Bangladesh. *Ocean & Coastal Management*, *50*(5-6), 411–427.

Hoq, M. E. (2020). Fisheries in the Sundarbans mangrove ecosystems of Bangladesh. *Aquaculture Asia*, *25*(4), 16–20.

Hoq, M. S., Uddin, M. T., Raha, S. K., & Hossain, M. I. (2024). Determinants of households' livelihood diversification strategies to adapt to natural hazards: Evidence from ecologically vulnerable haor region of Bangladesh. *Natural Hazards*, *114*(3), 3255–3291.

Hossain, P. R., Ludwig, F., & Leemans, R. (2018). Adaptation pathways to cope with salinization in south-west coastal region of Bangladesh. *Ecology and Society*, *23*(3), 40–60. https://doi.org/10.5751/ES-10215-230327

ISDR (International Strategy for Disaster Reduction). (2014). *Living with risk: A global review of disaster risk initiatives*. United Nations.

India Meteorological Department. (n.d.). *Cyclone e-Atlas; gridded rainfall and temperature data*. Government of India. https://www.imd.gov.in

Islam, M., & Wahab, M. (2020). A review on the present status and management of mangrove wetland habitat resources in Bangladesh with emphasis on mangrove fisheries and aquaculture. *Aquatic Biodiversity*, *30*(2), 65–190.

Islam, M. M., & Chuenpagdee, R. (2022). Towards a classification of vulnerability of small-scale fisheries. *Environmental Science & Policy*, *134*, 1–12. https://doi.org/10.1016/j.envsci.2022.04.003

Islam, M. M., Khan, M. I., Yeasmin, M. N., Barman, A., & Hannan, M. A. (2024). Socio-economic impacts of a poorly designed and managed Hilsa shad fish sanctuary in Bangladesh and learning for sustainability. *Environment, Development and Sustainability*, *26*(5), 12763–12787.

Islam, M. R. (2023). Environmental jeopardy and coping strategies of the small-scale fishers in the Bangladesh Sundarbans: The precedent of the world's largest mangrove. *Egyptian Journal Of Aquatic Biology And Fisheries*, *27*(6), 759–773.

Islam, M. S., & Haque, M. (2021). The mangrove-based coastal and nearshore fisheries of Bangladesh: Ecology, exploitation and management. *Reviews in Fish Biology and Fisheries*, *14*(2), 153–180. https://doi.org/10.1007/s11160-004-3769-8

Islam, S. N., & Gnauck, A. (2023). Mangrove wetland ecosystems in Ganges-Brahmaputra delta in Bangladesh. *Frontiers of Earth Science in China*, *20*(4), 439–448. https://doi.org/10.1007/s11707-008-0049-2

IWA Annual Report. (2008/2009). *Institute of Water. Administration. IWA*. Institute of Water Administration Inc.

Jabir, A. A., Hasan, G. J., & Anam, M. M. (2021). Correlation between temperature, sea level rise and land loss: An assessment along the Sundarbans coast. *Journal of King Saud University-Engineering Sciences*. https://doi.org/10.1016/j.jksues.2021.02.006

Jana, A., & Bhattacharya, A. K. (2013). Assessment of coastal erosion vulnerability around Midnapur-Balasore coast, eastern India using integrated remote sensing and GIS techniques. *Journal of the Indian Society of Remote Sensing*, *41*(3), 675–686.

Jana, H. K., Mitra, A., Zaman, S., Bose, R., & Raha, A. K. (2020). Will Avicennia alba thrive in climate change induced salinity rise? *International Journal of Scientific Research*, *3*(2), 459–461.

Jatoi, W. N., Mubeen, M., Ahmad, A., Cheema, M. A., Lin, Z., & Hashmi, M. Z. (2022). *Building climate resilience in agriculture*. Springer International Publishing. https://doi.org/10.1007/978-3-030-79408-8

Jayanthi, M., Duraisamy, M., Kabiraj, S., Thirumurthy, S., Samynathan, M., Panigrahi, A., & Muralidhar, M. (2022). Are the Sundarbans, the world's largest mangroves region under threat?—An ecosystem-based geospatial approach to assess changes past, present, and future in relation to natural and human-induced factors. *Land Degradation & Development*, *33*(13), 2137–2149. https://doi.org/10.1002/ldr.4294

Jentoft, S., Chuenpagdee, R., Barragán-Paladines, M. J., & Franz, N. (Eds.). (2017). *The small-scale fisheries guidelines: Global implementation* (Vol. 14). Springer. https://doi.org/10.1007/978-3-319-55074-9

Jevrejeva, S., Grinsted, A., & Moore, J. C. (2014). Upper limit for sea level projections by 2100. *Environmental Research Letters*, *9*(10), 104008.

Kais, S. M., & Islam, M. S. (2019). Perception of climate change in shrimp-farming communities in Bangladesh: A critical assessment. *International Journal of Environmental Research and Public Health*, *16*(4), 672–690. https://doi.org/10.3390/ijerph16040672

Kanan, A. H., Francesco, P., Mauro, M., & Rahman, M. M. (2023a). Mapping inundation from sea level rise and its interaction with land cover in the Sundarbans mangrove forest. *Climate Change*, *176*(2), 565–585. https://doi.org/10.1007/s10584-023-03557-6

Kanan, A.H., Pirotti, F., Masiero, M. et al. (2023b). Mapping inundation from sea level rise and its interaction with land cover in the Sundarbans mangrove forest. *Climatic Change* 176, 104. https://doi.org/10.1007/s10584-023-03574-5

Kanan, A. H., Srivastava, P. K., Islam, T., & Petropoulos, G. P. (2023). Mapping inundation from sea level rise and its interaction with land cover in the Sundarbans mangrove forest. *Climatic Change*, *176*, 147. https://doi.org/10.1007/s10584-023-03574-5

Karabi, D. (2018). Perils of women trafficking: A case study of Joynagar, Kultali administrative blocks, Sundarban, India. *International Journal of Education, Culture and Society*, *2*(2), 61–68. https://doi.org/10.11648/j.ijecs.20170202.12

Karim, M. F., & Mimura, N. (2008). Impacts of climate change and sea-level rise on cyclonic storm surge floods in Bangladesh. *Global Environmental Change*, *18*(3), 490–500. https://doi.org/10.1016/j.gloenvcha.2008.05.002

Katzenberger, A., Schewe, J., Pongratz, J., & Levermann, A. (2021). Robust increase of Indian monsoon rainfall and its variability under future warming in CMIP6 models. *Earth System Dynamics*, *12*(2), 367–386. https://doi.org/10.5194/esd-12-367-2021

Khan, A., Kemp, G. C., Currie-Alder, B., & Leone, M. (2018). Responding to uneven vulnerabilities: A synthesis of emerging insights from climate change hotspots. *Ocean & Coastal Management*, *85*(5), 80–110. https://doi.org/10.1016/j.ocecoaman.2018.01.025

Khan, M. R. (2013). The politics of international adaptation funding: Justice and division in the greenhouse. *Global Environmental Politics*, *13*(1), 29–48. https://doi.org/10.1162/GLEP_a_00153

Khan, N., Chakma, P., Uddin, M., & Hasan, C. (2024). Preparation and characterization of biodegradable plastic from red seaweed (Gracilaria sp.) from coastal areas in Bangladesh. *ChemistrySelect*, *9*(34), Article e202401990. https://doi.org/10.1002/slct.202401990

Kibria, A. S. M. G., Costanza, R., Groves, C., & Behie, A. M. (2018). The interactions between livelihood capitals and access of local communities to the forest provisioning services of the Sundarbans mangrove Forest, Bangladesh. *Ecosystem Services*, *32*, 41–49. https://doi.org/10.1016/j.ecoser.2018.05.003

Kienberger, S., Contreras, D., & Zeil, P. (2014). Spatial and holistic assessment of social, economic, and environmental vulnerability to floods-lessons from the Salzach River basin, Austria. In *Assessment of vulnerability to natural hazards: A European perspective* (pp. 53–73). Elsevier. https://doi.org/10.1016/B978-0-12-410528-7.00003-5

Kittinger, J. N., Finkbeiner, E. M., Ban, N. C., Broad, K., Carr, M. H., Cinner, J. E., . . . & Crowder, L. B. (2013). Emerging frontiers in social-ecological systems research for sustainability of small-scale fisheries. *Current Opinion in Environmental Sustainability*, *5*(3-4), 352–357.

Knowler, D., Philcox, N., Nathan, S., Delamare, W., Haider, W., & Gupta, K. (2009). Assessing prospects for shrimp culture in the Indian Sundarbans: A combined simulation modelling and choice experiment approach. *Marine Policy*, *33*(4), 613–623. https://doi.org/10.1016/j.marpol.2008.12.009

Kolding, J., Béné, C., & Bavinck, M. (2014). Small-scale fisheries: Importance, vulnerability and deficient knowledge. In S. M. Garcia, J. Rice, & A. Charles (Eds.), *Governance of marine fisheries and biodiversity conservation: Interaction and co-evolution* (pp. 317–331). Wiley-Blackwell.

Krishna, A. (2010). *One illness away: Why people become poor and how they escape poverty*. OUP Oxford.

Kumar, P. (2012). Impact of economic drivers on mangroves of Indian Sundarbans: An exploration of missing links. *Environment, Development and Sustainability*, *14*(6), 939–953. https://doi.org/10.1007/s10668-012-9361-9

Kumar, R., Sinha, R., Rakib, M. R. J., Padha, S., Ivy, N., Bhattacharya, S., . . . & Sharma, P. (2022). Microplastics pollution load in Sundarban delta of Bay of Bengal. *Journal of Hazardous Materials Advances*, *7*, 100099. https://doi.org/10.1016/j.hazadv.2022.100099

Kundu, S. K., & Santhanam, H. (2025). Assessing sustainability, livelihoods, and benefits of marine fishery advisory services in Odisha. *India. Academia Environmental Sciences and Sustainability*, *2*(4).

Lara, R. J., Neogi, S. B., Islam, M. S., Mahmud, Z. H., Yamasaki, S., & Nair, G. B. (2009). Influence of catastrophic climatic events and human waste on vibrio distribution in the Karnaphuli estuary, Bangladesh. *EcoHealth*, *6*(2), 279–286. https://doi.org/10.1007/s10393-009-0257-6

Lauria, V., Das, I., Hazra, S., Cazcarro, I., Arto, I., Kay, S., Ofori-Danson, P., Ahmed, M., Hossain, M. A. R., Barange, M., & Fernandes, J. A. (2018). Importance of fisheries for food security across three climate change vulnerable deltas. *Science of the Total Environment*, *640*, 1566–1577. https://doi.org/10.1016/j.scitotenv.2018.06.011

Lee, Y. J. (2014). Social vulnerability indicators as a sustainable planning tool. *Environmental Impact Assessment Review*, *44*, 31–42. https://doi.org/10.1016/j.eiar.2013.08.002

Loucks, C., Barber-Meyer, S., Hossain, M. A. A., Barlow, A., & Chowdhury, R. M. (2010). Sea level rise and tigers: Predicted impacts to Bangladesh's Sundarbans mangroves: A letter. *Climatic Change*, *98*(1), 291–298.

Lovelock, C. E., Feller, I. C., Reef, R., Hickey, S., & Ball, M. C. (2017a). Mangrove dieback during fluctuating sea levels. *Scientific Reports*, *7*(1), 1–9. https://doi.org/10.1038/s41598-017-01927-6

Lovelock, C. E., Feller, I. C., Reef, R., Hickey, S., & Ball, M. C. (2017b). Mangrove dieback during fluctuating sea levels. *Scientific Reports*, *7*(1), 1680.

Madhanagopal, D., & Pattanaik, S. (2020). Exploring fishermen's local knowledge and perceptions in the face of climate change: The case of coastal Tamil Nadu, India. *Environment, Development and Sustainability*, *22*, 3461–3489. https://doi.org/10.1007/s10668-019-00355-y

Majumdar, S. D., Pramanick, N., & Hazra, S. (2021). Biophysical vulnerability assessment of Indian Sundarbans mangrove. In *Sundarbans mangrove systems* (pp. 217–250). CRC Press.

Majumder, A. (2019). An empirical study on economic empowerment of women through self-help group of Indian Sunderban Delta of north 24 Parganas of West Bengal, India. In R. Das (Ed.), *Handbook of research on microfinancial impacts on women empowerment, poverty, and inequality* (pp. 129–164). IGI Global. https://doi.org/10.4018/978-1-5225-5240-6.ch007

Malakar, K., Mishra, T., & Patwardhan, A. (2018). Drivers of response to extreme weather warnings among marine fishermen. *Climatic Change*, *150*(3–4), 417–431. https://doi.org/10.1007/s10584-018-2287-y

Mallappa, H., & Shirur, M. (2021). *Climate change and resilient food systems*. Springer.

Mamun Siddiquee, M. A., Saha, S., Fatema, K., Uddin, M., & Rashid, H. (2022). Microplastics ingestion by fish and crab species in the Sundarbans mangroves waters. FSB Biennial Conference 2021. https://doi.org/10.13140/rg.2.2.22806.01601

Mandal, M. S. H., & Hosaka, T. (2020). Assessing cyclone disturbances (1988–2016) in the Sundarbans mangrove forests using Landsat and Google earth engine. *Natural Hazards*, *102*, 133–150. https://doi.org/10.1007/s11069-020-03914-z

Mandal, S., Hasan, I., Hawlader, H. H., Sultana, I., Rahman, M. M., & Majumder, M. S. I. (2017). Occupational health hazard and safety assessment of fishermen community in coastal zone of Bangladesh. *International Journal of Health Economics and Policy*, *2*(2), 63–71.

Mandal, U. K., Karim, F., Yu, Y., Ghosh, A., Zahan, T., Mallick, S., ... Mainuddin, M. (2025). Assessing vulnerability and climate risk to agriculture for developing resilient farming strategies in the Ganges Delta. *Climate Risk Management*, *47*, 100690.

Mastrojeni, G., & Pasini, A. (2017). *Greenhouse effect, war effect: Climate, conflicts, migrations: Italy in frontline*. Chiarelettere.

Mendelsohn, R., Emanuel, K., Chonabayashi, S., & Bakkensen, L. (2012). The impact of climate change on global tropical cyclone damage. *Nature Climate Change*, *2*(3), 205–209. https://doi.org/10.1038/nclimate1357

Micheli, F., Saenz-Arroyo, A., Aalto, E., Beas-Luna, R., Boch, C. A., Cardenas, J. C., ... Woodson, C. B. (2024). Social-ecological vulnerability to environmental extremes and adaptation pathways in small-scale fisheries of the southern California current. *Frontiers in Marine Science*, *11*, 1322108.

Miller, F., Osbahr, H., Boyd, E., Thomalla, F., Bharwani, S., Ziervogel, G., Walker, B., Van Der Leeuw, S., Hinkel, J., Downing, T., Folke, C., & Nelson, D. (2010). Resilience and vulnerability: Complementary or conflicting concepts? *Ecology and Society*, *15*(2), 45–65.

Mishra, M., Acharyya, T., Augusto, C., Santos, G., Richarde, M., Dipika, K., Kamal, A. H. M., & Susmita, R. (2021). Geo-ecological impact assessment of severe cyclonic storm Amphan on Sundarban mangrove forest using geospatial technology. *Estuarine, Coastal and Shelf Science*, *260*(3), 120–140. https://doi.org/10.1016/j.ecss.2021.107493

Mishu, R. A., Mallick, J., Akter, M. Y., Fattah, M. A., Kamruzzaman, M., Salam, M. A., Dhar, B., & Islam, A. R. M. T. (2025). Spatial and temporal variability of future extreme precipitation in Bangladesh using CMIP6 models. *Journal of Water and Climate Change*, *16*(4), 1281–1309. https://doi.org/10.2166/wcc.2025.415

Mitra, A. (2013). *Sensitivity of mangrove ecosystem to changing climate* (Vol. 62). Springer. https://doi.org/10.1007/978-81-322-1509-7

Mitra, A. (2017). Bioaccumulation of heavy metals by dominant oyster in Indian Sundarbans. *Journal of Science, Engineering and Health Management*, *1*(2), 34–39.

Mitra, A. (2018). Can species serve as proxy to climate change induced salinity alteration? *Journal of Marine Biology and Aquaculture*, *4*(1), 1–3.

Mitra A (2020) *Mangrove forests in India*. Cham: Springer International Publishing.

Mitra, A., & Barua, P. (2022a). Coping practices of women fisherfolk in responses to climate change at UNESCO declared world heritage site of Sundarbans. In U. Chatterjee, A. O. Akanwa, S. Kumar, S. K. Singh, & A. Dutta Roy (Eds.), *Ecological footprints of climate change*. Springer Climate.

Mitra, A., & Barua, P. (2022b). Coping practices of women fisherfolk in responses to climate change at UNESCO declared world heritage site of Sundarbans. In U. Chatterjee, A. O. Akanwa, S. Kumar, S. K. Singh, & A. Dutta Roy (Eds.), *Ecological footprints of climate change*. Springer.

Mitra, A., & Barua, P. (2023). Coping practices of women fisherfolk in responses to climate change at UNESCO declared world heritage site of sundarbans. In *Ecological footprints of climate change: Adaptive approaches and sustainability* (pp. 701–729). Cham: Springer International Publishing.

Mitra, A., Chowdhury, R., Sengupta, K., & Banerjee, K. (2020a). Impact of salinity on mangroves of Indian Sundarbans. *Journal of Coastal Environment*, *1*(1), 71–82.

Mitra, A., Dutta, J., Mitra, A., & Thakur, T. (2020b). Amphan Supercyclone: A death knell for Indian Sundarbans. *eJournal of Applied Forest Ecology*, *8*(1), 41–48.

Mitra, A., & Gati, R. C. (2015). Carbon census in the mangrove ecosystem of lower Gangetic delta. *Economol Journal*, *5*, 11–27.

Mitra, A., & Ghosh, R. (2014). Bioaccumulation pattern of heavy metals in commercially important fishes on and around Indian Sundarbans. *Global Journal of Animal Scientific Research*, *2*(1), 33–45.

Mitra, A., Nayak, B., Pramanick, P., Roy, M., Chakraborty, S., Banerjee, K., Amin, G., & Zaman, S. (2014). Heavy metal level in a mangrove associate species Ipomoea pes-caprae in and around the world heritage site of Indian Sundarbans. *Journal of Environmental Science, Computer Science and Engineering & Technology*, *4*(1), 1–7.

Mitra, A., Saha, A., Fernandes, M., Pramanick, P., Zaman, S., & Sengupta, G. (2019). Will non-conventional alternative livelihood scheme work in the framework of mangrove dominated Indian Sundarbans. *Annals of Marine Science*, *3*(1), 13–17. https://doi.org/10.17352/ams.000015

Mitra, A., & Zaman, S. (2016). *Basics of marine and estuarine ecology*. Springer. https://doi.org/10.1007/978-81-322-2707-6

Mitra, A., & Zaman, S. (2021). *Estuarine acidification*. Springer International Publishing.

Mitra, A., Zaman, S., & Pramanick, P. (2022). Blue economy in context to the mangrove dominated Indian Sundarbans. In *Blue economy in Indian Sundarbans: Exploring livelihood opportunities* (pp. 85–174). Cham: Springer International Publishing.

Mitra, A., Zaman, S., Pramanick, P., Bhattacharyya, S. B., & Biswas, P. (2024). Impacts of SLR induced erosion and salinity alteration on the most relevant sectors of blue economy pertinent to the framework of Sundarbans. *Environmental Science and Pollution Research*, *31*(2), 1–20. https://doi.org/10.1007/s11356-023-30003-y

Mohammad Raqibul Hasan, M. R., Hossain, M., & Rashid, A. M. (2023). The dilemma of prioritizing conservation over livelihoods: Assessing the impact of fishing restriction to the fishermen of the Sundarbans. *Trees, Forests and People*, *11*(3), 34–50.

Mohanty, B., Mohanty, S., Sahoo, J., & Sharma, A. (2010). Climate change: Impacts on fisheries and aquaculture. In S. Simard (Ed.), *Climate change and variability* (pp. 119–138). IntechOpen. https://doi.org/10.5772/9805

Molla, M. A. G., Islam, M. R., Islam, S., & Salam, M. A. (2020). Socio-economic status of crab collectors and fatteners in the southwest region of Bangladesh. *Journal of Bangladesh Agriculture University*, *7*(2), 411–419.

Mondal, B., Roy, A., & Saha, A. K. (2022). Vulnerability assessment of mangrove areas in coastal West Bengal, India. *Remote Sensing Applications: Society and Environment*, *25*, 100680.

Mondal, P., & Pal, S. (2023). Ecotoxicological risks of pesticide residues in the Hooghly-Matla estuarine complex. *Environmental Toxicology and Chemistry*, *42*(5), 1124–1135. https://doi.org/10.1002/etc.5573

Mondal, T., & Tripathy, B. (2022). A study on impact of bank erosion on socio-economic sectors of Sundarban region, West Bengal. *Madhya Bharti*, *82*(20), 178–189.

Mondal, T., & Tripathy, B. (2023a). Status of women empowerment in coastal area of India: A case study on Sundarban region, West Bengal. *The DNC Journal*, *2*(2), 20–45.

Mondal, T., & Tripathy, B. (2023b). Status of women empowerment in coastal area of India: A case study on Sundarban region. *West Bengal.The DNC journal.*, *2*(2), 1–21.

Moniruzzaman, M., Cottrell, A., King, D., & Islam, M. N. (2018). Environmental migrants in Bangladesh: A case study on climatic change hazards in the southwestern coastal area. *Springer Climate*, 99–117.

Moon, I. J., Kim, S. H., & Chan, J. C. (2019). Climate change and tropical cyclone trend. *Nature*, *570*(7759), E3–E5. https://doi.org/10.1038/s41586-019-1222-3

Mozumder, M. M. (2018). Coastal ecosystems services in the bay of Bengal and efforts to improve their management. *Indian Journal of Marine Sciences*, *47*(11), 60–75.

Mukherjee, A., & Ghosh, S. (2020). Pesticide pollution in Indian estuaries: A case study from the Sundarbans ecosystem. *Marine Pollution Bulletin*, *161*, Article 111770. https://doi.org/10.1016/j.marpolbul.2020.111770

Mukherjee, M. (2018). Empowerment of fisherwomen of Sundarbans through ornamental fish farming. *Asian Fisheries Sciences*, *40*(3), 125–150.

Mukherjee, P., Mitra, A., Zaman, S., & Mitra, A. (2021b). Impact of climate-change-induced salinity alteration on Ichthyoplankton diversity of Indian Sundarbans. In W. Leal Filho, J. Luetz, & D. Ayal (Eds.), *Handbook of climate change management: Research, leadership, transformation* (pp. 1–27). Springer International Publishing.

Mukherjee, P., Mitra, A., Zaman, S., & Mitra, A. (2022). Impact of climate-change-induced salinity alteration on Ichthyoplankton diversity of Indian Sundarbans. In W. L. Filho, et al. (Eds.), *Handbook of climate change management*. Springer Nature.

Mukherjee, P., Pramanick, P., Zaman, S., & Mitra, A. (2021a). Phytoremediation of heavy metals by the dominant mangrove associate species of Indian Sundarbans. *Journal of Environmental Engineering and Landscape Management*, *29*(4), 391–402.

Mukherjee, S., Chaudhuri, A., Sen, S., & Homechaudhuri, S. (2012). Effect of cyclone Aila on estuarine fish assemblages in the Matla River of the Indian Sundarbans. *Journal of Tropical Ecology*, *28*(4), 405–415. https://doi.org/10.1017/S0266467412000247

Mukhopadhyay, P. (2022). The 'fluid landscape'of the Sundarbans: Critically reviewing the 'managed retreat'discourse. *Conservation and Society*, *20*(3), 234–244.

Mukhopadhyay, U., Bhattacharya, S., Chouhan, P., Paul, S., Chowdhury, I. R., & Chatterjee, U. (Eds.). (2024). *Climate crisis, social responses and sustainability: Socio-ecological study on global perspectives*. Springer Nature Switzerland.

Murdiyarso, D., Purbopuspito, J., Kauffman, J. B., Warren, M. W., Sasmito, S. D., Donato, D. C., Manuri, S., Krisnawati, H., Taberima, S., & Kurnianto, S. (2015). The potential of Indonesian mangrove forests for global climate change mitigation. *Nature Climate Change*, *5*(12), 1089–1092. https://doi.org/10.1038/nclimate2734

Nandi, N. C., & Misra, A. (1987). Bibliography of the Indian Sundarbans with special reference to Fauna. Records of the Zoological Survey of India, occ. *Paper*, *97*, 1–135.

Nath, A., Samanta, S., Banerjee, S., Danda, A. A., & Hazra, S. (2021). Threat of arsenic contamination, salinity and water pollution in agricultural practices of Sundarban Delta, India, and mitigation strategies. *SN Applied Sciences*, *3*(5), 560.

Nautiyal, S., Rao, K. S., Kaechele, H., Raju, K. V., Schaldach, R., & Eds.). (2013). *Knowledge systems of societies for adaptation and mitigation of impacts of climate change*. Springer Science & Business Media.

Nawar, N., Rahman, M., Marzia, S., Ali, M., Akbor, M., Siddique, M., & Malafaia, G. (2022). Characterization of microplastic pollution in the Pasur River of the Sundarbans ecosystem (Bangladesh) with emphasis on water, sediments, and fish. *SSRN Electronic Journal*. https://doi.org/10.2139/ssrn.4290243

Neogi, S. B., Dey, M., Kabir, S. L., Masum, S. J. H., Kopprio, G., Yamasaki, S., & Lara, R. (2017). Sundarban mangroves: Diversity, ecosystem services and climate change impacts. *Asian Journal of Medical and Biological Research*, 2(4), 488–507.

Neogi, S. B., Dey, M., Lutful Kabir, S. M., Masum, S. J. H., Kopprio, G. A., Yamasaki, S., & Lara, R. J. (2016). Sundarban mangroves: Diversity, ecosystem services and climate change impacts. *Indian Journal of Marine Science*, *30*(3), 30–50.

Ni, W., Li, M., Ross, A. C., & Najjar, R. G. (2019). Large projected decline in dissolved oxygen in a eutrophic estuary due to climate change. *Journal of Geophysical Research: Oceans*, *124*, 8271–8289. https://doi.org/10.1029/2019JC015274

Nicholls, R. J., Adger, W. N., Hutton, C. W., & Hanson, S. E. (2020). *Deltas in the Anthropocene*. Springer Nature.

Nordhagen, A., Rizwan, A. A. M., Aakre, I., Moxness Reksten, A., Pincus, L. M., Bøkevoll, A., Mamun, A., Haraksingh Thilsted, S., Htut, T., Somasundaram, T., & Kjellevold, M. (2020). Nutrient composition of demersal, pelagic, and mesopelagic fish species sampled off the coast of Bangladesh and their potential contribution to food and nutrition security—The EAF-Nansen programme. *Food*, *9*(6), 730.

O'Brien, K., Eriksen, S., Nygaard, L. P., & Schjolden, A. (2007). Why different interpretations of vulnerability matter in climate change discourses. *Climate Policy*, *7*(1), 73–88. https://doi.org/10.1080/14693062.2007.9685639

Ortega, L., Celentano, E., Finkl, C., & Defeo, O. (2018). Effects of climate variability on the morphodynamics of Uruguayan sandy beaches. *Journal of Coastal Research*, *29*, 747–755. https://doi.org/10.2112/JCOASTRES-D-13-00003.1

Ortolano, L., Sánchez-Triana, E., & Ferdausi, S. A. (2019). Strategy for adapting to climate change and conserving biodiversity in the Bangladesh Sundarbans. *Climate and Development*, *9*(4), 325–336.

Ortolano, L., Sánchez-Triana, E., Pau, T., & Ferdausi, S. A. (2016). Managing the Sundarbans region: Opportunities for mutual gain by India and Bangladesh. *International Journal of Environment And Sustainable Development*, *15*(1), 16–31. https://doi.org/10.1504/IJESD.2016.073336

Osland, M. J., Feher, L. C., Griffith, K. T., Cavanaugh, K. C., Enwright, N. M., Day, R. H., . . . & Rogers, K. (2017). Climatic controls on the global distribution, abundance, and species richness of mangrove forests. *Ecological Monographs*, *87*(2), 341–359.

Ouheddou, M., Abelouah, M. R., Ben-Haddad, M., Hajji, S., Laaraj, N. E., Akhouchal, I., & Ait Alla, A. (2025). Microplastics in Morocco's most consumed fisheries: Chemical characterization, ecological traits, and implications for human health. *Marine Pollution Bulletin*, *210*, 117334. https://doi.org/10.1016/j.marpolbul.2024.117334

Pal, N., Zaman, S., & Mitra, A. (2017). Variation of survival rate of mangrove flora due to salinity in Indian Sundarbans. *Journal of Science, Engineering and Health Management*, *1*(4), 51–54.

Pal, S. C., Chatterjee, U., Saha, A., & Ruidas, D. (2025). Climate change: Conflict and resilience in the age of Anthropocene.

Partelow, S., Glaser, M., Solano Arce, S., Barboza, R., & Schlüter, A. (2018). Mangroves, fishers, and the struggle for adaptive comanagement: Applying the social-ecological systems framework to a marine extractive reserve (Resex) in Brazil. *Ecology and Society*, *23*(3), 45–65.

Pasini, A., Mastrojeni, G., & Tubiello, F. N. (2018). Climate actions in a changing world. *The Anthropocene Review*, *5*(3), 237–241.

Paszkowski, A., Goodbred, S., Jr., Borgomeo, E., Khan, M. S. A., & Hall, J. W. (2021). Geomorphic change in the Ganges-Brahmaputra-Meghna delta. *Nature Reviews Earth & Environment*, *2*, 763–780. https://doi.org/10.1038/s43017-021-00213-4

Paul, A. K., & Paul, A. (2023). Crisis on the coast and hinterland.

Payo, A., Mukhopadhyay, A., Hazra, S., Ghosh, T., Ghosh, S., Brown, S., . . . Haque, A. (2016). Projected changes in area of the Sundarban mangrove forest in Bangladesh due to SLR by 2100. *Climatic Change*, *139*(2), 279–291.

Perry, A. L., Low, P. J., Ellis, J. R., & Reynolds, J. D. (2005). Climate change and distribution shifts in marine fishes. *Science*, *308*(5730), 1912–1915.

Petrik, C. M., Stock, C. A., Andersen, K. H., van Denderen, P. D., & Watson, J. R. (2020). Large pelagic fish are most sensitive to climate change despite pelagification of ocean food webs. *Frontiers in Marine Science*, *7*, 588482.

Pinsky, M. L., Worm, B., Fogarty, M. J., Sarmiento, J. L., & Levin, S. A. (2013). Marine taxa track local climate velocities. *Science*, *341*(6151), 1239–1242.

Pitchaikani, J. S. (2020). Integrated coastal zone management practices for Sundarbans, India. *Indian Journal of Geo Marine Science*, *49*(3), 352–356.

Pramanick, P., Bera, D., Banerjee, K., Zaman, S., & Mitra, A. (2016). Seasonal variation of proximate composition of common seaweeds in Indian Sundarbans. *International Journal of Life Science and Scientific Research*, *2*(5), 570–578.

Pratchett, M. S., Bay, L. K., Gehrke, P. C., Koehn, J. D., Osborne, K., Pressey, R. L., . . . & Wachenfeld, D. (2011). Contribution of climate change to degradation and loss of critical fish habitats in Australian marine and freshwater environments. *Marine and Freshwater Research*, *62*(9), 1062–1081.

Rabbitt, S., Lilley, I., Albert, S., & Tibbetts, I. R. (2019). What's the catch in who fishes? Fisherwomen's contributions to fisheries and food security in Marovo lagoon, Solomon Islands. *Marine Policy*, *108*(3), 103–120.

Radenković, M., Zorić, K., & Petrović, A. (2023). Invasive macroinvertebrate and fish species and their impact on fish resources: A case study on fishing waters of Serbia. In V. Simić, S. Simić, & V. Pešić (Eds.), *Ecological sustainability of fish resources of inland waters of the Western Balkans in the Anthropocene*. Springer.

Raha, A., Das, S., Banerjee, K., & Mitra, A. (2012). Climate change impacts on Indian Sunderbans: A time series analysis (1924–2008). *Biodiversity and Conservation*, *21*, 1289–1307. https://doi.org/10.1007/s10531-012-0260-z

Rahman, M., Dustegir, M., Karim, R., Haque, A., Nicholls, R. J., Darby, S. E., . . . & Akter, M. (2018). Recent sediment flux to the Ganges-Brahmaputra-Meghna delta system. *Science of the Total Environment*, *643*, 1054–1064.

Rahman, M., Rahman, B., & Nazira, M. (2016). Observation on selectivity of fishing gears and ichthyofaunal diversity in the Paira river of southern Bangladesh. *International Journal of Fisheries and Aquatic Studies*, *4*(1), 95–100.

Rahman, M. A., Lee, S. G., Molla, M. H. R., Asare, O. E., Megwalu, F., Jahan, B., & Shaikh, M. M. (2018a). Fisheries management and governance in Bangladesh. *MOJ Ecology & Environmental Sciences*, *3*(6), 381–385.

Rahman, M. A., Lee, S. G., Molla, M. H. R., Asare, O. E., Megwalu, F., Jahan, B., & Shaikh, M. M. (2018b). Fisheries management and governance in Bangladesh. *MOJ Ecology & Environmental Sciences*, *3*(6), 381–385.

Rao, D. S., Rao, K. S., Iyer, C. S. P., & Chittibabu, P. (2008). Possible ecological consequences from the Sethu Samudram canal project, India. *Marine Pollution Bulletin*, *56*(2), 170–186.

Räsänen, A., Juhola, S., Nygren, A., Käkönen, M., Kallio, M., Monge Monge, A., & Kanninen, M. (2016). Climate change, multiple stressors and human vulnerability: A systematic review. *Regional Environmental Change*, *16*(8), 2291–2302. https://doi.org/10.1007/s10113-016-0974-7

Ray Chaudhuri, T., Fazli, P., Zaman, S., Pramanick, P., Bose, R., & Mitra, A. (2020). Impact of acidification on heavy metals in Hooghly estuary. *Journal of Harmonized Research in Applied Sciences*, *2*(2), 91–97.

Ray, S., et al. (2022). Agricultural runoff and chemical pollution: An emerging threat to the Indian Sundarbans. *Chemosphere*, *298*, 134332. https://doi.org/10.1016/j.chemosphere.2022.134332

Roy, A., De, A., Aftabuddin, M., Bera, A., Bayen, S., Ghosh, A., & Das, B. (2024). Analysis of health ailments and associated risk factors in small-scale fisherfolk community of Indian Sundarbans: A cross-sectional study. *Indian Journal of Community Medicine*, *49*(2), 360–366. https://doi.org/10.4103/ijcm.ijcm_906_22

Roy, A., Manna, R., Ghosh, S., Sahu, S., & Das, B. (2023c). Assessment of women's contribution in small-scale fisheries in Indian Sundarbans: Issues, strategies, and way forward for sustainability. *Fisheries Management and Ecology*, *30*(4), 378–391. https://doi.org/10.1111/fme.12630

Roy, A., Manna, R. K., Ghosh, S., Sahu, S. K., & Das, B. K. (2023b). Assessment of women's contribution in small-scale fisheries in Indian Sundarbans: Issues, strategies, and way forward for sustainability. *Fisheries Management and Ecology*, *30*(4), 378–391.

Roy, A., & Sen, A. (2024). Importance of indigenous small fish to boost livelihood and nutritional security: Evidences from Indian Sunderbans. In *Perspectives and applications of indigenous small fish in India: An introduction* (pp. 583–593). Springer Nature Singapore.

Roy, A., Sharma, A. P., Bhaumik, U., Pandit, A., Saha, P., & Mitra, A. (2023a). Socio-economic features of womenfolk of Indian Sunderbans involved in fish drying. *Indian Journal of Extension Education*, *53*(2), 142–146.

Roy, A., Sharma, A. P., Bhaumik, U., Pandit A., Singh, S. R. K., Saha, S. et al (2017) Socio-economic features of womenfolk of Indian Sundarbans involved in fish drying. *Indian Journal of Extension Education 53*(2):142–146

Roy, S., Kumar, V., Manna, R. K., & Suresh, V. R. (2016). Sundarbans mangrove deltaic system–an overview of its biodiversity with special reference to fish diversity. *Journal of Applied and Natural Science*, *8*(2), 1090.

Roychowdhury, R., Vyas, P., Zaman, S., Roy, A., & Mitra, A. (2019). Surface water pH: A proxy to acidification of estuarine water of Indian Sundarbans. *International Journal of Research and Analytical Reviews*, *6*(1), 1530–1535.

Rudra, A., & Chattopadhyay, A. (2019). Environmental change of coastal Sundarbans: Impact on livelihood and standard of living status of indigenous people. *Environmental Quality Management*, *29*(2), 77–84.

Sadrekarimi, H., Gardanova, Z. R., Bakhshesh, M., Ebrahimzadeh, F., Yaseri, A. F., Thangavelu, L., Hasanpoor, Z., Zadeh, F. A., & Kahrizi, M. S. (2022). Emerging role of human microbiome in cancer development and response to therapy: Special focus on intestinal microflora. *Journal of Translational Medicine*, *20*, 301. https://doi.org/10.1186/s12967-022-03488-3

Safonova, I., Lukjanova, I., & Zhadan, O. (2017). Untitled. *Journal of Aquaculture & Marine Biology*, *6*(1). https://doi.org/10.15406/jamb.6.1

Saha, A., Bera, B., Shit, P. K., & Bhattacharjee, S. (2025). Unraveling Sundarbans' erosion: How machine learning maps climate change impacts. *npj Climate Action*, *4*, 36. https://doi.org/10.1038/s44168-026-00336-8

Saha, S., & Dash, G. (2021). Climatic stress on cultured Scylla sp.: Impact on livelihood and public health in West Bengal. *International Journal of Researches in Biosciences, Agriculture & Technology*, *S17*, 331–344.

Saha, S., Halder, M., Mookerjee, S., & Palit, A. (2019). Seasonal influence, enteropathogenic microbial load and diarrhoeal enigma in the Gangetic Delta, India: Present scenario and health implications. *Journal of Infection and Public Health*, *12*, 540–548. https://doi.org/10.1016/j.jiph.2019.01.066

Sahana, M., Hong, H., Ahmed, R., Patel, P. P., Bhakat, P., & Sajjad, H. (2019). Assessing coastal island vulnerability in the Sundarban biosphere reserve, India, using geospatial technology. *Environmental Earth Sciences*, *78*(10), 1–22.

Sahana, M., Rehman, S., Paul, A. K., & Sajjad, H. (2021). Assessing socio-economic vulnerability to climate change-induced disasters: Evidence from Sundarban biosphere reserve, India. *Geology, Ecology, and Landscapes*, *5*(1), 40–52. https://doi.org/10.1080/24749508.2019.1700670

Sahana, M., & Sajjad, H. (2019). Vulnerability to storm surge flood using remote sensing and GIS techniques: A study on Sundarban biosphere reserve, India. *Remote Sensing Applications: Society and Environment*, *13*, 106–120.

Sahoo, B., & Bhaskaran, P. K. (2018). Multi-hazard risk assessment of coastal vulnerability from tropical cyclones–a GIS based approach for the Odisha coast. *Journal of Environmental Management*, *206*, 1166–1178. https://doi.org/10.1016/j.jenvman.2017.10.075

Salagrama, V. (2012). Climate change and fisheries: Perspectives from small-scale fishing communities in India on measures to protect life and livelihood. *Asian Fisheries Science*, *40*(4), 130–150.

Salehin, M., Chowdhury, M. M. A., Clarke, D., Mondal, S., Nowreen, S., Jahiruddin, M., & Haque, A. (2018). Mechanisms and drivers of soil salinity in coastal Bangladesh. In R. Nicholls, C. Hutton, W. Adger, S. Hanson, M. Rahman, & M. Salehin (Eds.), *Ecosystem services for well-being in deltas* (pp. 331–347). Palgrave Macmillan.

Salunke, M., Kalyankar, A., Khedkar, C. D., Shingare, M., & Khedkar, G. D. (2020). A review on shrimp aquaculture in India: Historical perspective, constraints, status and future implications for impacts on aquatic ecosystem and biodiversity. *Reviews in Fisheries Science & Aquaculture*, 1–20.

Samanta, B. (2018). Population growth in Sundarban region - a spatio-temporal analysis. *Scholars Journal of Arts, Humanities and Social Science*, *20*(2), 45–65.

Samanta, S., Hazra, S., Mondal, P. P., Chanda, A., Magsi, H., & Abdullah-Al-Wadud, M. (2021). Assessment and attribution of mangrove forest changes in the Indian Sundarbans from 2000 to 2020. *Remote Sensing*, *13*(24), 4957. https://doi.org/10.3390/rs13244957

Sardar, P., & Samadder, S. R. (2023). Long-term ecological vulnerability assessment of Indian Sundarban region under present and future climatic conditions under CMIP6 model. *Ecological Informatics*, *76*(2), 102–125.

Sardar, P., & Samadder, S. R. (2023). Long-term ecological vulnerability assessment of Indian Sundarban region under present and future climatic conditions under CMIP6 model, *Ecological Informatics*, *76*, 102140, ISSN 1574-9541. https://doi.org/10.1016/j.ecoinf.2023.102140

Sarkar, B., Islam, A., & Majumder, A. (2021). Seawater intrusion into groundwater and its impact on irrigation and agriculture: Evidence from the coastal region of West Bengal. *India. Regional Studies in Marine Science*, *44*, 101751.

Sarkar, M. (2011). Secondary students' environmental attitudes: The CASE of environmental education in Bangladesh. *International Journal of Academic Research in Business and Social Sciences*, *1*, 106.

Sarkar, S., et al. (2022). Pesticide bioaccumulation in estuarine fishes: Implications for food safety and ecosystem health. *Ecotoxicology and Environmental Safety*, *240*, 113690. https://doi.org/10.1016/j.ecoenv.2022.113690

Sarker, S., Huda, A., Niloy, M., & Chowdhury, G. (2022). Trophic transfer of microplastics in the aquatic ecosystem of Sundarbans mangrove forest, Bangladesh. *Science of the Total Environment*, *838*, 155896. https://doi.org/10.1016/j.scitotenv.2022.155896

Sen, S., & Homechaudhuri, H. (n.d.). Population characteristics and trends in artisanal fishery of Scylla serrata (Forsskål, 1775) in Indian Sundarban: Implications on future managements. *Ocean & Coastal Management*, *143*(2), 105–114.

Sen, S., & Homechaudhuri, S. (2017a). Population characteristics and trends in artisanal fishery of Scylla serrata (Forsskål, 1775) in Indian Sundarban: Implications on future managements. *Ocean & Coastal Management*, *143*, 105–114.

Sen, S., & Homechaudhuri, S. (2017b). Population characteristics and trends in artisanal fishery of Scylla serrata (Forsskål, 1775) in Indian Sundarban: Implications on future managements. *Ocean & Coastal Management*, *143*, 105–114.

Sengupta, T., & Mitra, A. (2020). Spatial variation of the microbial diversity in the mangrove dominated Sundarban Forest of India. In S. De Mandal & P. Bhatt (Eds.), *Recent advancements in microbial diversity* (pp. 333–350). Academic Press.

Shaffril, H. A. M., Abu Samah, A., & D'Silva, J. L. (2017). Climate change: Social adaptation strategies for fishermen. *Marine Policy*, *81*, 256–261.

Siason, I. M. (2001). Women in fisheries in the Philippines. In *International symposium on women in Asian fisheries, Malaysia: ICLARM*. World Fish Center. It is Siason, *2001*.

Siddique, M. R., Hossain, M., Rashid, A. M., Khan, N. A., Shuvo, S. N., & Hassan, M. Z. (2024). Evaluating co-management in the Sundarbans mangrove forest of Bangladesh: Success and limitations from local forest users' perspectives. *Ecology and Society*, *29*(2).

Sidik, F., Supriyanto, B., Krisnawati, H., & Muttaqin, M. Z. (2018). Mangrove conservation for climate change mitigation in Indonesia. *Wiley Interdisciplinary Reviews: Climate Change*, *9*(5), e529. https://doi.org/10.1002/wcc.529

Simard, M., Fatoyinbo, L., Smetanka, C., Rivera-Monroy, V. H., Castañeda-Moya, E., Thomas, N., & Van der Stocken, T. (2019). Mangrove canopy height globally related to precipitation, temperature and cyclone frequency. *Nature Geoscience*, *12*(1), 40–45.

Simić, V. M., Simić, S. B., Stojković Piperac, M., Petrović, A., & Milošević, D. (2020). Commercial fish species of inland waters: A model for sustainability assessment and management. *Science of the Total Environment*, *497*, 642–650. https://doi.org/10.1016/j.scitotenv.2019.07.371

Singh, A., Bhattacharya, P., Vyas, P., & Roy, S. (2017). Contribution of NTFPs in the livelihood of mangrove forest dwellers of Sundarban. *Journal of Human Ecology*.

Sinha, M. (1998). *Fisheries in coastal areas of West Bengal and required conservation*. Barrackpore, India: CIFRI.

Smallhorn-West, P., Cohen, P., Phillips, M., Jupiter, S., Govan, H., & Pressey, R. (2022, e13970). Linking small-scale fisheries co-management to U.N. Sustainable development goals. *Conservation Biology*, *36*(6). https://doi.org/10.1111/cobi.13977

Smederevac-Lalić, M., Špelić, I., Đug, S., Pengal, P., Çobani, M., Mrdak, D., & Piria, M. (2023). Political and socio-economic aspects of fisheries in inland and coastal waters of the Western Balkan. In V. Simić, S. Simić, & V. Pešić (Eds.), *Ecological sustainability of fish resources of inland waters of the Western Balkans in the Anthropocene*. Springer.

Sowrav, S. F. F., Debsarma, S. K., Das, M. K., Ibtehal, K. M., Rahman, M., Hridita, N. T., & Hoque, M. S. A. (2025). Developing a semi-automated technique of surface water quality analysis using GEE and machine learning: A case study for Sundarbans. *Heliyon*, *11*(3), e42404.

Sumaila, R. U., Cheung, W. W. L., Lam, V. W. Y., Pauly, D., & Herrick, S. (2011). Climate change impacts on the biophysics and economics of world fisheries. *Nature Climate Change*, *120*(2), 55–75. https://doi.org/10.1038/nclimate1301

Sumaila, U. R., Palacios-Abrantes, J., & Cheung, W. W. (2020). Climate change, shifting threat points, and the management of transboundary fish stocks. *Ecology and Society*, *25*(4).

Szabo, S., Brondizio, E., Renaud, F. G., Hetrick, S., Nicholls, R. J., Matthews, Z., … & Dearing, J. A. (2016). Population dynamics, delta vulnerability and environmental change: comparison of the Mekong, Ganges–Brahmaputra and Amazon delta regions. *Sustainability Science*, *11*(4), 539–554.

Taher, T. B., Althaus, C. E., Tranter, P. J., & Evans, M. C. (2023). Impacts of shrimp aquaculture on the local communities and conservation of the world's largest protected mangrove forest. *Environmental Science & Policy*, *147*, 351–360. https://doi.org/10.1016/j.envsci.2023.03.014

Thomas, A., Mangubhai, S., Fox, M., Meo, S., Miller, K., Naisilisili, W., Veitayaki, J., & Waqairatu, S. (2021). Why they must be counted: Significant contributions of Fijian women fishers to food security and livelihoods. *Ocean & Coastal Management*, *205*, 105571.

Thomas, N., Lucas, R., Bunting, P., Hardy, A., Rosenqvist, A., & Simard, M. (2017). Distribution and drivers of global mangrove forest change, 1996–2010. *PLoS One*, *12*(6).

Thoya, P., Horigue, V., Möllmann, C., Maina, J., & Schiele, K. S. (2022). Policy gaps in the east African blue economy: Perspectives of small-scale fishers on port development in Kenya and Tanzania. *Frontiers in Marine Science*, *9*, 933111. https://doi.org/10.3389/fmars.2022.933111

Tierney, J. E., Oppo, D. W., Rosenthal, Y., Russell, J. M., & Linsley, B. K. (2010). Coordinated hydrological regimes in the Indo-Pacific region during the past two millennia. *Paleoceanography*, *25*(1).

Titumir, R. A. M. (2022). *Sundarbans and its ecosystem services: Traditional knowledge, customary sustainable use and community based innovation*. Palgrave Macmillan.

Trivedi, S., Zaman, S., Ray Chaudhuri, T., Pramanick, P., Fazli, P., Amin, G., & Mitra, A. (2016). Inter-annual variation of salinity in Indian Sundarbans. *Indian Journal of Geo-Marine Sciences*, *45*(3), 410–415.

Turner, B. L., Matson, P. A., McCarthy, J. J., Corell, R. W., Christensen, L., Eckley, N., Hovelsrud-Broda, G. K., Kasperson, J. X., Kasperson, R. E., Luers, A., Martello, M. L., Mathiesen, S., Naylor, R., Polsky, C., Pulsipher, A., Schiller, A., Selin, H., & Tyler, N. (2003). Illustrating the coupled human-environment system for vulnerability analysis: Three case studies. *Proceedings of the National Academy of Sciences of the United States of America*, *100*, 8080–8085. https://doi.org/10.1073/pnas.1231335100

Uddin, M. S., Steveninck, E. R., Stuip, M., & Shah, M. A. R. (2013). Economic valuation of provisioning and cultural services of a protected mangrove ecosystem: A case study on Sundarbans reserve Forest, Bangladesh. *Ecosystem Services*, *5*, 88–93. https://doi.org/10.1016/j.ecoser.2013.07.002

UNDRR. (2023). *United Nations Office for Disaster Risk Reduction Annual Report*, 2023.

Villagran, J. C. (2006). *Vulnerability: A conceptual and methodological review (studies of the university: Research, counsel, education - Nr. 4)*. UNU Institute for Environment and Human Security.

Visser, T. A. M., Valbi-Jorgensen, J., & Chomo, V. (2021). *Good practices guidelines for data collection systems to support sustainable inland and recreational fisheries in the Western Balkans region*. FAO Fisheries and Aquaculture.

Vivekananda, J., Schilling, J., Mitra, S., & Pandey, N. (2018). On shrimp, salt and security: Livelihood risks and responses in South Bangladesh and East India. *Environment, Development and Sustainability*, *16*(6), 1141–1161.

Walsh, K. J., McBride, J. L., Klotzbach, P. J., Balachandran, S., Camargo, S. J., Holland, G., Knutson, T. R., Kossin, J. P., Lee, T., Sobel, A., & Sugi, M. (2016). Tropical cyclones and climate change. *Wiley Interdisciplinary Reviews: Climate Change*, *7*(1), 65–89. https://doi.org/10.1002/wcc.371

Wang, X., Jiang, D., & Lang, X. (2017). Future extreme climate changes linked to global warming intensity. *Science Bulletin*, *62*(24), 1673–1680.

Ward, R. D., Friess, D. A., Day, R. H., & Mackenzie, R. A. (2016). Impacts of climate change on mangrove ecosystems: A region by region overview. *Ecosystem Health and Sustainability*, *2*(4), e01211. https://doi.org/10.1002/ehs2.1211

Wisner, B., Oxley, M., Budihardjo, P. H., Copen, K., Castillo, G., Cannon, T., . . . Bonduelle, S. (2014). "Down home, it's all the same": Building synergisms between community-based disaster risk reduction and community-based climate change adaptation. In L. Schipper, J. Ayers, H. Reid, S. Huq, & A. Rahman (Eds.), *Community-based adaptation to climate change: Scaling it up* (pp. 172–191). London: Earthscan.

Woodruff, J. D., Irish, J. L., & Camargo, S. J. (2013). Coastal flooding by tropical cyclones and sea-level rise. *Nature*, *504*(7478), 44–52.

World Bank. (2020). *Bangladesh: Climate change and sustainable development (report no. 21104)*. World Bank.

World Bank Group. (2014). *Doing business 2015: Going beyond efficiency: Comparing business regulations for domestic firms in 189 economies: A world bank group flagship report*. World Bank Publications.

Xie, D., Schwarz, C., Kleinhans, M. G., Zhou, Z., & van Maanen, B. (2022). Implications of coastal conditions and sea-level rise on mangrove vulnerability: A bio-morphodynamic-modeling study. *Journal of Geophysical Research: Earth Surface*, *127*, e2021JF006301. https://doi.org/10.1029/2021JF006301

Yadav, S. K., & Majumdar, K. (2020). Mangrove associated seaweeds in Sundarban biosphere reserve, West Bengal, India. *International Journal of Advanced Research in Biological Sciences*, *7*(12), 53–62.

Yoon, D. K. (2012). Assessment of social vulnerability to natural disasters: A comparative study. *Natural Hazards*, *63*, 823–843. https://doi.org/10.1007/s11069-012-0189-2

Zacharia, P. U., &Najmudeen, T. M. (2012). Marine finfish resources of India: Distribution, commercial exploitation, utilization pattern and trade. In *World trade agreement and indian fisheries paradigms: A policy outlook*. [Conference paper presented at World Trade Agreement and Indian Fisheries Paradigms, September 17–26, 2012, Kochi].

4 Navigating the Tides: *The Enduring Resilience of Fisherfolk*

4.1 INTRODUCTION

Fish, as renewable resources, possess two primary characteristics: (1) the "utilization" of a unit of fish resource results in its complete and irreversible destruction (Sumaila and Munro, 2009), and (2) the fish population may be replenished to ensure sustained availability over time. Fisheries economics examines the temporal distribution of resources, reconciling present utilization with future accessibility. Although capture fisheries often incur modest ongoing expenses post-initial investment, their profitability may lead to overexploitation, especially in open-access or common-pool contexts. Fish stocks, being substantial resources, yield economic advantages over time, and their sustainable management requires the use of less than their natural growth. Effective fisheries management aims to maximize the long-term net economic advantages to society over time rather than pursuing short-term profits (Sin et al., 2019; Sumaila & Munro, 2009). Gibson's (2020) "Affordances theory" defines resources as accessible elements (such as people, institutions, or objects) in an environment that can be interacted with in various ways and offer benefits depending on individual needs (Sulikowski et al., 2020; Sun et al., 2016). The comprehensive interpretation of resources pertains explicitly to how an element in the environment might be transformed into a resource (Schultz, 2017; Tagliolatto et al., 2020). The community's resilience is linked to its capacity to identify and exploit environmental affordances, as this recognition and utilization facilitate adaptive responses to challenges and stressors. Certain groups possess a significant potential to endure, acquire knowledge, and adjust to change (Sepulveda et al., 2020; Weber et al., 2019; Zhou et al., 2019), while others exhibit lesser capability. Understanding the factors that enable communities to adapt (Saithong et al., 2022) and prepare for change is essential for mitigating vulnerability and strengthening resilience (Cinner et al., 2018; Murali & Sundar, 2017; Schwartzberg et al., 2021).

4.2 CHALLENGES OF SYSTEM BOUNDARIES AND MULTISCALE INTERACTIONS: RETHINKING RESILIENCE, ADAPTABILITY, AND ADAPTIVE CAPACITY

Resilience, adaptability, and adaptive capacity are three distinct but interrelated concepts (Spalding et al., 2023). All three concepts are crucial for the adaptability and sustainability of a socioecological system (Perissi et al., 2017). Adaptation and

DOI: 10.1201/9781003729990-4

adaptive capability are critical components to consider in the discussion of resilience. Adaptation refers to the strategies and measures enacted to maintain the capacity to navigate changes within the socioecological system while safeguarding essential structure and function (Liles et al., 2018; Madsen et al., 2021; Mohamed Shaffril et al., 2021). Adaptive capacity is an essential component of resilience, as the ability of individual systems to adjust and recover from shocks enhances the overall resilience of interconnected systems when facing complex, multiscale challenges (Malakar et al., 2023). Studies on the limitations of adaptation often investigate human understanding and actions in reaction to environmental changes, partially reflecting their capacity to deal with future disruptions (Alessa et al., 2017; Eagaard et al., 2018; Kane et al., 2020; Lai et al., 2021), which may either limit or increase their adaptive strategies (Hadjicostis, 2020; Zhang, 2021). Havice et al. (2012) illustrated the complex relations between psychological factors and the social, technical, and regulatory frameworks within the human impact assessment model. Kyle et al. (2020) later recognized these human sociopsychological attributes (e.g., values, norms, and attitudes) as latent variables (Saithong et al., 2022) within the complex interface of the socioecological system. When dealing with open systems, system boundaries should reflect relevant temporal, geographical, and institutional scales. The principal issue outlined pertains to regions and communities experiencing multiple crises that undermine societal functioning and restrict the capacity of governments to protect their populations (Mahfuzul Haque et al., 2023; Forrestall et al. 2019). A further challenge is to address significant disparities in wealth, income, power, and ecological impact observed at various scales. While these phenomena are interconnected, each remains intricate, dynamic, and subject to change across diverse scales and time frames (Ward & Langlands 2019). In many systems evaluations, spatial and temporal dimensions are routinely considered, though the scales related to jurisdiction, knowledge hierarchies, and institutional frameworks are often overlooked (Islam, 2018). Accordingly, there is a need for a multisystemic approach that transcends conventional boundaries and promotes innovative research and stakeholder engagement (Sarker et al., 2021). The principles and indicators of adaptation have been tabulated in Table 4.1. Despite the widespread notion that the processes that produce resilience are still constrained, the understanding of resilience is increasingly aligned across several disciplines. Alam

TABLE 4.1
Principles and Indicators for Adaptation

Principles	Indicators	Literature Sources
1. Flexible management structure	– Adaptive management strategies – Explicit planning concepts – Encourage optimal resource management techniques – Community-oriented governance	Collins et al., (2009); Doswald and Osti (2011); Watson et al. (2013); Watson et al. (2012); Mercer et al. (2012); Eslamian et al. (2024)

(Continued)

Table 4.1 (Continued)

Principles	Indicators	Literature Sources
2. Knowledge-based adaptation	– Cultivate understanding and consciousness – Collaborations between research and management – Optimal scientific data and indigenous knowledge – Culturally sensitive	Doswald and Osti (2011); McCarthy (2012); Mercer et al. (2012)
3. Maximum stakeholder involvement	– Involve the greatest number of stakeholders – Engage local communities – Collaborative plan involving many partners – Cooperation and reliance	Collins et al. (2009); Doswald and Osti (2011); Watson et al. (2012); Eslamian et al. (2024)
4. Variety	– Mitigate ambiguities – Prioritize the effects of climate change – Investigate various adaptation alternatives – Comprehend trade-offs	Doswald and Osti (2011); Watson et al. (2012); Jones et al. (2012); Mercer et al. (2012)
5. Multiscale operation	– Incorporate into developmental techniques – Support for sectoral adaptation planning – Multisectoral methodologies – Various geographic scales	Collins et al. (2009); Doswald and Osti (2011); Travers et al. (2012); Mercer et al. (2012)
6. Ensuring governance	– Responsibility – Lucid decision-making – Gender equity – Equity – Methodical assessment and appraisal	Doswald and Osti (2011); Jones et al. (2012); Walsh (2020)
7. Resilience building	– Resilience against resistance – Regulate climatic fluctuations – Address long-term climate change – Mitigate disaster risk reduction vulnerability – Alleviate nonclimatic stressors	Collins et al. (2009); Watson et al. (2012); McCarthy (2012); Elsamian et al. (2024); Abramson et al. (2010)
8. Maintaining ecosystem	– Foster ecosystem resilience – Preserve ecosystem services – Augment biodiversity – Conserve resources – Prevent maladaptation	Collins et al. (2009); Watson et al. (2013); Doswald and Osti (2011); McCarthy (2012); Travers et al. (2012); Mercer et al. (2012); Nalau et al. (2018)
9. Integrating development	– Deliver social, economic, and environmental advantages – Augment livelihood assistance – Guarantee cost-efficiency	Collins et al. (2009); Pérez et al. (2010); Nalau et al. (2018)

et al. (2020), a prominent developmental psychologist focused on resilience, define it as "the capacity of a dynamic system to adapt successfully to disturbances that threaten system function, viability, or development." This concept strongly aligns with the work of notable systems ecologist Smith et al. (2021), who define resilience as "the capacity of a system to absorb disturbance and reorganise while changing, thereby maintaining essentially the same function, structure, feedback, and consequently, identity." Both definitions emphasize a shift from disarray and dysfunction to recovery, adaptation, or systemic transformation before, during, and after facing adversity (Kabir & Hossain, 2021) and highlight the importance of examining how human and nonhuman systems adeptly navigate shocks and disturbances to thrive in the face of shocks and disturbances (O'Donnell & Wodon, 2015). The current perspective of multisystemic resilience presents a novel, transdisciplinary framework for understanding interactions within the socioecological system (SES). Instead of focusing on isolated ecological or social elements, it emphasizes that resilience arises from processes (Ahmed & Khan, 2023; Sammonds et al., 2021; Younus & Kabir, 2018). The use of a network analysis method facilitates to assess the critical system characteristics through the investigation of the network's topology. Networks are instrumental in the study of community-level adaptation as it reveals the relationships at the individual level and the properties of the socioecological system (Nomura et al., 2022).

A substantial corpus of research investigates the livelihoods linked to fisheries through diverse network methodologies (Beaudreau et al., 2018; Blasiak et al., 2017; Dee et al., 2017; Fisher et al., 2021). The socioecological network described by Cinner et al. (2018) and Cline et al. (2018) is based on trophic interactions and fishing methods. By establishing networks of cross-fisheries engagement across different "métiers," or fishing methods characterized by combinations of species compositions and gear types, future research explores flexibility (Cota-Nieto et al., 2018). Comparable research has established networks based on the interconnections among fishers through various permits (Kroetz et al., 2019) or species landings (Levin, 2019), facilitating the analysis of fisheries spillover and the potential redistribution of fishing effort across different fisheries. Dwiartama and Rosin (2014) contend that actor-network theory expands the notion of agency to encompass nonhumans, hence promoting a more integrated view of a socioecological system. The actor-network theory is related to vulnerability and resilience. The authors argue that emphasizing agency, not only in individual actions but also in connections and networks, as shared competencies may explain socioecological dichotomies within the SES paradigm (Christmann et al., 2012). The attributes of complexity are essential for analyzing systemic resilience in several dimensions. First, recognizing the flexibility and ambiguity of system boundaries makes it easier to incorporate and update analysis to incorporate potentially marginal agents and participants. It emphasizes that, in the delivery of humanitarian aid, recipients are not merely passive victims, but individuals and communities equipped with local resources and the capacity to tackle future challenges. There are currently no methods for documenting these attributes or tools to measure the frequently intangible traits and unclear processes that determine whether recovery, adaptation, and transformation efforts succeed or fail following a major disaster.

Global threats to the resilience of socioecological systems, whether immediate or distant, necessitate a parallel resilience in legal frameworks. The subsequent five principles are apparent in the works of professionals in catastrophe resilience.

- Principle 1: The essential capacity of resilience is to "absorb disturbance" and "reorganize into a fully functional system" (Cutter et al., 2008). The United Nations employed the phrase "build back better" to illustrate the ideal outcomes of a resilient community's recovery, rehabilitation, and reconstruction following a disaster (United Nations International Strategy for Disaster Reduction, 2017). The principal focus is on "build better" rather than "build back."
- Principle 2: Resilience should be regarded as a process rather than an outcome. Resilience is defined by adaptation rather than stability (Norris et al., 2011). The ongoing process encompasses perception, anticipation, learning, and adaptation (Park et al., 2012).
- Principle 3: Disasters affect nearly all socioecological situations, encompassing the economic, built, cultural, and political facets of society. These elements are tightly interconnected and profoundly influence one another.
- Principle 4: Since a catastrophe is a cyclical phenomenon, the postdisaster response, reconstruction, and recovery from a specific disaster serve as predisaster preparedness for the next event, which is vital for communities in hazard-prone regions (Wu & Hou, 2019). Promoting resilience is an ongoing effort necessitating persistent engagement between individuals and their communities.
- Principle 5: Fostering resilience necessitates compliance with "the principles of equity, fairness, and access to resources" (Cutter, 2016). Since resilience is a collective characteristic of a system, improving resilience does not prioritize one component over another. It encapsulates and advocates for social equality (Tierney, 2006).

Fuller et al. (2007) analyze actor-network theory in relation to vulnerability and resilience, arguing that focusing on distributed agency in human actions and networks can clarify and bridge socioecological divides within the SES framework. The characteristics of complexity are vital for analyzing systemic resilience in various dimensions. Recognizing the openness and indeterminacy of system boundaries facilitates integrating and adapting analysis to include actors and stakeholders that may appear marginal. It underscores that in providing humanitarian aid, beneficiaries are not merely helpless victims but rather persons and communities endowed with local resources and the ability to confront future challenges. Hence, we require methodologies to discern these properties and instruments to measure the frequently intangible characteristics and ambiguous processes that influence the success or failure of recovery, adaptation, and transformation following a significant catastrophe (Giron-Nava et al., 2020).

An examination of resilience principles across many academic disciplines (Ojeda-Ruiz et al., 2019) identified seven comparable principles that elucidate the functioning of resilience in stressed systems (human, constructed, or natural). The elements comprise (1) resilience in adverse contexts; (2) resilience as a process; (3) trade-offs among resilient systems; (4) a resilient system characterized by openness, dynamism, and complexity; (5) a resilient system promote connectivity; (6) a resilient system exemplifying experimentation and learning; and (7) a resilient system incorporating diversity, redundancy, and participation (Oken et al., 2020). Therefore, process resilience encompasses environmental perception and alteration (e.g., sensing [cells, tissues], perception [organisms], communication [populations], and calibration), reactions (actions), and outcomes. With each cycle, the environment undergoes changes that affect the initial and most crucial phase (perception of environmental change) (Stoll et al., 2017).

A multisystemic approach to resilience recognizes the intricate interplay between inequality and sustainability, functioning through complex mechanisms and pathways across various scales—from psychological to environmental—and encompassing both gradual and rapid variables and feedback loops (White et al., 2020) (Figure 4.1). This statement points to the existence of numerous conditions where interventions may have been introduced, but their success largely depends on the timing and the process used. However, if fundamental moral frameworks and cultural norms are not changed in tandem with user behavior modification, efforts may not be effective (Troell et al., 2019; Vázquez-Delfín et al., 2019).

From a livelihood standpoint, adaptive capacity posits that diversity is crucial for enhancing options and flexibility in responding to disruptions and sustaining income (Bobel et al., 2018; Cramer & Krueger, 2020). Diversification and turnover in fisheries have historically been regarded as essential elements of socioecological resilience (Cutuli & Herbers, 2018; Hutchison & Sterbenz, 2018). Historically, diversification and turnover in the fisheries have been considered as the two main factors that maintain resilience in the socioecological system (Cutuli & Herbers, 2018; Hutchison & Sterbenz, 2018). It has been shown that diversifying a harvest portfolio, or the assemblage of captured species, can not only avoid the occurrence of economic troubles but also increase fishers' capacity of risk management in a particular set of fisheries (for instance, fishery closures, lowered demand, and seasonality) (Mishna et al., 2016; Olsson et al., 2020; Oshri et al., 2018). However, the development of specialization can bring some financial benefits, yet it is accompanied by significant risks that endanger the possibility of adaptation in the event of a fisheries collapse (Baumann & Ylinen, 2020; Quinlan et al., 2016). Boksa et al. (2018), Younus and Kabir (2018), and Fanian et al. (2019) assert that a diverse array of species with uncorrelated or disassociated abundances across time may mitigate a fisherman's risk by ensuring more stable and less variable income. Understanding the variety and organization of fishing portfolios is crucial for assessing the adaptive capacity of fisheries. Adaptive ability can differ among individuals, families, communities, governments, and nations (Siders, 2019). Thus, defining the portfolios of fishing communities or port associations is beneficial for reinforcing regional strategies designed to improve the adaptive capacity of fishery livelihoods.

The socioeconomic condition of fishermen is intricately dependent on climate change (Sarker et al., 2019), since unstable or inadequate income from diminishing fish stocks may hinder their capacity to address challenges (Xu & Kajikawa, 2017). Recent evaluations of marine ecosystems and fisheries' responses to climate change

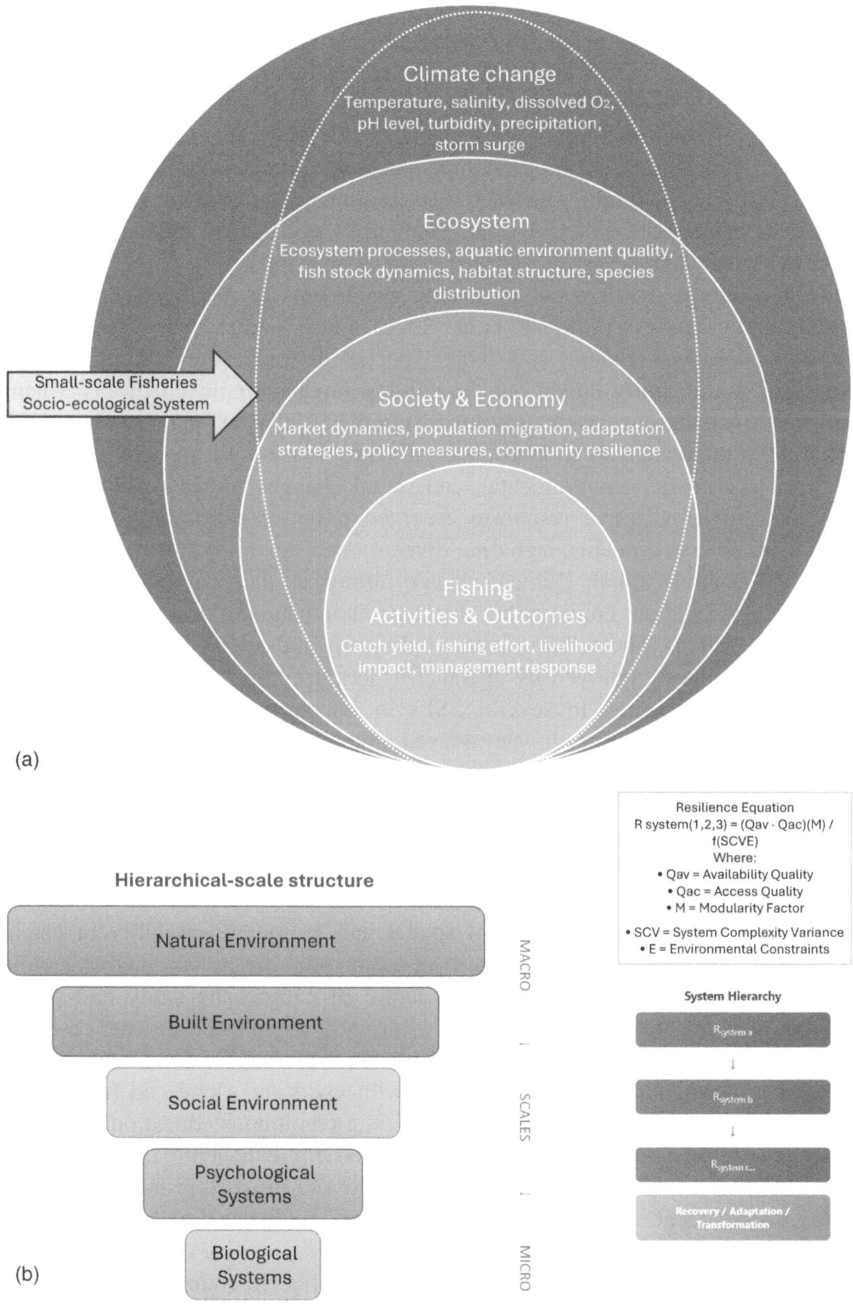

FIGURE 4.1 (a) Socioecological system of small-scale fisheries of Sundarbans comprised of interacting scales. (b) Hierarchical organization of interacting scales of a system. An expression of resilience (in contexts where a population is exposed to above-normal levels of adversity).
(Modified after Ungar (2021)).

highlight the lack of conclusive evidence of specific adaptation tactics and methodologies (Folke et al., 2018). In this context, SSFs are predominantly neglected in the scientific literature (Beckett et al., 2018). Different studies have examined adaptation to climate change in SSFs. Cutter (2019) investigated the constraints and challenges associated with the limitation of fishing activity due to cyclones in Bangladesh. The researchers concluded that the limitations encompass meteorological and maritime physical attributes, including the increased frequency and duration of cyclones and concealed sandbars. Conversely, the challenges encompass inadequate weather forecasts, substandard vessels, limited market and financial access, low-income levels, inadequate enforcement of fishing regulations, and a lack of sustainable livelihood alternatives (Villasante et al., 2022). Cutuli and Herbers (2018) reported that socioecological change and target species diversification, along with gear ownership, significantly influenced the cultural models employed by fishermen in response to alterations in the resource system. Hutchison and Sterbenz (2022) performed a meta-analysis to assess the influence of adaptive capacity domains on reactions to climatic, environmental, and social changes in SSFs in 20 countries. Community-level adaptive responses emerged only when communities accessed assets from diverse domains, including diversity and flexibility, learning and knowledge, and natural capital. Climate change influences the responses and adaptive strategies of indigenous communities that depend on fisheries(Leal Filho et al., 2020; Olsson et al., 2015). The authors identified diversity and coadaptive capability as the primary techniques employed by artisanal fishers in indigenous communities. The ongoing data deficiency in several SSFs in impoverished nations occasionally obstructs the assessment of climate-induced alterations (Lauria et al., 2018; Woodcraft et al. 2020). The study integrated scientific and indigenous ecological knowledge, illustrating that a partnership of self-directed and government-led adaptations is crucial for addressing the challenges of climate change (Scherer et al., 2019; Ungar, 2018). The Adaptive Core Capacities of Coastal Communities have been tabulated in Table 4.2.

These fisheries exhibit significant spatial and temporal variability in catches, attributable to substantial fluctuations in environmental parameters such as temperature and salinity. The interdisciplinary nature of catastrophe study and practice requires balancing all social elements concurrently to enhance disaster resistance. Disaster resilience in the rehabilitation of human settlements necessitates the amalgamation of swift physical reconstruction with enduring social and humanitarian recovery. Encouraging catastrophe resilience is a continuous educational endeavor. Community participation is an effective method for enhancing the resilience of individuals, families, and communities. Research and practice in hazard and catastrophe management require interdisciplinary and multistakeholder collaboration. Effective leadership is essential for the collaborative involvement of experts and stakeholders, aimed at enhancing the resilience of affected communities.

Rights-based regulatory frameworks governing shellfishing (Pita et al., 2019) exhibit considerable resilience to the impacts of climate change (Ojea et al., 2017). Fluctuations in temperature and/or salinity, even within the tolerable thresholds for each species, can interfere with various physiological processes, including growth and

TABLE 4.2
List of Adaptive Core Capacities of Coastal Communities

Nine Core Adaptive Capacities Required for Community Resilience	Elements of the Capacity
Local knowledge	Comprehending the community's vulnerabilities, providing training and public disaster education, and fostering collective effectiveness and empowerment
Community networks and relationships	Interconnectedness, unity, trustworthiness, common ideals, robust relationships
Communication	Robust communication networks, varied modes and contexts of communication, risk communication, crisis communication
Health	Existing community health, comprehension and mitigation of health risks, accessibility to health services, and quality treatment for physical and mental health concerns
Governance and leadership	Infrastructure and services, public engagement and assistance, local involvement in planning, reaction, and recovery
Resources	Provisions, hydration, medical supplies, accommodation, conveyance, critical equipment, monetary and communal resources
Economic investment	Allocation of financial resources, economic planning, cost-efficient interventions, economic advancement of postdisaster infrastructure
Preparedness	Risk evaluation, exercises, strategizing, alleviation strategies
Mental outlook	Perspectives on uncertainty, optimism, and flexibility

reproduction (Domínguez et al., 2020, 2021; Macho et al., 2016; Peteiro et al., 2018; Vázquez-Delfín et al., 2019), as well as behavior (Woodin et al., 2020). The related restrictions may result in recruitment failures in adult populations or prolong the time needed for the species to reach commercial levels.

Geographic diversification and species diversification (changing the target species) are two significant worldwide diversification techniques in SSFs. The latter pertains to seasonal mobility, internal migration, trans-local livelihoods (Islam & Herbeck, 2013), and/or the geographical displacement of fisheries participants, enabling them to function at regional scales outside their local fishing regions or communities. Consequently, spatial diversity necessitates that participants transition between alternative fishing areas or alter their conventional landing locations. It may also encompass fishing entities whose extraction activities are distributed over extensive geographical areas. Spatial diversification in mobility has been identified as an adaptive strategy in several SSF systems and is a significant trait of numerous populations dependent on marine resources (Badjeck et al., 2010; Njock & Westlund, 2010; Overå, 2005). Participants in fisheries may possess varied motivations and competencies that either limit or facilitate their mobility (Islam & Herbeck,

2013; Sievanen, 2014; Wanyonyi et al., 2016), while regulations and institutions significantly impact diversification processes (Nunan, 2010; Nunan et al., 2012). Diversification has consistently been crucial to SSF initiatives, and its significance is anticipated to escalate as climate change advances (Young et al., 2019).

Spatial diversity presents distinct governance problems, occasionally resulting in disputes among stakeholders (Crona & Rosendo, 2011; Dubik et al., 2019), and is frequently linked to overexploitation (Binet et al., 2012; Wilson, 2017). The occurrences of recurrent exploitation and systematic depletion of fishing grounds have been documented in numerous global case studies (Turner & Berkes, 2006). Proposed regulations linking access rights to geographically defined boundaries aim to mitigate overexploitation; however, these management strategies often overlook spatial diversification as an adaptive approach and fail to address the fundamental socioeconomic, political, and environmental factors (Finkbeiner, 2015; Finkbeiner et al., 2017). These strategies may result in unintended outcomes, such as effort displacement (Abbott & Haynie, 2013; Kroetz et al., 2019) or diminished diversification skills (Kasperski & Holland, 2013), while simultaneously heightening individual vulnerability and rule infractions (Stoll et al., 2016). A comprehensive understanding of geographical diversification and their temporal variations is crucial for enhancing fisheries management across extensive scales (Cudney-Bueno & Basurto, 2009; Nunan et al., 2012). Socioecological resilience (Folke et al., 2018) defines the ability of the socio-ecosystem to endure the effects of "novel, unforeseen events and extreme shocks," such as climatic extremes or disturbances. It involves mitigating their severity, autonomously returning to original SES patterns and functions, or transitioning to alternative socioecological states. A resilient SES can restore ecosystem functionality and the provision of ecological services (ES), as well as rehabilitate impacted sectors, infrastructure, and institutions. Conversely, a tilted SES must traverse through a modified ES spectrum and potential failures across several sectors, infrastructure, and institutions, necessitating substantial modifications in management and institutional frameworks. The idea of Society for Ecological Restoration (SER) has evolved from the concept of robustness (Anderies et al., 2004) and the impact of resource utilization on social and ecological resilience (Adger, 2000), but still encounters implementation problems. Hence, it necessitates the amalgamation of ecological resilience (Gunderson, 2000; Hodgson et al., 2015) and societal transformation processes (Carpenter et al., 2012; Folke et al., 2005). However, the exact delineation of pertinent SES features has yet to be established, and considerable obstacles remain in implementing the SER concept. Comprehending and improving socioecological adaptation capability to climatic extremes presents a considerable challenge.

Artisanal fisheries frequently encounter the consequences of climate change and may benefit from strategies that enhance adaptation and strengthen resilience (Bossier et al., 2025). Study indicates that SSFs may respond to change through three pathways: adaptation, response, and coping. The diversity and adaptability are crucial elements for adaptive capability and that fishermen can modify their practices even in the absence of financial resources. Fishers frequently modify their techniques in response to variations in weather, particularly oceanic conditions influenced by climate change. Fishers predict changes by reflecting on their connection with nature, which could relate to conservation initiatives. Fishers may adapt by switching between species or multispecies exploitation on a seasonal or annual basis in response to market fluctuations or fisheries closures. The notion of resilience has developed through the interplay of ecological, social, and technological factors (Bush et al., 2010), impacting community well-being. Recognizing the roles and contributions of men, women, youth, elders, and other stakeholders in marine resource economies is essential for bolstering resilience in coastal communities facing environmental and economic changes on a global scale (Ravera et al., 2016). Resilience evaluation and quantification are essential avenues for explaining complex systems, uncovering overarching patterns, and identifying emergent characteristics (Quinlan et al., 2016) to facilitate decision-making. The "all affected principle" posits that individuals, regardless of their physical or temporal proximity, should be incorporated into decision-making processes, even when their representation poses challenges. Future generations, if not formally acknowledged in representative democratic systems, face a limitation that results in short-termism (Brown et al., 2019). In systemic resilience, the absence or nonobservability of agents in open systems poses a problem for their incorporation and for methodologies addressing the influence of indirect and indeterminate events on these systems (Frawley et al. 2021). Understanding the role of women and gender within these frameworks is essential for creating effective policies and programs that balance the sustainability of fisheries resources with the viability of fishing communities (Berkes, 2015; Jess Grunblatt & Alessa 2017). Human-centered approaches that recognize the interdependent dynamics of social and ecological systems are becoming more prevalent in fisheries governance and policy (Bennett, 2019). The collection and dissemination of sex-disaggregated data (Harper et al., 2013) are essential to understand the resource utilization patterns for ecosystem-based fisheries management (de la Torre-Castro et al., 2017). A gender perspective is essential for understanding the implications of diverse management strategies, evaluating trade-offs, and enhancing the efficacy of fisheries management initiatives when men and women pursue different species, utilize distinct gear, and fish in various habitats (Harper et al., 2020; Kleiber et al., 2015). While modernization in the 1950s improved small-scale fishermen's access to resources, markets, and livelihoods, inherent logical fallacies in the process rendered them more susceptible to climate change. Resilience thinking significantly influences legislation and its ability to facilitate adaptive governance. The law is a complex adaptive system of organizations and mechanisms operating at various levels, from international to local, governing multiple facets of human behavior, either directly or indirectly. However, the law does not operate autonomously from other social systems; instead, it coevolves with them within a network of interconnected systems. The legal system, as a subsystem, possesses resilience characteristics that may be defined by the general resilience theory exhibiting reliability, efficiency, scalability, modularity, and flexibility. "Adaptive governance" is a conceptual framework employed to regulate the resilience of systems toward socially desired outcomes. When effectively implemented, the resilience of the legal system generates the necessary adaptive legislation to facilitate responsive governance.

4.3 NAVIGATING CHALLENGES AND PROSPERITY: ADAPTIVE STRATEGIES AND RESILIENT LIVELIHOODS IN THE SUNDARBANS SMALL-SCALE FISHERIES

Fishing in the Sundarbans generally requires a crew of at least three individuals: one to operate the net, another to navigate the vessel, and a third to monitor tiger activity, thereby enhancing efficiency and protecting against tiger attacks and piracy. The fisherman is typically joined by family members, such as sons, brothers, or fathers, who remain with them in the event of any misfortune. This confidence and devotion constitute the primary safeguard for their fishing ventures and enhance profitability. In reaction to seasonal, interannual, and multidecadal climate variability and disasters, fishers are adopting various strategies, including reducing and modifying household consumption, diversifying income sources, restricting the use of specific gear types, changing target species, relocating to less productive areas to enhance catch, engaging in vessel buybacks, participating in illicit fishing, and utilizing mutual insurance and social networks in Sundarbans. To ensure the sustainability of small-scale fishing in developing nations, Evans and Andrew (2011) emphasized the necessity of establishing social networks that strengthen resilience and empower local resource users. Social units possess varied coping mechanisms that allow them to address recorded damage and mitigate possible risks linked to a hazard. Entitlement and endowment are associated with these capabilities, and both concepts can be broadened to encompass additional social safety nets that enhance resilience function (Turner et al., 2003). The social structure, particularly through social networks, influences economic outcomes for three primary reasons. Social networks primarily influence the dissemination and quality of information. An abundance of information is intricate, nuanced, and challenging to authenticate, which leads individuals to distrust impersonal sources and instead rely on personal connections. Second, social networks serve as a significant source of reward and punishment, with their impact magnified when originating from individuals with whom one has close familiarity. Third, trust, defined as the assurance that individuals will act ethically despite opposing incentives, arises within the framework of a social network (Granovetter, 2005). Trust, reciprocity, and kinship collectively form the essential enduring mechanisms that reinforce familial bonds and social networks.

The customary practice and the belief in the forest goddess Bon Bibi confer psychological resilience upon individuals, enabling them with the psychological fortitude to undertake hazardous endeavors in the wilderness (Grafton, 2010; Islam & Chuenpagdee, 2013; Kazim, 2011; Morato et al., 2006; OECD, 2010). Reports indicate a collaboration between boat owners and dealers (aratdars) to enhance profits. In response to the heightened wave activity, gillnetters have elevated their vessels by one foot to a total height of 2.5 feet and expanded their width from 12 feet to 17 feet, enhancing their power from 20 to 25 horsepower to over 100 horsepower. This modification aims to enhance efficiency and extend the range of fishing expeditions, thereby increasing capacity and augmenting rivalry with other vessel proprietors (Salagrama, 2012). Occupational divergence is seen as the principal strategy for addressing livelihood difficulties. The fishing season commences in mid-July and persists until October. Every month during the peak season includes a phase (Gon) of abundant catch, followed by a phase of diminished catch. However, the catch fluctuates daily throughout the Gon season. The prime fishing season in the Bay of Bengal adjacent to the Sundarbans forest extends from December to March; yet this period may occasionally be deemed inconsequential as fishery resource collectors infrequently participate in fishing activities in the Bay. Fish output in the ocean from mid-April to mid-July significantly declines due to increased salinity in coastal waters and the necessity for net construction or repair, vessel upkeep, and financial planning for the upcoming peak season (Islam, 2018; Keith, 2003). During the lean season or alongside fishing, fishermen participate in rice cultivation, serve as crew members, engage in shrimp farming, perform daily labor, harvest Golpata (*Nypa* sp.), collect firewood, operate boats, repair boats and nets, and pull rickshaws. The rural livelihood of Sundarbans is intrinsically linked to seasonality, and the fisherwomen have devised several techniques to sustain their livelihoods in response to these cyclical variations. The fisherwomen consistently gather fish seeds throughout the year. However, seed availability is constrained throughout winter, necessitating reliance on alternative livelihood techniques, such as fish drying. Fisherwomen frequently process and dry fish from mid-October until late January or mid-February. During the rainy season, women also capture fish from small water channels due to the increased availability of fish. From June to October, several individuals engage in various horticultural activities in their own gardens or as workers. During winter, the women in the Sundarbans gather timber for firewood from the forest. Women in the Sundarbans frequently engage in fish drying for four months due to the scarcity of alternative economic options during that period. Women have a vital role in post-harvest operations. Sundarbans fisherwomen engage in drying and preparing fish from October to February to provide financial assistance for their families (Roy and Bhaumik, 2014; Kundu et al., 2025). The "secret savings" of women, acquired through net crafting and mending, shrimp-seed harvesting, fish commerce, aquaculture employment, and the production and sale of desiccated fish, act as financial safeguards for their families. With the introduction of commercially produced nets, net-making transitioned from being predominantly a female occupation. Traditional female fishermen are broadening their responsibilities beyond home and industrial labor and are also managing small enterprises in adjacent cities to sustain their families. Neglecting women's contributions exacerbates the marginalization of small-scale fishermen and SSFs, which receive significantly less governmental support and management focus than industrial sectors (Jacquet & Pauly, 2008; Schuhbauer et al., 2017). Nontargeted and endangered species are harvested out of necessity during scarcity and for additional income during peak seasons. The fry collectors and fishermen corrupt local law enforcement officials to facilitate illegal and off-season harvesting. Hilsa is the primary catch in Bangladesh, with fishermen predominantly using small-meshed set bag nets to capture juvenile fish. Annually, around 463 million Jatka, equating to 3,707 tons, are illicitly taken with monofilament nets during the spawning season. To reduce reliance on capturing a single species, Bombay Duck, sardines, tuna, and specific catfish species that were previously overlooked are now being utilized. Consequently, the multispecies fisheries of Sundarbans offer distinctive options for fishermen to adjust their target species (Islam, 2018; Salagrama, 2012). The fishermen sought advice from family, neighbors, local officials, esteemed village elders, friends, fisheries extension agents, and NGO representatives to address issues arising from ecological changes and their professional ramifications. In general, the chief (Sardar) of each fishing hamlet possessed the primary authority to negotiate with other parties regarding solutions to prevailing issues. Ecological information is mostly imparted through seminars and hands-on experiences, augmented by mentorship and teaching the youth. The predominant age demographic of fishermen is between 40 and 49 years, possessing 10–15 years of fishing experience. Although fishermen prefer their children to receive sufficient education for employment outside the Sundarbans, 87% of fishing families choose fishing as the main income, while the remaining 13% engage in fishing as daily labor (Chacraverti, 2014). Crab and prawn seed harvesting has become highly lucrative, particularly for landless individuals, as it requires minimal labor (only two persons) and presents a reduced risk of tiger assaults. Consequently, fishermen employ diversity to mitigate risks linked to their biological and economic contexts (Minnegal & Dwyer, 2008). In numerous impoverished nations, comprehension of the determinants affecting success in small-scale fishing remains rudimentary. The preponderance of documented success stories originates from affluent nations where SSFs have minimal significance (Isaacs, 2013; Ratner et al., 2012).

4.4 CONCLUSION

Government environmental policies have frequently overlooked the unique socio-ecological dynamics and indigenous knowledge systems, focusing instead on technology-driven techniques that have proven mainly ineffective in managing the environment. Adopting local involvement and honoring the socioecological frameworks of communities might reduce vulnerability and improve environmental governance. Recognizing that local vulnerabilities are deeply intertwined with ecological systems, two options emerge to support tribal communities: one is allowing natural selection to adopt resilience. Simultaneously, the latter involves intentional governmental measures to foster community investment and local habitation. Population growth in these communities can boost social resilience, facilitating the development of locally informed conservation efforts. Investing in local communities improves societal well-being and environmental conservation, creating a mutually beneficial relationship. Fisheries policies and adaptation programs informed by this understanding would more efficiently address resource decline and enhance local capacities.

BIBLIOGRAPHY

Abbott, J. K., & Haynie, A. C. (2013). What are we protecting? Fisher behavior and the unintended consequences of spatial closures as a fishery management tool. *Ecological Applications*, *23*(4), 762–777.

Abramson, D. M., Stehling-Ariza, T., Park, Y. S., Walsh, L., & Culp, D. (2010). Measuring individual disaster recovery: A socioecological framework. *Disaster Medicine and Public Health Preparedness*, *4*(Suppl. 1), S46–S54. https://doi.org/10.1001/dmp.2010.12

Adger, W. N. (2000). Social and ecological resilience: Are they related? *Progress in Human Geography*, *24*(3), 347–364. https://doi.org/10.1191/030913200701540465

Ahmed, S., & Khan, M. A. (2023). Spatial overview of climate change impacts in Bangladesh: A systematic review. *Climate and Development*, *15*(2), 132–147. https://doi.org/10.1080/17565529.2022.2062284

Alam, M. S., Mikhailovna, R. O., & Vasilevich, G. V. (2020). The aftermath of natural disasters in Bangladesh and a proposal to minimise casualty. *ICTACT Journal of Management Studies*, *6*, 1166–1169.

Alessa, A., Aleid, D., Almutairi, S., AlGhamdi, R., Huaidi, N., Almansour, E., & Youns, S. (2017). Awareness of polycystic ovarian syndrome among Saudi females. *International Journal of Medical Science and Public Health*, *6*(6), 1013–1020. https://doi.org/10.5455/ijmsph.2017.0202507022017

Anderies, J. M., Janssen, M. A., & Ostrom, E. (2004). A framework to analyze the robustness of social-ecological systems from an institutional perspective. *Ecology and Society*, *9*(1), 18. https://doi.org/10.5751/ES-00610-090118

Badjeck, M.-C., Allison, E. H., Halls, A. S., & Dulvy, N. K. (2010). Impacts of climate variability and change on fishery-based livelihoods. *Marine Policy*, *34*(3), 375–383. https://doi.org/10.1016/j.marpol.2009.08.007

Baumann, L. J., & Ylinen, A. (2020). Prevention: Primary, secondary, tertiary. In M. Gellman (Ed.), *Encyclopedia of behavioral medicine* (pp. 1–3). Springer. https://doi.org/10.1007/978-3-030-39903-0_215

Beaudreau, A. H., Levin, P. S., Norman, K. C., & Mantua, N. J. (2018). Climate change and the adaptive capacity of Pacific northwest fisheries: A framework for management response. *Fisheries*, *43*(10), 452–463.

Beaudreau, A. H., Ward, E. J., Brenner, R. E., Shelton, A. O., Watson, J. T., Womack, J. C., Anderson, S. C., Haynie, A. C., Koehn, K. A., Levin, P. S., Marshak, A. R., Soldevilla, M. S., & Williams, B. C. (2019). Thirty years of change and the future of Alaskan fisheries: Shifts in fishing participation and diversification in response to environmental, regulatory and economic pressures. *Fish and Fisheries*, *20*(3), 601–619. https://doi.org/10.1111/faf.12364

Beckett, C., Maughan, B., Rutter, M., Castle, J., Colvert, E., Groothues, C., & Sonuga-Barke, E. J. S. (2018). Do the effects of early severe deprivation on cognition persist into early adolescence? Findings from the English and Romanian adoptees study. *Child Development*, *77*(3), 696–711. https://doi.org/10.1111/j.1467-8624.2006.00898.x

Bennett, D. (2019). Meeting society's expectations of graduates: Education for the public good. In J. Higgs, G. Crisp, & W. Letts (Eds.), *Education for employability I: The employability agenda* (pp. 35–48). Rotterdam, The Netherlands: Brill-Sense Publishers.

Berkes, F. (2015). *Coasts for people: Interdisciplinary approaches to coastal and marine resource management* (1st ed.). Routledge. https://doi.org/10.4324/9781315771038

Binet, T., Failler, P., & Thorpe, A. (2012). Migration of Senegalese fishers : A case for regional approach to management. *Maritime Studies*, *11*(1), 1–14.

Blasiak, R., Spijkers, J., Tokunaga, K., Pittman, J., Yagi, N., & Österblom, H. (2017). Climate change and marine fisheries: Least developed countries top global index of vulnerability. *PLoS One*, *12*(2), 80–95. https://doi.org/10.1371/journal.pone.0179632

Bobel, T. S., Hackl, S. B., Langgartner, D., Jarczok, M. N., Rohleder, N., Rook, G. A., & Reber, S. O. (2018). Less immune activation following social stress in rural vs. urban participants raised with regular or no animal contact, respectively. *PNAS*, *115*(20), 5259–5264. https://doi.org/10.1073/pnas.1719866115

Boksa, P., Joober, R., & Kirmayer, L. J. (2018). Mental wellness in Canada's aboriginal communities: Striving toward reconciliation. *Journal of Psychiatry & Neuroscience*, *40*(6), 363–365. https://doi.org/10.1503/jpn.150309

Bossier, S., Ota, Y., Pozas-Franco, A. L., & Cisneros-Montemayor, A. M. (2025). How much time and who will do it? Organizing the toolbox of climate adaptations for small-scale fisheries. *Frontiers in Marine Science*, *12*, Article 1521526. https://doi.org/10.3389/fmars.2025.1521526

Brown, P., Bocken, N., & Balkenende, R. (2019). Why do companies pursue collaborative circular oriented innovation? *Sustainability*, *11*(3), 635.

Bush, T., Bell, L., & Middlewood, D. (2010). Introduction: New directions in educational leadership. In T. Bush, L. Bell, & D. Middlewood (Eds.), *The principles of educational leadership and management* (2nd ed., pp. 3–12). Sage.

Carpenter, S. R., Folke, C., Norström, A., Olsson, O., Schultz, L., Agarwal, B., Balvanera, P., Campbell, B., Castilla, J. C., Cramer, W., DeFries, R., Eyzaguirre, P., Hughes, T. P., Polasky, S., Sanusi, Z., Scholes, R., & Spierenburg, M. (2012). Program on ecosystem change and society: An international research strategy for integrated social–ecological systems. *Current Opinion in Environmental Sustainability*, *4*(1), 134–138. https://doi.org/10.1016/j.cosust.2012.01.001

Chacraverti, S. (2014). *The Sundarbans fishers coping in an overly stressed mangrove estuary*.

Christmann, A., Knutti, R., & Zscheischler, J. (2012). Framing the way to relate climate extremes to climate change. *Climatic Change*, 115(2), 283–290.https://doi.org/10.1007/s10584-012-0441-5

Cinner, J. E., Adger, W. N., Allison, E. H., Barnes, M. L., Brown, K., Cohen, P. J., Gelcich, S., Hicks, C. C., Hughes, T. P., Lau, J., Marshall, N. A., & Morrison, T. H. (2018). Building adaptive capacity to climate change in tropical coastal communities. *Nature Climate Change*, *8*(2), 117–123. https://doi.org/10.1038/s41558-017-0065-x

Cline, T. J., Schindler, D. E., & Hilborn, R. (2018). Fisheries portfolio diversification and turnover buffer Alaskan fishing communities from abrupt resource and market changes. *Nature Communications*, *8*(1), 1–7. https://doi.org/10.1038/ncomms14042

Collins, J. A., Blackman, D. K., Harris, A., & Carlson, R. L. (2009). Seismic and drilling constraints on velocity structure and reflectivity near IODP Hole U1309D on the central dome of Atlantis massif, mid-Atlantic ridge 30°N. *Geochemistry, Geophysics, Geosystems*, *10*. https://doi.org/10.1029/2008GC002121

Cota-Nieto, J. J., Erisman, B., Aburto-Oropeza, O., Moreno-Báez, M., Hinojosa-Arango, G., & Johnson, A. F. (2018). Participatory management in a small-scale coastal fishery—Punta Abreojos, Pacific coast of Baja California Sur, Mexico. *Regional Studies in Marine Science*, *18*, 68–79. https://doi.org/10.1016/j.rsma.2017.12.014

Cramer, J., & Krueger, A. B. (2020). Disruptive change in the taxi business: The case of Uber. *American Economic Review*, *106*(5), 177–182. https://doi.org/10.1257/aer.p20161002

Crona, B. I., & Rosendo, S. (2011). Outside the law? Analyzing policy gaps in addressing fishers' migration in East Africa. *Marine Policy*, *35*(3), 379–388. https://doi.org/10.1016/j.marpol.2010.11.003

Cudney-Bueno, R., & Basurto, X. (2009). Lack of cross-scale linkages reduces robustness of community-based fisheries management. *PLoS One*, *4*(7), e6253. https://doi.org/10.1371/journal.pone.0006253

Cutter, S., Barnes, L., Berry, M., Burton, C., Evans, E., Tate, E., & Webb, J. (2008). A place-based model for understanding community resilience to natural disasters. *Global Environmental Change*, *18*(4), 598–606. https://doi.org/10.1016/j.gloenvcha.2008.07.013

Cutter, S. L. (2016). Resilience to what? Resilience for whom? *The Geographical Journal*, 182(2), 110–113.

Cutter, S. L. (2019). The landscape of disaster resilience indicators in the USA. *Natural Hazards*, *80*(2), 741–758. https://doi.org/10.1007/s11069-015-1993-2

Cutuli, J. J., & Herbers, J. E. (2018). Resilience in the context of development: Introduction to the special issue. *Journal of Early Adolescence*, *38*(9), 1205–1214. https://doi.org/10.1177/0272431618757680

de la Torre-Castro, M., Fröcklin, S., Börjesson, S., Okupnik, J., & Jiddawi, N. S. (2017). Gender analysis for better coastal management-increasing our understanding of social-ecological seascapes. *Marine Policy*, *83*, 62–74. https://doi.org/10.1016/j.marpol.2017.05.015

Dee, L. E., De Lara, M., Costello, C., & Gaines, S. D. (2017). To what extent can ecosystem services motivate protecting biodiversity? *Ecology Letters*, *20*(8), 935–946. https://doi.org/10.1111/ele.12790

Domínguez, R., Olabarria, C., Woodin, S. A., Wethey, D. S., Peteiro, L. G., Macho, G., & Vázquez, E. (2021). Contrasting responsiveness of four ecologically and economically important bivalves to simulated heat waves. *Marine Environmental Research*, *164*, 105229. https://doi.org/10.1016/j.marenvres.2020.105229

Domínguez-Castro, F., Reig, F., Vicente-Serrano, S. M., Aguilar, E., Peña-Angulo, D., Noguera, I., Revuelto, J., van der Schrier, G., & El Kenawy, A. M. (2020). A multidecadal assessment of climate indices over Europe. *Scientific Data*, *7*, 125. https://doi.org/10.1038/s41597-020-0464-0

Doswald, N., & Osti, M. (2011). *Ecosystem-based approaches to adaptation and mitigation: Good practice examples and lessons learned in Europe*. Federal Agency for Nature Conservation (BfN).

Dubik, B. A., Clark, E. C., Young, T., Jones Zigler, S. B., Provost, M. M., Pinsky, M. L., & St. Martin, K. (2019). Governing fisheries in the face of change: Social responses to long-term geographic shifts in a U.S. fishery. *Marine Policy*, *99*, 243–251. https://doi.org/10.1016/j.marpol.2018.10.032

Dwiartama, A., & Rosin, C. (2014). Exploring agency beyond humans: The compatibility of actor-network theory (ANT) and resilience thinking. *Ecology and Society*, *19*(3), 28. https://doi.org/10.5751/ES-06805-190328

Eagaard, O. R., Marchal, P., Gislason, H., & Rijnsdorp, A. D. (2018). Technological development and fisheries management. *Reviews in Fisheries Science and Aquaculture*, *22*(2), 156–174. https://doi.org/10.1080/23308249.2014.899557

Eslamian, S., Huda, M. B., Rather, N. A., & Eslamian, F. A. (Eds.). (2024). *Handbook of climate change impacts on River Basin management: Technology, innovations and management practices* (1st ed.). CRC Press. https://doi.org/10.1201/9781003473374

Evans, J. P. (2011). Resilience, ecology and adaptation in the experimental city. *Transactions of the Institute of British Geographers*, *36*(2), 223–237. https://doi.org/10.1111/j.1475-5661.2010.00420.x

Fanian, S., Young, S. K., Mantla, M., Daniels, A., & Chatwood, S. (2019). Evaluation of the kǫts'iihtła ("we light the fire") project: Building resiliency and connections through strengths-based creative arts programming for indigenous youth. *International Journal of Circumpolar Health*, *74*, Article 27672. https://doi.org/10.3402/ijch.v74.27672

Finkbeiner, E. M. (2015). The role of diversification in dynamic small-scale fisheries: Lessons from Baja California Sur, Mexico. *Global Environmental Change*, *32*, 139–152. https://doi.org/10.1016/j.gloenvcha.2015.03.009

Finkbeiner, S. D., Fishman, D. A., Osorio, D., & Briscoe, A. D. (2017). Ultraviolet and yellow reflectance but not fluorescence is important for visual discrimination of conspecifics by *Heliconius erato*. *Journal of Experimental Biology*, *220*(7), 1267–1276. https://doi.org/10.1242/jeb.153593

Fisher, M. C., Moore, S. K., Jardine, S., Watson, J., & Samhouri, J. F. (2021). Climate shock effects and mediation in fisheries. *Proceedings of the National Academy of Sciences*, *118*(2), 700–720. https://doi.org/10.1073/pnas.2014379117

Folke, C., Biggs, R., Norström, A. V., Reyers, B., & Rockström, J. (2018). Social-ecological resilience and biosphere-based sustainability science. *Ecology and Society*, *21*(3), Article 41. https://doi.org/10.5751/ES-08748-210341

Folke, C., Hahn, T., Olsson, P., & Norberg, J. (2005). Adaptive governance of social-ecological systems. *Annual Review of Environment and Resources*, *30*, 441–473. https://doi.org/10.1146/annurev.energy.30.050504.144511

Forrestall, F. C., Schirripa, M., Goodyear, C. P., Arrizabalaga, H., Babcock, E. A., Coelho, R., Ingram, W., Lauretta, M., Ortiz, M., Sharma, R., & Walter, J. (2019). Testing robustness of CPUE standardization and inclusion of environmental variables with simulated longline catch datasets. *Fisheries Research*, *210*, 1–13. https://doi.org/10.1016/j.fishres.2018.09.025

Frawley, T. H., Briscoe, D. K., Daniel, P. C., Britten, G. L., Crowder, L. B., Robinson, C. J., & Gilly, W. F. (2021). Impacts of a shift to a warm-water regime in the Gulf of California on jumbo squid (Dosidicus gigas). *ICES Journal of Marine Science*, *76*, 2413–2426. https://doi.org/10.1093/icesjms/fsz133

Fuller, R. A., Irvine, K. N., Devine-Wright, P., Warren, P. H., & Gaston, K. J. (2007). Psychological benefits of greenspace increase with biodiversity. *Biology Letters*, *3*, 390–394. https://doi.org/10.1098/rsbl.2007.0149

Gibson, J. J. (2020). *The ecological approach to visual perception*. Houghton Mifflin.

Giron-Nava, A., Johnson, A. F., Cisneros-Montemayor, A. M., & Aburto-Oropeza, O. (2020). Managing at maximum sustainable yield does not ensure economic well-being for artisanal fishers. *Fish and Fisheries*, *20*(2), 214–223. https://doi.org/10.1111/faf.12332

Grafton, R. Q. (2010). Adaptation to climate change in marine capture fisheries. *Marine Policy*, *34*(3), 606–615. https://doi.org/10.1016/j.marpol.2009.11.011

Granovetter, M. (2005). The impact of social structure on economic outcomes. *Journal of Economic Perspectives*, *19*(1), 33–50. https://doi.org/10.1257/0895330053147958

Gunderson, L. H. (2000). Ecological resilience—In theory and application. *Annual Review of Ecology and Systematics*, *31*, 425–439. http://dx.doi.org/10.1146/annurev.ecolsys.31.1.425

Hadjicostis, C. N. (2020). *Estimation and inference in discrete event systems communications and control engineering*. Springer Nature.

Harper, L., Kalfa, N., Beckers, G. M. A., Kaefer, M., Nieuwhof-Leppink, A. J., Fossum, M., Herbst, K. W., Bagli, D., & ESPU Research Committee. (2020). The impact of COVID-19 on research. *Journal of Pediatric Urology*, *16*(5), 715–716. https://doi.org/10.1016/j.jpurol.2020.07.002

Harper, S., Zeller, D., Hauzer, M., Pauly, D., & Sumaila, U. R. (2013). Women and fisheries: Contribution to food security and local economies. *Marine Policy*, *39*, 56–63. https://doi.org/10.1016/j.marpol.2012.10.018

Havice, E., Reed, K., Havice, M., & Reed, K. (2012). Fishing for development? Tuna resource access and industrial change in Papua New Guinea. *Journal of Agrarian Change*, *12*, 413–435. https://doi.org/10.1111/j.1471-0366.2011.00351.x

Hodgson, D., McDonald, J. L., & Hosken, D. J. (2015). What do you mean, 'resilient'? *Trends in Ecology & Evolution*, *30*(9), 503–506. https://doi.org/10.1016/j.tree.2015.06.010

Hutchison, D., & Sterbenz, J. P. G. (2018). Architecture and design for resilient networked systems, computer communications. *Computer Communications*, *131*, 13–21. https://doi.org/10.1016/j.comcom.2018.07.028

Hutchison, D., & Sterbenz, J. P. G. (2022). Architecture and design for resilient networked systems, computer communications. *Computer Communications*, *131*, 13–21. https://doi.org/10.1016/j.comcom.2022.07.028

Isaacs, M. (2013). *Small-scale fisheries governance and understanding the snoek (Thyrsites atun) supply chain in the ocean view fishing community, Western cape, South Africa. Ecology and Society*, *18*(4), 17. https://doi.org/10.5751/ES-05863-180417

Islam, M. M., & Herbeck, J. (2013). Migration and translocal livelihoods of coastal small-scale fishers in Bangladesh. *The Journal of Development Studies*, *49*(6), 832–845. https://doi.org/10.1080/00220388.2013.766719

Islam, M. R. (2018). Climate change, natural disasters, and socioeconomic livelihood vulnerabilities: Migration decision among the char land people in Bangladesh. *Social Indicators Research*, *136*, 575–593. https://doi.org/10.1007/s11205-017-1563-y

Jacquet, J., & Pauly, D. (2008). Funding priorities: Big barries to small-scale fisheries. *Conservation Biology*, *22*, 832–835.https://doi.org/10.1111/j.1523-1739.2008.00978.x

Jess Grunblatt, J., & Alessa, L. (2017). Role of perception in determining adaptive capacity: Communities adapting to environmental change. *Sustainability Sciences*, *12*, 3–13. https://doi.org/10.1007/s11625-016-0394-0

Jones, H. P., Hole, D. G., & Zavaleta, E. S. (2012). Harnessing nature to help people adapt to climate change. *Nature Climate Change*, *2*(7), 504–509. https://doi.org/10.1038/nclimate1463

Kabir, M. H., & Hossain, T. (2021). Assessment on social vulnerability and response towards natural disaster in a disaster-prone coastal village: An example from Bangladesh. *International Journal of Disaster Management*, *4*(1), 39–60.

Kane, E. A., Ball, A. C., & Brehmer, P. (2020). Total admissible catches and experimental fishing quotas in the pink lobster (Palinurus mauritanicus, Gruvel 1911): A new opportunity for economic development in West Africa? *Iranian Journal of Fisheries Sciences*, *19*(4), 2053–2061.

Kasperski, S., & Holland, D. S. (2013). Income diversification and risk for fishermen. *Proceedings of the National Academy of Sciences of the United States of America*, *110*(6), 2076–2081. https://doi.org/10.1073/pnas.1212278110

Kazim, A. (2011). The impact of climate change on relative humidity trends in Iraq. *Journal of Wasit College of Education*, *1*(10).

Keith, P. (2003). Biology and ecology of amphidromous Gobiidae of the indo-Pacific and the Caribbean regions. *Journal of Fish Biology*, *63*(4), 831–847. https://doi.org/10.1046/j.1095-8649.2003.00197.x

Kleiber, D., Harris, L. M., & Vincent, A. C. J. (2015). Gender and small-scale fisheries: A case for counting women and beyond. *Fish and Fisheries*, *16*(4), 547–562. https://doi.org/10.1111/faf.12075

Kroetz, K., Reimer, M. N., Sanchirico, J. N., Lew, D. K., & Huetteman, J. (2019). Defining the economic scope for ecosystem-based fishery management. *Proceedings of the National Academy of Sciences of the United States of America*, *116*, 4188–4193. https://doi.org/10.1073/pnas.1816545116

Kundu, S., Shukla, S., Banerjee, B., & Mohanta, A. (2025). Advanced bibliometric analysis of global research trends in drought monitoring and assessment. *Reprint Version of Research Square*. https://doi.org/10.21203/Rs.3.Rs-6125428/V1

Kyle, M. A., Aveling, E.-L., & Singer, S. J. (2020). Establishing high-performing teams: Lessons from health care. *MIT Sloan Management Review*, *61*(3), 14–18.

Lai, A., Lahaye, S., & Giua, A. (2021). Verification of detectability for unambiguous weighted automata. *IEEE Transactions on Automatic Control*, *66*(3), 1437–1444. https://doi.org/10.1109/TAC.2020.2995173.

Lauria, F., Tebaldi, T., Bernabò, P., Groen, E. J. N., Gillingwater, T. H., & Viero, G. (2018). riboWaltz: Optimization of ribosome P-site positioning in ribosome profiling data. *PLoS Computational Biology*, *14*(8), e1006169. https://doi.org/10.1371/journal.pcbi.1006169

Leal Filho, W., Azul, A. M., Brandli, L., Özuyar, P. G., & Wall, T. (Eds.). (2020). *Climate action*. Springer International Publishing. https://doi.org/10.1007/978-3-319-71063-1

Levin, S. A. (2019). The architecture of robustness. In *Global challenges, governance, and complexity: Applications and frontiers* (pp. 16–23). Edward Elgar Publishing Ltd.. https://doi.org/10.4337/9781788115421.00008

Liles, M. J., Gaos, A. R., Bolaños, A. D., Lopez, W. A., Arauz, R., Gadea, V., Urteaga, J., Yañez, I. L., Pacheco, C. M., Seminoff, J. A., & Peterson, M. J. (2018). Survival on the rocks: High bycatch in lobster gillnet fisheries threatens hawksbill turtles on rocky reefs along the eastern Pacific coast of Central America. *Latin American Journal of Aquatic Research*, *45*(3), 521–539. https://doi.org/10.3856/vol45-issue3-fulltext-7

Macho, G., Woodin, S. A., Wethey, D. S., & Vázquez, E. (2016). Impacts of sublethal and lethal high temperatures on clams exploited in European fisheries. *Journal of Shellfish Research*, *35*(2), 405–419. https://doi.org/10.2983/035.035.0215

Madsen, J. K., Bailey, R., Carrella, E., & Koralus, P. (2021). From reactive towards anticipatory fishing agents. *Journal of Simulation*, *15*(1–2), 23–37. https://doi.org/10.1080/17477778.2020.1742588

Mahfuzul Haque, M., Jannat Hossain, N., Moula, N., & Eslamian, S. (2023). Linkage between social and hydrological systems to support resilience: A case of freshwater wetland in Bangladesh. In S. Eslamian & F. Eslamian (Eds.), *Disaster risk reduction for resilience: Climate change and disaster risk adaptation* (pp. 3–35). Springer Nature Switzerland.

Malakar, A., Ray, C., D'Alessio, M., Shields, J., Adams, C., Stange, M., Weber, K. A., & Snow, D. D. (2023). Interplay of legacy irrigation and nitrogen fertilizer inputs to spatial variability of arsenic and uranium within the deep vadose zone. *Science of the Total Environment*, 897, 165299. https://doi.org/10.1016/j.scitotenv.2023.165299

McCarthy, J. (2012). The financial crisis and environmental governance "after" neoliberalism. *Tijdschrift voor Economische en Sociale Geografie*, *103*(2), 180–195. https://doi.org/10.1111/j.1467-9663.2012.00711.x

Mercer, J., Kelman, I., Taranis, L., & Suchet-Pearson, S. (2012). Framework for integrating indigenous and scientific knowledge for disaster risk reduction. *Disasters*, 36(3), 430–448.

Minnegal, M., & Dwyer, P. D. (2008). Mixed messages: Buying back Australia's fishing industry. *Marine Policy*, *32*(6), 1063–1071.

Mishna, F., Khoury-Kassabri, M., Schwan, K., Wiener, J., Craig, W., Beran, T., Pepler, D., & Daciuk, J. (2016). The contribution of social support to children and adolescents' self-perception: The mediating role of bullying victimization. *Children and Youth Services Review*, *63*(C), 120–127. https://doi.org/10.1016/j.childyouth.2016.02.013

Mohamed Shaffril, H. A., Samsuddin, S. F., & Abu Samah, A. (2021). The ABC of systematic literature review: The basic methodological guidance for beginners. *Quality and Quantity*, *55*(4), 1319–1346. https://doi.org/10.1007/s11135-020-01059-6

Morato, T., Cheung, W. W. L., & Pitcher, T. J. (2006). Vulnerability of seamount fish to fishing: Fuzzy analysis of life-history attributes. *Journal of Fish Biology*, *68*(1), 209–221. https://doi.org/10.1111/j.0022-1112.2006.00894.x

Murali, K., & Sundar, V. (2017). Reassessment of tidal energy potential in India and a decision-making tool for tidal energy technology selection. *International Journal of Ocean and Climate System*, *8*(2), 85–97. https://doi.org/10.1177/1759313117694629

Nalau, J., Becken, S., Schliephack, J., Parsons, M., Brown, C., & Mackey, B. (2018). *The role of indigenous and traditional knowledge in ecosystem-based adaptation: A review of the literature and case studies from the Pacific Islands. Weather, Climate, and Society*, *10*(4), 851–865. https://doi.org/10.1175/WCAS-D-18-0032.1

Njock, J.-C., & Westlund, L. (2010). Migration, resource management and global change: Experiences from fishing communities in west and Central Africa. *Marine Policy*, *34*(4), 752–760. https://doi.org/10.1016/j.marpol.2010.01.020

Nomura, Y., Newcorn, J. H., & Hurd, Y. L. (2022). Response to uncertainty in developmental origins of health and disease research. *Journal of Child Psychology*, *65*(7), 989–990. https://doi.org/10.1111/jcpp.13546

Norris, F. H., Sherrieb, K., & Pfefferbaum, B. (2011). Community resilience: Concepts, assessment, and implications for intervention. In S. M. Southwick & B. T. Litz (Eds.),

Resilience and mental health: Challenges across the lifespan (pp. 162–175). Cambridge University Press. https://doi.org/10.1017/CBO9780511994791.013
Nunan, F. (2010). Mobility and fisherfolk livelihoods on Lake Victoria: Implications for vulnerability and risk. *Geoforum, 41*(5), 776–785. https://doi.org/10.1016/j.geoforum.2010.04.009
Nunan, F., Luomba, J., Lwenya, C. A., Yongo, E., Odongkara, K., & Ntambi, B. (2012). Finding space for participation: Fisherfolk mobility and co-management of Lake Victoria fisheries. *Environmental Management, 50*(2), 204–216. https://doi.org/10.1007/s00267-012-9881-y
O'Donnell, A., & Wodon, Q. (Eds.) (2015). *Climate change adaptation and social resilience in the sundarbans* (1st ed.). Routledge. https://doi.org/10.4324/9781315767802
OECD (2010). *Globalisation in fisheries and aquaculture: Opportunities and challenges.* OECD Publishing. https://doi.org/10.1787/9789264074927-en
Ojea, E., Pearlman, I., Gaines, S. D., & Lester, S. E. (2017). Fisheries regulatory regimes and resilience to climate change. *Ambio*, 46, 399–412. https://doi.org/10.1007/s13280-016-0850-1
Ojeda-Ruiz, M. A., Marín-Monroy, E. A., Galindo-De la Cruz, A. A., & Cota-Nieto, J. J. (2019). Analysis and management of multispecies fisheries: Small-scale finfish fishery at Bahía Magdalena-Almejas, Baja California Sur, Mexico. *Ocean and Coastal Management, 178*, Article 104835. https://doi.org/10.1016/j.ocecoaman.2019.104835
Oken, K. L., Punt, A. E., & Holland, D. S. (2020). The effects of population synchrony, life history, and access constraints on benefits from fishing portfolios. *Ecotoxicology, 20*(2), 45–65.
Olsson, L., Jerneck, A., Thoren, H., Persson, J., & O'Byrne, D. (2015). Why resilience is unappealing to social science: Theoretical and empirical investigations of the scientific use of resilience. *Science Advances, 1*(4), 56–65. https://doi.org/10.1126/sciadv.1400217
Olsson, L., Jerneck, A., Thoren, H., Persson, J., & O'Byrne, D. (2020). Why resilience is unappealing to social science: Theoretical and empirical investigations of the scientific use of resilience. *Science Advances, 1*(4), 50–65. https://doi.org/10.1126/sciadv.1400217
Oshri, A., Duprey, E. B., Kogan, S. M., Carlson, M. W., & Liu, S. (2018). Growth patterns of future orientation among maltreated youth: A prospective examination of the emergence of resilience. *Developmental Psychology, 54*(8), 1456–1471. https://doi.org/10.1037/dev0000528
Overå, R. (2005). Institutions, mobility and resilience in the Fante migratory fisheries of West Africa. *Transactions of the Historical Society of Ghana, 9*, 103–123.
Park, S. E., Marshall, N. A., Jakku, E., Dowd, A.-M., Howden, S. M., Mendham, E., & Fleming, A. (2012). Informing adaptation responses to climate change through theories of transformation. *Global Environmental Change, 22*(1), 115–126. https://doi.org/10.1016/j.gloenvcha.2011.10.003
Pérez, A. A., Fernández, B. H., & Gatti, R. C. (2010). *Building resilience to climate change: Ecosystem-based adaptation and lessons from the field.* IUCN.
Perissi, I., Bardi, U., El, T., & Lavacchi, A. (2017). Dynamic patterns of overexploitation in fisheries. *Ecological Modelling, 359*, 285–292. https://doi.org/10.1016/j.ecolmodel.2017.06.009
Peteiro, L. G., Woodin, S. A., Wethey, D. S., Costas-Costas, D., Martínez-Casal, A., Olabarria, C., & Vázquez, E. (2018). Responses to salinity stress in bivalves: Evidence of ontogenetic changes in energetic physiology on *Cerastoderma edule*. *Scientific Reports, 8*, 8329. https://doi.org/10.1038/s41598-018-26706-9

Pita, P., Fernandez-Marquez, D., Antelo, M., Macho, G., Villasante, S., 2019. Socioecological changes in data-poor S-fisheries: A hidden shellfisheries crisis in Galicia (NW Spain). *Marine Policy*, *101*, 208–224. https://doi.org/10.1016/j.marpol.2018.09.018

Quinlan, A., Berbes-Blazquez, M., Haider, J. L., & Peterson, G. D. (2016). Measuring and assessing resilience: Broadening understanding through multiple disciplinary perspectives. *Journal of Applied Ecology*, *53*(3), 677–687. https://doi.org/10.1111/1365-2664.12550

Ratner, H., George, E., & Iveson, C. (2012). *Solution focused brief therapy: 100 key points and techniques*. Routledge. https://doi.org/10.4324/9780203116562

Ravera, S., Castroflorio, T., Garino, F., Daher, S., Cugliari, G., & Deregibus, A. (2016). Maxillary molar distalization with aligners in adult patients: A multicenter retrospective study. *Progress in Orthodontics*, *17*, 12. https://doi.org/10.1186/s40510-016-0126-0

Roy, D. A., & Bhaumik, U. (2014). *Participation of rural women of sunderban in decision making related to fishery activities*. SSRN.

Saithong, A., Satumanatpan, S., Kanongdate, K., Piyawongnarat, T., & Srisantear, P. (2022). Adaptiveness to enhance the sustainability of freshwater-aquaculture farmers to environmental changes. *Environment and Natural Resources Journal*, 20(3), 288–296. https://doi.org/10.32526/ennrj/20/202100217

Salagrama, V. (2012). *Climate change and fisheries: Perspectives from small-scale fishing communities in India on measures to protect life and livelihood*. Chennai: International Collective in Support of Fishworkers.

Sammonds, P., Shamsudduha, M., & Ahmed, B. (2021). Climate change driven disaster risks in Bangladesh and its journey towards resilience. *Journal of the British Academy*, *9*(8), 55–77. https://doi.org/10.5871/jba/009s8.055

Sarker, M. N. I., Wu, M., Alam, G. M., & Shouse, R. C. (2019). Livelihood vulnerability of riverine-island dwellers in the face of natural disasters in Bangladesh. *Sustainability*, *11*(6), Article 1623. https://doi.org/10.3390/su11061623

Sarker, M. N. I., Wu, M., Alam, G. M., & Shouse, R. C. (2021). Livelihood vulnerability of riverine-island dwellers in the face of natural disasters in Bangladesh. *Sustainability*, *11*(6), Article 1623. https://doi.org/10.3390/su11061623

Scherer, L. D., Maynard, A., Dolinoy, D. C., Fagerlin, A., & Zikmund-Fisher, B. J. (2019). The psychology of "regrettable substitutions": Examining consumer judgements of bisphenol A and its alternatives. *Health, Risk & Society*, *16*(7-8), 649–666. https://doi.org/10.1080/13698575.2014.969687

Schuhbauer, A., Chuenpagdee, R., Cheung, W. W. L., Greer, K., & Sumaila, U. R. (2017). How subsidies affect the economic viability of small-scale fisheries. *Marine Policy*, *82*, 114–121. https://doi.org/10.1016/j.marpol.2017.05.013

Schultz, O. J. (2017). The micro-politics of small-scale fisheries governance in South Africa: A case study of community-based political representation as a form of brokerage. *Maritime Studies*, *16*(1), 45–65. https://doi.org/10.1186/s40152-017-0060-x

Schwartzberg, J. E., Alam, M., Calkins, P. B., Spear, T. G. P., Dikshit, K. R., Allchin, F. R., Subrahmanyam, S., Wolpert, S. A., Srivastava, A. L., Thapar, R., & Champakalakshmi, R. (2021). India. In *Encyclopedia Britannica*. https://www.britannica.com/place/India

Sepulveda, C. A., Aalbers, S. A., Wang, M., Kneebone, J., & Bernal, D. (2020). Postrelease survivorship of Pacific bluefin tuna (Thunnus orientalis) captured in the California recreational fishery. *Fisheries Research*, *223*(2), 70–90. https://doi.org/10.1016/j.fishres.2019.105432

Siders, A. R. (2019). Social justice implications of US managed retreat buyout programs. *Climatic Change*, *152*(2), 239–257. https://doi.org/10.1007/s10584-018-2272-5

Sievanen, L. (2014). How do small-scale fishers adapt to environmental variability? Lessons from Baja California, Sur, Mexico. *Maritime Studies*, *13*(1), 9.

Sin, M. S., Yew, T. S., & Noh, K. M. (2019). Effort dynamics and alternative management policies: The case of the west coast zone B trawl fishery in peninsular Malaysia. *Marine Resource Economics*, *34*(1), 77–103. https://doi.org/10.1086/701995

Smith, A. C., Tasnim, T., Irfanullah, H. M., Turner, B., Chausson, A., & Seddon, N. (2021). Nature-based solutions in Bangladesh: Evidence of effectiveness for addressing climate change and other sustainable development goals. *Frontiers in Environmental Science*, *50*(2), 456–467. https://doi.org/10.3389/fenvs.2021.737659

Spalding, A. K., Grorud-Colvert, K., Allison, E. H., et al. (2023). Engaging the tropical majority to make ocean governance and science more equitable and effective. *npj Ocean Sustainability*, 2, Article 8. https://doi.org/10.1038/s44183-023-00015-9

Stoll, J. S., Beitl, C. M., & Wilson, J. A. (2016). How access to Maine's fisheries has changed over a quarter century: The cumulative effects of licensing on resilience. *Global Environmental Change*, *37*, 79–91. http://dx.doi.org/10.1016/j.gloenvcha.2016.01.005

Stoll, J. S., Fuller, E., & Crona, B. I. (2017). Uneven adaptive capacity among fishers in a sea of change. *PLoS One*, *12*(6), e0178266. https://doi.org/10.1371/journal.pone.0178266

Sulikowski, J. A., Golet, W., Hoffmayer, E. R., Driggers, W. B., Natanson, L. J., Carlson, A., & Sweezey, B. B. (2020). Observing post-release mortality for dusky sharks, Carcharhinus obscurus, captured in the U.S. pelagic longline fishery. *Fisheries Research*, *20*(2), 150–170. https://doi.org/10.1016/j.fishres.2019.105342

Sumaila, U. R., & Munro, H. C. (2009). Fisheries economics. In J. H. Steele (Ed.), *Encyclopedia of ocean sciences* (2nd ed., pp. 491–498). Academic Press.

Sun, J., Hinton, M. G., & Webster, D. G. (2016). Modelling the spatial dynamics of international tuna fleets. *PLoS One*, *11*(8), 1–28. https://doi.org/10.1371/journal.pone.0159626

Tagliolatto, A. B., Goldberg, D. W., Godfrey, M. H., Monteiro-Neto, C., & Itabapoana, F. D. (2020). Spatio-temporal distribution of sea turtle strandings and factors contributing to their mortality in south-eastern Brazil study site. *Aquatic Conservation Marine and Freshwater Ecosystem*, *30*(2), 331–350. https://doi.org/10.1002/aqc.3244

Tierney, K. (2006). Social inequality, hazards and disasters. In H. K. Ronald Daniels and Donald Kettl (Eds.), *On risk and disaster: Lessons from Hurrican Katrina*. Pennsylvania: University of Pennsylvania Press.

Travers, A. A., Vaillant, C., Arneodo, A., & Muskhelishvili, G. (2012). *DNA structure, nucleosome placement and chromatin remodelling: A perspective. Biochemical Society Transactions*, *40*(2), 335–340. https://doi.org/10.1042/BST20110768

Troell, M., Jonell, M., & Crona, B. (2019). *The role of seafood in sustainable and healthy diets: The EAT-lancet commission report through a blue lens*. Beijer Institute.

Turner, B. L., Kasperson, R. E., Matson, P. A., McCarthy, J. J., Corell, R. W., Christensen, L., Eckley, N., Kasperson, J. X., Luers, A., Martello, M. L., Polsky, C., Pulsipher, A., & Schiller, A. (2003). A framework for vulnerability analysis in sustainability science. *Proceedings of the National Academy of Sciences*, *100*(14), 8074–8079. https://doi.org/10.1073/pnas.1231335100

Turner, N. J., & Berkes, F. (2006). Developing resource management and conservation. *Human Ecology*, *34*, 475–478. https://doi.org/10.1007/s10745-006-9060-y

Ungar, M. (2018). Systemic resilience: Principles and processes for a science of change in contexts of adversity. *Ecology and Society*, *23*(4), Article 34. https://doi.org/10.5751/ES-10385-230434

Ungar, M. (Ed.) (2021). *Multisystemic resilience: Adaptation and transformation in contexts of change*. Oxford University Press. https://doi.org/10.1093/oso/9780190095888.001.0001

United Nations Office for Disaster Risk Reduction (UNDRR). 2017. *The Sendai Framework terminology on disaster risk reduction*. https://www.undrr.org/terminology/disaster-risk-reduction

Vázquez-Delfín, E., Freile-Pelegrín, Y., Pliego-Cortés, H., & Robledo, D. (2019). Seaweed resources of Mexico: Current knowledge and future perspectives. *Botanica Marina*, *62*(3), 275–289. https://doi.org/10.1515/bot-2018-0070

Villasante, S., Macho, G., Silva, M. R., Lopes, P. F., Pita, P., Simón, A., & Calvo, N. (2022). Resilience and social adaptation to climate change impacts in small-scale fisheries. *Frontiers in Marine Science*, *9*, Article 802762. https://doi.org/10.3389/fmars.2022.802762

Walsh, F. (2020). Facilitating family resilience: Relational resources for positive youth development in conditions of adversity. In M. Ungar (Ed.), *The social ecology of resilience: A handbook of theory and practice* (pp. 173–185). Springer. https://doi.org/10.1007/978-1-4614-0586-3_15

Wanyonyi, I. N.,Wamukota, A., Tuda, P., Mwakha, V. A., & Nguti, L. M. (2016). Migrant fishers of Pemba: Drivers, impacts and mediating factors. *Marine Policy*, *71*, 242–255. https://doi.org/10.1016/j.marpol.2016.06.009

Ward, T., & Langlands, R. L. (2019). Repairing the rupture: Restorative justice and the rehabilitation of offenders. *Aggression and Violent Behavior*, *14*(3), 205–214. https://doi.org/10.1016/j.avb.2009.03.001

Watson, J. E., Iwamura, T., & Butt, N. (2013). Mapping vulnerability and conservation adaptation strategies under climate change. *Nature Climate Change*, *3*(11), 989–994.

Watson, P. A. G., Karoly, D. J., Allen, M. R., Faull, N., & Lee, D. S. (2012). *Quantifying uncertainty in future southern hemisphere circulation trends. Geophysical Research Letters*, *39*. https://doi.org/10.1029/2012GL054158

Weber, C. T., Borit, M., & Aschan, M. (2019). An interdisciplinary insight into the human dimension in fisheries models. A systematic literature review in a European Union context. Frontiers in Marine *Science*, *6*, Article 369. https://doi.org/10.3389/fmars.2019.00369

White, E. R., Froehlich, H. E., Gephart, J. A., Cottrell, R. S., Branch, T. A., Bejarano, R. A., & Baum, J. K. (2020). Early effects of COVID-19 on US fisheries and seafood consumption. *Fish and Fisheries*, *22*(1), 232–239. https://doi.org/10.1111/faf.12525

Wilson, J. (2017). Learning, adaptation, and the complexity of human and natural interactions in the ocean. *Ecology and Society*, *22*. https://doi.org/10.5751/ES-09356-220243

Woodcraft, S., Bacon, N., Caistor-Arendar, L., & Hackett, T. (2020). *Design for social sustainability: A framework for creating thriving new communities*. Social Life/Young Foundation.

Woodin, S. A., Wethey, D. S., Olabarria, C., Vazquez, E., Domínguez, R., Macho, G., & Peteiro, L. (2020). Behavioral responses of three venerid bivalves to fluctuating salinity stress. *Journal of Experimental Marine Biology and Ecology*, *522*, 151256.

Wu, H., & Hou, C. (2019). Utilizing co-design approach to identify various stakeholders' roles in the protection of intangible place-making heritage: The case of Guchengping Village. *Disaster Prevention and Management: An International Journal*, *29* (1), 22–35. https://doi.org/10.1108/DPM-09-2018-0291

Xu, L., & Kajikawa, Y. (2017). An integrated framework for resilience research: A systematic review based on citation network analysis. *Sustainability Science*, *13*(1), 235–254. https://doi.org/10.1007/s11625-017-0487-4

Young, T., Fuller, E. C., Provost, M. M., Coleman, K. E., St. Martin, K., McCay, B. J., & Pinsky, M. L. (2019). Adaptation strategies of coastal fishing communities as species shift poleward. *ICES Journal of Marine Science*, *76*, 93–103. https://doi.org/10.1093/icesjms/fsy140

Younus, M. A. F., & Kabir, M. A. (2018). Climate change vulnerability assessment and adaptation of Bangladesh: Mechanisms, notions, and solutions. *Sustainability*, *10*(11), Article 4286. https://doi.org/10.3390/su10114286

Zhang, K. (2021). State-based opacity of real-time automata. In A. Castillo-Ramirez, P. Guillon, & K. Perrot (Eds.), *27th IFIP WG 1.5 international workshop on cellular automata and discrete complex systems (AUTOMATA 2021)* (Vol. 90, pp. 12:1–12:15). Open Access Series in Informatics (OASIcs). https://doi.org/10.4230/OASIcs.AUTOMATA.2021.12

Zhou, C., Jiao, Y., & Browder, J. (2019). How much do we know about seabird bycatch in pelagic longline fisheries? A simulation study on the potential bias caused by the usually unobserved portion of seabird bycatch. *PLoS One*, *14*(8), Article e0220797. https://doi.org/10.1371/journal.pone.0220797

5 Empowering Sundarbans Small-Scale Fisheries through Sustainable Stewardship and Strategic Policy Innovation

5.1 INTRODUCTION

Sustainable development pertains to the advancement of human potential by augmenting the substantive freedoms and capacities of the current generation without jeopardizing the capabilities of future generations (Noor Islam et al., 2014). This principle closely aligns with the foundational criteria articulated by Pearce et al. (1989), who argued that sustainability requires "that the next generation should inherit a stock of wealth, comprising man-made assets and environmental assets, no less than the stock inherited by the previous generation," and more stringently, "that the next generation should inherit a stock of environmental assets no less than the stock inherited by the previous generation." These conditions highlight the imperative of intergenerational equity at the core of sustainable development, particularly in ecosystems where natural capital underpins livelihoods and economic stability.

Comprehending the trade-off dynamics among ecological, economic, and social goals is crucial for formulating strategies aimed at managing or rehabilitating ecosystems (Liberati et al., 2018; Oliver et al., 2019; Paniagua & Rayamajhee, 2024). Human interventions to (1) tame the dynamic character of the landscape, (2) exploit natural resources, and (3) reduce exposure to prevailing hazards have led to massive infrastructure development. These critical areas are experiencing current and anticipated global change effects, characterized by high population density and significant ecological and economic relevance for sustainable development (Haegenlocher et al., 2018). However, the "paradox of plenty," or "natural resource curse" (Karl, 1999), describes the negative effects that can result from an abundance of natural resources. This curse arises from two primary factors: first, policy mismanagement, where easy access to resource revenues fosters rent-seeking conduct. Secondly, this surplus resource diminishes the urgency for critical economic changes, eventually obstructing long-term development and stability.

 DOI: 10.1201/9781003729990-5

Within marine and fisheries systems, these sustainability challenges are further complicated by governance failures. As Botsford et al. (1997) observed

Traditionally, open-access problems in the sea have been approached with management schemes based on effort controls, catch quotas, gear restrictions, and various types of area/ season closures. The strategic/legal framework for marine property rights (whether private, communal, or collective) has been seldom addressed. The challenge for the next century lies more on the crafting of local and regional institutions, than in the filling of scientific gaps or on the implementation of novel management measures.

This comprehension remains highly relevant, particularly for intricate frameworks like small-scale fishing (SSF) systems wherein ecological and socioeconomic methodologies are interdependently linked (Younus & Kabir, 2018). The socioeconomic problems of small-scale fishermen are currently attracting increased attention due to: (1) the realization that small-scale fisheries are not a transitory feature of fisheries development, (2) a growing focus on enhancing the socioeconomic conditions of low-income populations, and (3) the new opportunities for local fisheries arising from the declaration of extended fisheries jurisdictions. In response, governments in developing nations are contemplating development aid initiatives to enhance SSF. However, such efforts may be unsatisfying without a thorough understanding of the factors responsible for their current depressed situation and the existing potential for further development (Chuenpagdee & Jentoft, 2019). The success of small-scale fisheries in some developing nations may be attributed to many factors: (1) favorable ocean conditions in recent decades, (2) a single, accountable management agency, and (3) a well-established program of limited entry to the fishery (Azam et al., 2021; Islam et al., 2021) (4) The biocomplexity of the fish stock structure (Mitra et al., 2023). In addition, nations such as India and Bangladesh exhibit significantly divergent governance capabilities and resources allocated to fisheries management, functioning as essentially "self-governed" entities. Historically, the primary paradigm of fisheries governance in the Asia-Pacific has been characterized as "resource nationalism" (Tesfahunegn & Gebru, 2021; Tesfahunegn et al., 2021, 2019; Myers et al., 2018; Koch & Perreault, 2019), whereby governments have deliberately sought to enhance fisheries productivity and trade. As Helgeson et al. (2012) point out in their insightful historical analysis of national seafood production systems, "national seafood systems promoted volume [of extraction] to ensure the reproduction of domestic capital, sustain new industrial societies by providing cheap food for workers and their families, and extend[ed] geopolitical influence."

This phenomenon was evident throughout the extensive industrial growth over most of Asia in the postwar era (Murali & Sundar, 2021). Numerous developing nations in the Asia-Pacific region have adopted the concept of "comanagement" (Faridah-Hanum et al., 2014), which emphasizes shared responsibility for resource management among the state, resource users, and other stakeholders, including civil society groups (Chakraborty et al., 2020; Mollick et al., 2022; Paszkowski et al., 2021). In practice, comanagement models vary on a continuum from centralized, where government undertakes most functions, through consultative and collaborative, to delegated models, where fishers undertake most governance functions (Hale et al., 2019; Salehin et al., 2018). Spijker et al. (2022) conducted a systematic investigation revealing that Asian governance systems have a strong inclination toward data collection and public accessibility. It is anticipated that it helps to assess fish populations and collect in-port data at multiple locations by compiling and validating self-reported catch and effort data in critical fisheries. Additionally, several Asian countries make their national fishing and vessel registers publicly accessible, encouraging transparency in fisheries management. There has been a recent shift toward "private" or "market-based governance," which is based on the idea that market participants should play a primary role in environmental sustainability governance (Bush & Oosterveer, 2019; Groeneveld et al., 2017). The state's legal authority is minimal, and access to fishery resources is regulated by customary and/or informal institutions that intersect with other social institutions.

A society may be seen as a hierarchical system, including smaller social units integrated into larger structures. These smaller units operate autonomously under well-recognized norms (Paniagua & Rayamajhee, 2024). Nested arrangements facilitate cooperative systems, whereby each unit functions autonomously until its problems impact adjacent jurisdictions. As issues escalate, they generate "nested externalities," necessitating elevated cooperation. Smaller issues can be handled locally, but larger challenges necessitate cooperation between units, aligning objectives and rules for mutual benefit. However, when many parties are involved, collective decision-making becomes more difficult and costly due to diverging interests (Buchanan, 1973). Thus, it often requires more complex, multi-center governance arrangements (Paniagua & Rayamajhee, 2024). Since the 1980s, Neoliberalism has become a unique form of economic governance. The 1950s vision of oceans as commons with inherent problems of overfishing and overcapacity gave rise to neoliberalism in marine (O'Sullivan et al., 2019). Hence, human nature to overexploit shared resources and to compete for fish has led to the development of innovative technology, causing both overfishing and overcapacity (Schlueter et al., 2018). It is believed that converting ocean commons to private property and then allocating access to the resource through market mechanisms will maximize economic efficiency and harness the profit motive to accomplish conservation goals (Brown et al., 2019; Manson et al., 2020). Thus, policymaking, which is a technical process, is possible to attain the right outcomes if only the right governance institutions are put in place. Yet, policies are always implemented in contexts—they interact with preexisting forms of governance, and with the social and political-economic contexts (Fabinyi & Barclay, 2022).

5.2 RESILIENT SHORES: SUSTAINABLE STRATEGIES FOR FISHERIES AND FISHERFOLK IN THE CLIMATE-AFFECTED SUNDARBANS

The fisheries in the Sundarbans, located along the shared border of two nations, encounter a distinctive and multi-layered array of challenges due to the fluctuating nature of fish stocks, the complex ecosystems they occupy, and the substantial expenses associated with monitoring and managing these resources (Paniagua & Rayamajhee, 2024). The dynamic characteristics of fish populations, affected by tides, salinity, and seasonal variations, hinder sustainable management of these resources. The interdependence of the Sundarbans' ecology may have extensive impacts across the whole system, influencing not only fish populations but also the residents that depend on these resources for their survival. High monitoring costs, particularly in remote and inaccessible areas, exacerbate governance issues, making it more difficult to enforce sustainable fishing methods and prevent illegal fishing. Local communities, governmental bodies, nongovernmental organizations (NGOs), and international organizations (Galappaththi et al., 2022; Sunny et al., 2022) urgently need to work together extensively (Sarkar et al., 2023). Coordinated, multi-tiered governance initiatives are essential to safeguard the Sundarbans' distinctive ecology and ensure the sustainability of its SSF communities. To ensure the stable economic viability of fisherfolk communities, a resource-based strategy, environmentally friendly solutions, and organizational or strategic interventions based on climate change adaptation and mitigation measures must be followed (Siam et al., 2020; Soga & Gaston, 2020). Integrated environmental management facilitates the protection, restoration, and management of economically relevant fish habitats while promoting diverse fishing techniques and key fish species. Simultaneously, adaptations in the built environment include the design of coastal infrastructure to accommodate and manage increasing sea levels, recurrent cyclones, and flooding.

This section discusses potential recommendations for improving the conditions of small-scale fishers in the Sundarbans, emphasizing the necessity for policy-based practices to incorporate community-driven, climate-responsive strategies to ensure the sustainability of fisheries within existing social governance and economic frameworks (Lam et al., 2020; Moosavi, 2017; Mutombo & Ölçer, 2017) (Figure 5.1).

5.2.1 Ecosystem-Based Adaptation (EbA)

Ecosystem-based adaptation (EbA) is a climate change adaptation approach that incorporates environmental services to enhance human well-being (Girot et al., 2012). Vignola et al. (2009) and Mercer et al. (2012) emphasize that EbA enhances both local and external ecological information to determine effective climate adaptation strategies, acknowledging the different requirements of diverse populations and facilitating local adaptation and ecosystem management. According to Girot et al. (2012), EbA prioritizes the conservation of natural resources while strengthening the efficiency of adaptation initiatives. By emphasizing environmental services that enhance human resilience, EbA reduces climate vulnerability and promotes community resilience by focusing on environmental services that improve human resilience (Pérez et al., 2010). The strategy is based on the idea that conserving ecosystems may facilitate climate adaptation by maintaining the provision of vital ecosystem services. The Convention on Biological Diversity (CBD, 2009) states that EbA integrates ecosystem services and biodiversity into a comprehensive adaptation strategy, promotes biodiversity conservation, and yields economic, social, and cultural benefits. Robust ecosystems are vital for EbA, as they exhibit greater resilience and consistently provide key services compared to degraded systems (Colls et al., 2009), the comprehensive knowledge of EbA continues to develop globally (Huq et al., 2013).

Small-scale fisheries, although locally orientated, may have a global influence akin to larger fisheries, since they use common fish stocks and may extend beyond local or national seas into deep-sea and remote regions, sometimes prompted by

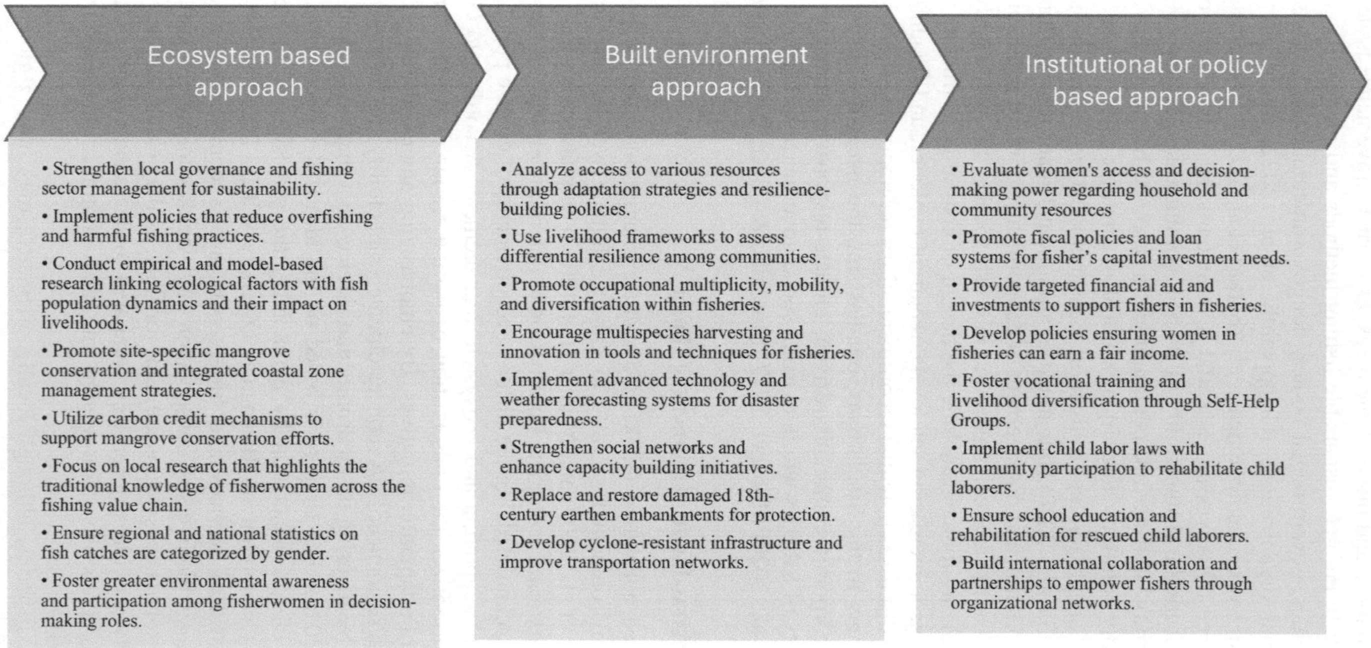

FIGURE 5.1 Recommendations for the enhancement of the livelihoods of small-scale fisherfolk in the Sundarbans under climate change challenges.

localized resource depletion. This methodology adopts a comprehensive strategy that affects fishing at both the community and environmental levels (Lei et al., 2021). The long-term sustainability of livelihoods reliant on small-scale fisheries is challenged by various processes and factors. Policies aimed at reducing fishing effort, restoring overexploited or depleted fish stocks, promoting coastal development, and restricting destructive fishing practices and pollution can enhance the reproductive capacity of fish stocks and limit climate-induced impact on fishery output. Effective management in the context of rapidly changing climatic conditions requires empirical or model-derived studies that link different ecological factors with fish population dynamics and their effects on fishing practices and the livelihoods of fishermen (Caihong et al., 2019; Ghassen et al., 2021). Strategic spatial and temporal periodic closures enhance environmental resilience and reduce ecological and societal vulnerability to climate change. Still, it often overlooks how restrictions affect women's access to resources and income among communities of fishermen.

5.2.1.1 Fisheries Conservation and Management

Prioritizing the diversity and resilience of fisheries via adaptive regulations, while enhancing the resilience of fisherfolk communities via sustainable livelihood strategies, is important in addressing climate change. This would connect socioeconomic vulnerability with sustainable ecological health (Ashrafuzzaman, 2022; Rakib et al., 2018; Sammonds et al., 2021; Smith et al., 2021). The focus in both rich and poor nations has transitioned toward Environment-Based Fisheries Management (EBFM), which seeks to replace traditional fisheries research, centered only on target species, with an awareness of a fishery's impact on the wider environment (Haque et al., 2022; Nalau et al. 2018). The government needs to advocate for the ecosystem approach to fisheries (EAF) to use current aquatic resources sustainably. Fish sanctuaries are seen as vital instruments for the protection and conservation of fishery resources, accompanied by a policy change toward community involvement (Dutta et al., 2020; Ellison, 2019). The evaluation of spatially explicit alternatives, including reproductive reserves, rotation, and various forms of spatially variable effort allocation, as well as productivity manipulation (enhancement, predator control, habitat modification), and experimental management utilizing spatially defined experimental units (Funari et al., 2019), does not equate "data" (in the narrow sense implied by the requirements of conventional stock assessment models) with information. Qualitative spatially explicit information may be as pertinent as traditional quantitative data, as in S-fisheries[1] management, the "where" often outweighs the "how much." Management strategies for S-fisheries are often limited by spatial demands from conflicting stakeholders; Geographical Information System (GIS) platforms are particularly beneficial in this regard. Comprehending the rationale behind the geographical distribution of fishing efforts (spatial allocation of effort) is essential for formulating successful strategies. The use of abundant empirical data (both documented and undocumented) (Ellison, 2019) and the information necessary for spatially explicit management choices is not always available due to financial and logistical limitations, which can only be gathered with the involvement of fishermen. Observing abundance trends across multiple locations can provide clearer insights into the dynamics of harvested resources compared to global stock assessments (Füssel & Klein, 2019). Additionally, the use of simple feedback decision rules driven by data (as opposed to mediated by assessment models) to adjust harvest regulations in response to monitoring indices. Such rules could include different sorts of spatially explicit regulations. The approach used for testing management procedures in conventional fisheries (Dutta et al., 2020) is a promising option for the design of stock-specific harvesting strategies for S-fisheries (Hazra & Ghosh, 2019).

Currently, India lacks any Marine Protected Areas (MPAs) near the Sundarbans (Donato et al., 2011), and the conservation plan focuses on species rather than adopting an EbA. Fish sanctuaries must be designated, particularly in shared rivers such as the Raimangal. Fishing restrictions should be implemented based on a thorough study of the natural life cycle of dominant fish species. He et al. (2022) suggested that the establishment of MPAs and Gear-exclusion Areas can effectively alleviate the competition between user groups or increase the opportunities for certain users, thus resolving the conflicts of artisanal fishermen (Ghosh 2021).

Fishing restrictions in the Sundarbans must prioritize the bio-ecology of fish resources. Spatial and temporal limitations and gear restrictions must be devised and enforced to prevent the adverse environmental consequences of wild-fry and fish congregation. Mosquito nets and offshore set bag nets must be outlawed to prevent fishermen from collecting juveniles, sub-adults, juvenile shrimps, and fish to ensure recruitment and ecological balance. All gears driven by fixed engine boats should be permanently outlawed, and the permissible mesh size for all gill nets, lift nets, shore seines, and set bag nets should be 5 cm. Severe overexploitation of *Pangasius* and L. *calcarifer*, a five-year moratorium on their harvest is warranted. Fishing of *T. ilisha* should be prohibited from November to April, with a minimum size restriction of 23 cm. Mud crab harvesting should be limited winter season to males with a minimum carapace width of 10 cm. The collection of *M. rosenbergii*, *Plotosus* spp., & *Mugil cephalus* should be prohibited from May to June. Capturing or possessing female mud crabs, female *M. rosenbergii* (Hoq et al., 2007), and live gigantic oysters should be prohibited by law (Alongi, 2020). An immediate need exists to develop strategies for implementing the seasonal and geographical prohibition and to diminish the fishing capability of commercial prawn trawlers to restore the stock. The current regulation governing the trawler fleet, which prohibits bottom trawling within 12 nautical miles of the shoreline and mechanized fishing within 15 kilometers of the shoreline, should be reinforced to mitigate bycatch. Implementing operational modifications, such as reducing trawling speeds and durations, as well as avoiding high bycatch zones, alongside reevaluating methods to enhance the efficacy of bycatch reduction technologies, such as turtle exclusion devices, will contribute to ecological sustainability.

Diverse habitats in river ecosystems facilitate different life stages of fish, which emphasizes the need for connections between these habitats to enable migration. Habitat fragmentation can severely disrupt connectivity, particularly in rivers, due to their unique dendritic structure, resulting in more pronounced consequences than in terrestrial landscapes (Fagan, 2002). The loss of connection has resulted in a decline in migratory species, while human activities like dam building and water diversion intensify habitat fragmentation and degradation (Bunn & Arthington, 2002; Dynesius & Nilsson, 1994). Conservation efforts must also include flood regimes and other natural flow patterns essential for sustaining biodiversity and fish populations in highly seasonal river systems (Welcomme, 1985; Winemiller, 2004). Conventional protected areas are inadequate for migratory species navigating different habitats; however, integrated strategies can enhance conservation results by focusing on habitat connectivity and ecological processes across freshwater, estuarine, and marine ecosystems (Galat & Zweimuller, 2001). Research and conservation initiatives thus aimed at the conservation of freshwater fishes on a regional scale are becoming more vital owing to the rising environmental concerns, including the overexploitation of water resources and global environmental changes. Extending research on ecosystem functions in riverine habitats, including meta-analyses and modeling, like Macro-ecological models (Oberdorff et al., 1995), will improve comprehension of fish population dynamics and guide long-term conservation and fisheries management (Walters, 1997; Winemiller, 2004). As the genetic makeup of a fish population directly influences its ecological roles, survival, and adaptation to the environment. Ecological pressures, in turn, drive genetic changes within a population over time. Hence, genetic study and knowledge on systematics of fish taxa aid in establishing Evolutionary Significant Units (ESUs) of different population structure, geographic distribution, and connectedness of fish species (Talarico et al.2023; Hogan, 2013; So et al., 2006). This mapping is crucial for identifying threats to genetic diversity, including unidirectional gene flow, and the formulation of protection strategies to conserve the integrity of fish populations (Jager et al., 2001; Matsubara et al., 2001; Mayden & Wood, 1995). In addition, the large-scale worldwide declines of freshwater and marine fish diversity and population abundance severely endanger ecosystem sustainability as well as human livelihoods. Environmental deoxyribonucleic acid (DNA) (eDNA)-based biomonitoring offers a powerful, efficient, and budget-friendly method to identify species occurrences and population trends in various aquatic habitats. Hence, it is a promising tool for understanding biodiversity responses and supporting fish conservation and management. However, the numerous technical aspects and rapid changes in the eDNA field that are happening concurrently may overwhelm researchers (Yao et al., 2022). Hence, decision-making tools for river conservation must be pragmatic, ecologically viable, and pertinent to management challenges to guarantee the effectiveness and sustainability of conservation initiatives (Abell et al., 2007).

5.2.1.2 Mangrove Conservation Efforts

The mangrove environment is closely linked to fish habitats and nurseries; therefore, the active involvement of fishermen in safeguarding this unique ecosystem through Joint Mangrove Management (JMM) is fundamental to sustainable resource management. Members of the Forest Protection Committee/Eco Development Committee should be motivated to establish a Community Fund to support developmental initiatives in the villages (Raha et al., 2013). The Sundarbans mangroves are a typical example of high carbon sink ecosystems, evident from their total blue carbon stocks of 26.62 Tg and carbon sequestration rates of 4.71–6.54 Mg C ha^{-1} $year^{-1}$ (Bera et al., 2022). The enormous carbon potential of Sundarbans wetlands can be turned into promising carbon credit projects that can create the financing required both for conservation and community development in a sustainable way. Strategic plantation

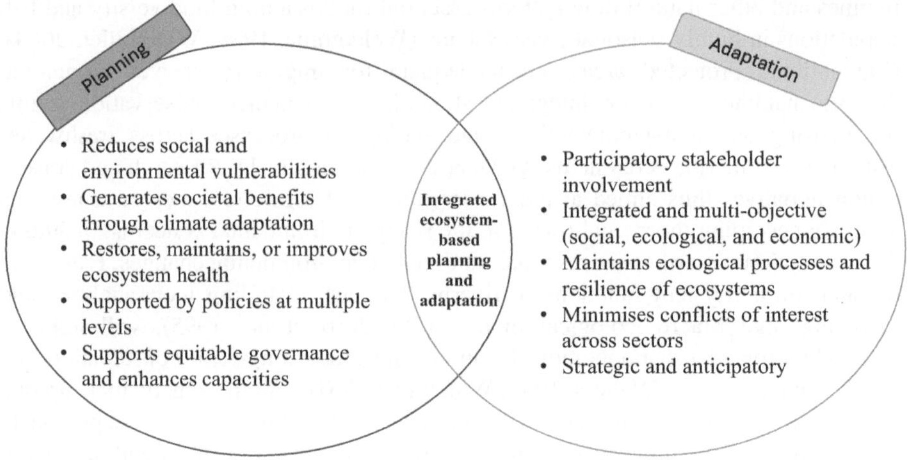

FIGURE 5.2 Integrated ecosystem-based planning and adaptation.

sites should prioritize areas with high restoration potential and proximity to fishing grounds to maximize ecological and economic benefits. Native species, including *Heritiera fomes*, *Rhizophora mucronata*, and *Avicennia marina*, should be selected based on site-specific salinity gradients and soil conditions (Rehman et al., 2019). Integrated Mangroves Fishery Farming System (IMFFS) offers innovative solutions combining mangrove restoration with sustainable aquaculture, enhancing productivity while providing carbon sequestration benefits (Krishnan et al., 2019). Community-based management ensures equitable benefit distribution from carbon credit revenues while maintaining traditional fishing practices. Carbon credit mechanisms can generate \$10–50 per ton CO_2 equivalent, providing substantial revenue streams for restoration activities and alternative livelihoods. However, the recipe for success includes strong monitoring protocols, community engagement, and coordination with existing coastal zone management frameworks (Figure 5.2).

5.2.2 Built Environment Approach

5.2.2.1 Infrastructure Development

The built environment has five essential capital assets: social, financial, natural, physical, and human, which are accessed and used via various adaptation techniques to enhance resilience in response to emerging threats (Jabir et al., 2021; Joa et al., 2018). The capacity of a constructed environment is often referred to as the inherent resilience of a community. Therefore, the interdependence of the natural and built environments must be recognized during the formulation of policies aimed at preserving ecosystem services and enhancing the livelihoods of women fishers. Therefore, the interdependence of the natural and constructed environments must be recognized during the formulation of policies aimed at preserving ecosystem services and enhancing the livelihoods of women fishers (Jonah et al., 2017).

The livelihood framework techniques may elucidate the varying resilience of fisherwomen in managing crises (Karmakar, 2021). The durable fishery-dependent life and livelihoods include the occupational multiplicity of the women fisherfolk, the professional mobility and spread of the outside of fisheries, zoo-geographical variation, expansion with the various species, multi-gears approach, and new harvesting tools and techniques (Mondol et al., 2023). In addition to traditional and indigenous knowledge for formulating adaptation strategies, technological advancements and predictive modeling for disaster forecasting or establishing social cohesiveness and networks enhance cooperation and may augment the adaptive ability of fisherfolk families (Newell et al., 2019). But in the case of the women fisherfolk, as they play an important role in their household, it becomes difficult or nearly impossible for them. In this context, an alternate livelihood can help them (Widmalm, 2021). An urgent need exists to restore and repair the breached earthen embankments from the 18th century to safeguard the fishing community from sea level rise and flooding. Appropriate infrastructure for transit and cyclone-resistant architectural design is essential. Education and technology may empower women fishermen to expand their options. Women fishermen need supplementary investment to adapt to climate change, including both direct and indirect gains, whether modest or substantial, that provide a return on investment and create a mutually beneficial scenario (Shankar et al., 2021). Integrated Coastal Zone Management (ICZM) aims to adapt to natural hazards, which is crucial for the survival of coastal inhabitants and ecosystems (Pitchaikani, 2020) by promoting the development or relocation of dwellings for those displaced by flood susceptibility. This will diminish the vulnerability of the fisherfolk (Barua & Mitra, 2024; Roy et al., 2022; Uddin et al., 2023).

5.2.2.2 Technology and Innovation

The strategic application of information and communication technologies (ICTs) is a promising instrument (Kaur & Sood, 2020) in the management of small-scale fisheries and operational value chains (Mallalieu, 2015). Numerous Asian nations employ various ICT techniques to better the precision of weather forecasts, augment catch through improved accuracy, provide multi-hazard early alerts, conduct environmental impact assessments, and promote information exchange and commerce. The Sundarbans continues to fall behind in the adoption of such tactics (Hoq, 2007) due to inadequate ICT infrastructure, with internet and satellite mobile phone networks in remote regions being prohibitively expensive and inaccessible (Cespedes, 2012; Levy, 2008; Banerjee, 2008). Recently, in certain rural regions of Sundarbans, NGOs and government entities launched mobile van services to market their products; however, this initiative faltered due to the limited interest of the fishing community. The fishermen encounter difficulties in obtaining information due to their exceedingly low literacy rates and insufficient ICT-based understanding. Implementing e-commerce initiatives will enhance growth and enable local populations to access services. ICT-related authorities ought to proactively offer ICT resources and training to fishermen, thereby enhancing their earnings through direct engagement with the market and clients for product sales (Omar & Chhachhar, 2012). By implementing a comprehensive and progressive fisheries policy, the Sundarbans may optimize its fisheries resources while ensuring ecological sustainability, thus enhancing the economic empowerment of its small-scale fishers. Enhancing access to regional climate variability information and using early warning strategies would reduce vulnerability (Neogi et al., 2016). Automatic Weather Stations (AWSs) can be essential for anticipating climate variability and its effects on the community (Nayak & Takemi, 2022).

5.2.2.3 Access to Resources and Market Infrastructure

The remote rural regions, or "less-favored areas," of the Sundarbans are marked by persistent poverty, elevated transaction costs, and ineffective transportation systems—particularly during the rainy season when transport is primarily reliant on boats. These areas suffer from limited or nonexistent market access and inadequate government services and infrastructure, especially in health, education, and credit facilities (including banks and microcredit institutions), resulting in the sector's undercapitalization (Bird et al., 2002; Cochrane et al., 2023; Hulme et al., 2001; Jalan & Ravallion, 1999; Ruben, 2005). Strategic collaboration with extension field officials from the Department of Fisheries, coupled with the provision of policy support for essential services in traditional fishing communities—encompassing health, education, literacy, digital inclusion, nutrition, potable water, energy, housing, home repair assistance, transportation, sanitation facilities, and sufficient sewage systems—can also alleviate water body pollution. Policy reforms primarily aim to rectify the malfunction of environmental, economic, and sociopolitical systems within societies. They focus on amending historical mistakes in institutional frameworks. Grindle and Thomas (1991) perceive policy reform as a calculated governmental initiative to rectify these discrepancies. Schneider and Heredia (2003) identified four principal elements affecting reform: political perspectives, economic pressures, strategic decisions by state reformers, and international influence. Both viewpoints indicate that policy reforms are influenced by external and internal causes. In the case of Sundarbans, global concerns to preserve the historic Delta and the economic prospects of ecotourism may have influenced the Government's strategy of fisheries governance. Such reforms can profoundly rewrite the power dynamics among stakeholders, resulting in the emergence of winners and losers, contingent upon the reforms' trajectory (Mosepele et al., 2017).

Potential initiatives for enhancing fisheries markets may involve establishing marketing collectives of fishermen and connecting with fish entrepreneurs. Integrated infrastructure development should primarily focus on multimodal transportation networks, hub-and-spoke market systems with all-inclusive facilities including ice plants and cold storage, and digital connectivity for sharing market information in real-time. Establishing several fish landing and market centers locally on the landside of the Sundarbans, along with ice producers through public–private partnerships, the cycle of indebtedness can be disrupted. With appropriate training, intermediaries can serve as "conservation agents," adept at merging indigenous knowledge from artisanal fishers with the regulatory frameworks of a complicated management system. As the fisheries service remains inadequate in the Sundarbans, construction of hatcheries or backyard hatcheries for wild shrimp fry (Hondula et al., 2020) can be widely adopted to reduce the pressure on nature and to promote alternative livelihood. Community-centered design principles naturally involve participatory planning and cultural sensitivity (Ahamed et al., 2012). The phased development, focused on foundation infrastructure, market integration, and value addition, can maximize impact. Public–private partnerships and community participation are the two main pillars for sustainability. On the one hand, well-planned infrastructure investments can cut postharvest losses by 30–40%, and conversely, fisher incomes can go up by 25–35% due to better market access and product quality (Béné et al., 2016). Strategic development of shrimp processing hubs in coastal Bangladesh is a good example of how clustered infrastructure development can create economic opportunities while ensuring quality product for export markets.

5.2.3 Institutional/Policy-Based Approach

5.2.3.1 Governance Frameworks

Government is not supposed to be a profit-making venture. (Cochrane et al., 2023)

A primary goal of government policy: to improve people's lives. The economic dimension and the social dimension ("people") encompass everything else connected with human aspirations: equity (translated as income distribution), inclusion (commonly operationalized as employment), and health (expressed by an indicator such as life expectancy or access to medical services) (Tesfahunegn et al., 2019).

Responsible governance of land, fisheries, and forests in small-scale fisheries is essential to human rights, food security, poverty eradication, sustainable livelihoods, social stability, housing security, economic growth, and rural and social development (Hutton et al., 2021). Emphasis was placed on women's tenure rights. It is essential to implement suitable mechanisms to identify, document, and uphold the rights of legal tenure rights holders (FAO, 2022). The recognition, respect, and protection of local customs and practices, as well as customary or otherwise preferential access to fishery resources and land by SSF communities, including ethnic minorities and indigenous peoples, should be in accordance with international human rights law (Yawson et al., 2020). Robust governance frameworks that include the socioecological intricacies of the Sundarbans are necessary. It is crucial to focus on decentralized and local government systems involved in governance and development initiatives, as well as SSF communities, including the area of research. In the Sundarbans, where the formal regulatory framework is lacking, the patterns of SSF production and commercialization are often shaped by a variety of self-governing systems (Chuenpagdee, 2011; Frawley et al., 2019). While cooperative fishermen enter into formal agreements with others to outline roles and responsibilities, collaborating as part of collectively owned and democratically organized entities (such as fishing cooperatives), those who do not cooperate (for instance, in patron-client dynamics) form informal arrangements with fisheries patrons, exchanging their catch for favors, loans, or access to markets (Basurto et al., 2013; Johnson, 2010). Therefore, well-defined property rights, inclusive decision-making processes, and the development of cost-effective monitoring technologies are essential for the sustainable management of fisheries. (Figure 5.3).

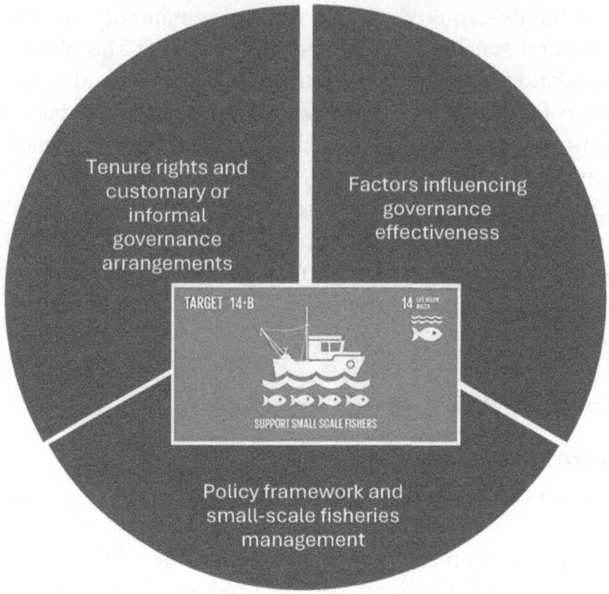

FIGURE 5.3 Steps of effective governance.

- Transparency: Ensures clear communication and accessibility of information.
- Accountability: Decision-makers are responsible to the public and institutions
- Participation: Encouraging community participation in the process of making decisions.
- Rule of Law: Equitable and unbiased enforcement of legislation.
- Responsiveness: Institutions must effectively serve all stakeholders
- Equity and Inclusiveness: All individuals should feel included, and their interests acknowledged
- Effectiveness and Efficiency: Processes and institutions provide outcomes that satisfy requirement
- Consensus-Oriented: Facilitation of divergent interests to achieve widespread agreement
- Strategic Vision: Leaders possess a long-term perspective on government and growth.

The establishment and implementation of exclusive zones for SSF should be considered. Highly migratory species, including tuna, billfish, and sharks, traverse the high seas and the exclusive economic zones (EEZs) that have been established by different authorities (Leal Filho, 2018), rendering them particularly vulnerable to overexploitation (Zimmermann et al., 2018). Such species may be influenced by the policies and activities of governments and business entities from far jurisdictions. Even the choices made by secondary actors—such as processors, marketers, supporting industries, and consumers—have implications for the overall sustainability of fisheries (Vachette, 2017; UNODRR, 2021). Hence, Governments should promote the involvement of fishermen in Biodiversity Management Committees to develop estuarine and marine biodiversity registries for conservation and protection purposes (Banerjee et al., 2017). India lacks distinct inland fisheries policy frameworks, and if such frameworks do exist, they are often subordinate to national marine fisheries policies (Ashrafuzzaman & Furini, 2019). In September 2003, the Government of Bangladesh implemented a total prohibition on the collection of prawn fry to mitigate the ecological consequences of fry fishing. The prohibition mostly criminalized poor individuals without options, had little effect on fry capture, and allegedly fostered minor corruption as authorities accepted bribes to ignore violations (Dubey et al., 2017). The West Bengal Inland Fisheries Act of 1984, along with its amendments in 1993 and 1997, imposed limits on the capture of "specified size, group, or species" and regulated "the erection or use of fixed engines" for fishing certain fish groups. However, in the absence of governmental support, fishers find it challenging to acquire various types of nets or equipment for different species due to their considerable expense. To prevent fish poisoning among fishermen, particularly during the lean season, enhanced and consistent patrols inside the forest, along with livelihood assistance, would eliminate this malpractice. The implementation of fishing limitations and safety measures should be grounded in democratic engagement with experts and stakeholders, since Sundarbans fishers operate collaboratively (Chatterjee et al., 2015; Dasgupta et al., 2020).

Artisanal fishermen are required to be listed in a national registry; they are often permitted to operate within the area of their registration (Ahmed & Haq, 2017). A "virtual" fishing community, defined as the functional equivalent of a geographically defined community in which the members of the community may not all reside in the same area. It may be thought of as a community of interest, as distinct from a community of place. (Akhtar et al., 2020)

In this context, "multi-level comanagement" has gained significance beyond traditional definitions of comanagement in recent years, explicitly highlighting the principles of power devolution, cooperative partnerships, democratic participation, polycentric governance networks, and subsidiarity (Finkbeiner & Basurto, 2015). Rights-based fisheries management (RBM) is a fisheries management strategy (Parsaulian et al., 2024) that emphasizes the establishment of rights over resource access, rather than implementing effective management and legislation to address overcapacity and/or overfishing within a fishery. In fisheries management, neoliberal privatization and market mechanisms have been brought together in the form of individual transferable quotas (ITQs) (Fabinyi & Barclay, 2022; Haider et al., 2018; Klinsky et al., 2017). The implementation of ITQs would provide stewardship incentives to craftsmen, thereby reducing rivalry among fishers, capital investment, and labor, while specifically restricting the quantity of fish a fleet may catch and enhancing individual profitability. Instead, this is debatable in prawn fisheries, characterized by fluctuating stocks, the presence of multi-cohort populations, and the practice of discarding over-quota harvest. The implementation of ITQs for Sundarbans fishermen may lead to a capacity cascade displacement or spillover, given their propensity to transition across fisheries (Ward et al., 2004). Equity issues may be alleviated by distributing ITQs to "communities" as community transferable quotas (CTQs) or by instituting Territorial Use Rights for Fisheries (TURF), which grant exclusive access to fish or shellfish harvesting within designated geographical regions to a specified group of fishers, characterized by nontransferability, temporal limitations, or seasonality (Wingard, 2000). To establish a balanced rights-based framework, it is imperative to incorporate monitoring, control, and surveillance to enhance social equity rather than facilitate monopolistic practices. Ensuring equity in interventions across regions of the Sundarbans where formal regulatory capacity is weak and natural resource use is regulated from the bottom-up requires explicit attention to how different self-governance forms influence vulnerability and response to extreme events and increasing environmental variability (Frawley et al., 2024).

Ecological and environmental quotas are essential for the effective "operation of Individual Transferable Quota (ITQ) systems" (Cochrane & Garcia, 2009) within the framework of ecosystem-based fisheries management (Jentoft & Chuenpagdee, 2015; Sumailla et al., 2000). The TURF system must be expanded to include the whole of the fisheries. Study found that the combination of TURFs and an open-access fishery was a (1) "recipe for disaster;" (2) the implementation procedure was too complex. The system had to be designed so that it could be efficiently implemented in hundreds of TURFs; (3) the data collection system was considered unnecessarily demanding and too costly, and its purpose was not clearly understood by the fishers. Too much emphasis was placed on the collection of ecological data, which, although scientifically interesting, was not needed for management. The advice was to simplify the information required, and (4) in general, implementation of the system was viewed as too "top-down." The advice was to empower fishers' organizations. TURFs with a "certificate of origin" marking. A notable advancement was the heightened focus and investment in vigilance, with several communities conducting round-the-clock patrols of their territories. The development of a TURF and a fishers' organization exhibits positive feedback, which is a considerable advantage (Payne & Castilla, 1994; Stotz, 1997; plus Avilez, 2003; Barría, 2003; Barrios-Garrido et al., 2003; Oresanz et al., 2005). A Framework for comanagement for the Sundarbans small-scale fishery involving fisherfolk has been proposed in Figure 5.4.

The implementation of eco-labeling and seafood wallet card initiatives in the small-scale fisheries of Sundarbans could generate market-driven incentives for environmentally sustainable products and enhance the management of fisheries production processes through life-cycle assessment, thereby promoting long-term stewardship for national economic welfare (Bridgespan Group, 2005). Consumers evaluate product features in relation to environmental and ecological goals, alongside economic and social purposes, such as fair trade, assistance for small farmers, the prevention of child labor, and the protection of human rights among Sundarbans fishermen. This will provide one of the least coercive market-based strategies to enhance conservation results and create new prospects for capital investment and joint ventures in the Sundarbans, serving as a crucial factor for accessing new premium green markets (Titumir and Shah Paran 2022).

To improve the governance of small-scale fisheries in the Sundarbans, enhanced cooperation and tangible links among governmental bodies, researchers, NGOs,

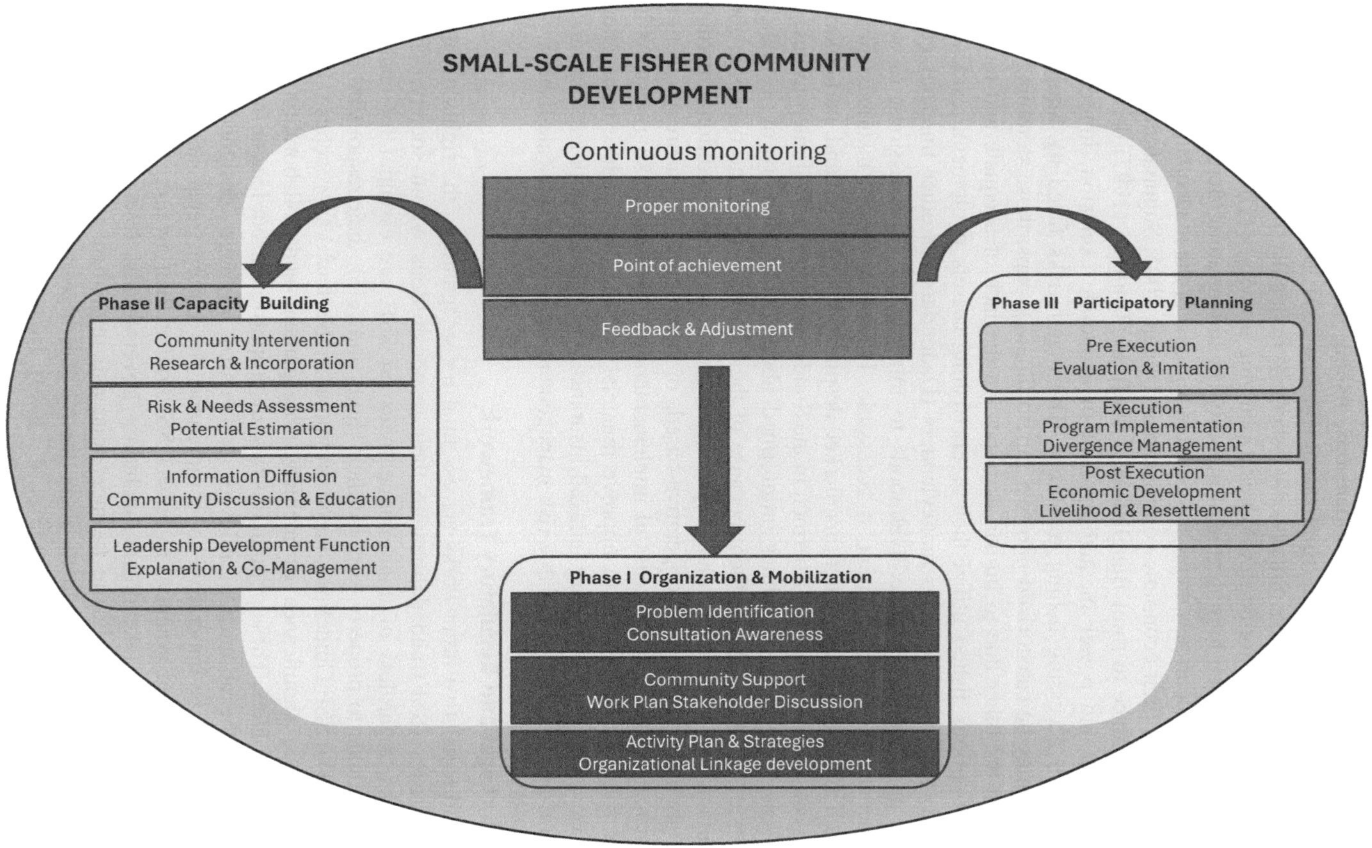

FIGURE 5.4 Proposed framework for comanagement for Sundarbans small-scale fishery involving fisherfolk. **(Modified after Mitra & Barua (2022)).**

fisheries administrators, and stakeholders are essential. Managers of small-scale fisheries may implement stringent fishing regulations, such as restrictions on the number of fishers, boats, boat dimensions, fishing days, or seasons, to mitigate overexploitation and semi-industrialization (Berkes, 2007; Bertram, 1995; Pomeroy, 2012; Purcell et al., 2013). Small-scale fisherfolk frequently find themselves marginalized in macroeconomic development planning processes (Thorpe, 2004). This can be because of their high degree of mobility (including unregistered international migrants), residence in peripheral and isolated regions, or the inadequate recognition and valuation of their economic contributions (Dugan, 2005; Sugunan et al., 2007). Historical reference to rural Bangladesh, a study by Toufique (1997) depicted that systemic bias and power inequality result in economic exclusion, hindering the poorest from accessing fishing resources. The major risks faced by Sundarbans fishermen may be categorized into two superordinate groups: those stemming from nature and the other due to human activities. In the lack of adequate and rational insurance options, fishers utilize their assets (fishing vessels, equipment, residences, and land) and incur debts to moneylenders (Dadondars) or other financial institutions, or pursue alternative livelihoods to mitigate the repercussions, or seek assistance from the Government, if accessible. However, private insurance firms are much more reluctant to enter this market, despite significant demand, particularly among cooperative artisanal fishermen in the Sundarbans region. In numerous cases, small-scale fishermen are either uninformed about the availability of insurance services or do not prioritize them because of their limited cash. The fishers of the Sundarbans can be offered various types of insurance, including traditional, partner agent, mutual, community/cooperative-based, or parametric, to ensure protection against natural disasters. Successful implementation of mobile banking systems in remote fishing communities to improve financial inclusion and reduce transaction costs. Development of weather-indexed insurance and fishing equipment insurance programs may provide affordable risk management tools for small-scale fishers.

5.2.3.2 Legal and Regulatory Framework

Legal and legislative frameworks are essential for safeguarding the rights of small scale fisheries (SSF) and delineating their responsibilities within society. These frameworks, regardless of their unique focus on SSF, are essential for enabling SSF individuals to properly exercise their rights and fulfill their responsibilities as fishermen. The SSF Guidelines, based on human rights-based approaches (HRBA), underscore the significance of inclusive engagement and collaboration among diverse stakeholders. These populations bear disproportionate burdens or insufficient protections in the absence of rules or preferential privileges designed for SSF. Consequently, it is imperative to comprehend the current legal and regulatory frameworks and their potential to bolster SSF, rather than awaiting the enactment of new legislation. Stakeholders, including governments, researchers, and SSF communities, must collaborate to understand and implement these tools (Joydeb, 2016). Engaging SSF people to claim their rights and enforce duties is key in advancing the SSF Guidelines (Shi et al., 2019). Eight key legal issues and their potential legal effects in small-scale fisheries have been depicted in Table 5.1. Given that the Sundarbans straddle the border of two nations, it is crucial to address the

TABLE 5.1
Eight Key Legal Issues, Potential Legal Effects, and Innovative Approaches in Small-Scale Fisheries

Legal Issue	Potential Legal Effects	Innovative Approaches
1. Legal Definition	• Promote, empower, and enhance awareness of the SSF sector • Acknowledge and confirm the rights of SSF individuals and/or confer preferred privileges • Ensure responsibilities are commensurate with their capabilities	• Digital ID systems: Blockchain-based registries and smart ID cards for SSF verification • Mobile apps: Digital platforms for fishers to prove legal status and access benefits • Participatory definition processes: Community involvement in defining SSF criteria • Flexible categorization: Tiered definitions recognizing different scales within SSF
2. Tenure and/or Labor Rights	• Establish a registry or alternative mechanism to document these rights • Exempt SSF from fees or obligations that may obstruct documentation of rights • Classify violations of these rights as an offence with corresponding penalties • Mandate fisheries officers and inspectors to uphold and safeguard SSF tenure and labor rights • Confer professional identity to SSF and enable access to social security benefits • Facilitate streamlined administrative and/or judicial processes for SSF individuals to assert their rights	• Blockchain tenure registries: Immutable, transparent documentation of fishing rights • Paralegal programs: Training fishers to understand and advocate for their rights • Legal aid clinics: Community-based simplified legal procedures • TURFs (Territorial Use Rights): Community-managed fishing territories with legal recognition • Mobile registration: App-based systems for documenting rights without bureaucratic barriers • Community legal observers: Peer networks monitoring rights violations
3. Participation in Fisheries Management and Conservation	• Confer authority to SSF individuals in planning, drafting, negotiating, and adopting fisheries management measures (e.g., quotas, closed areas, and seasons) and management plans • Impose responsibilities regarding conservation measures (e.g., use of permitted fishing gear, prohibition of fishing protected species, and prevention of bycatch and gear loss) • Ensure legal recognition of SSF knowledge and practices as integral to management and conservation parameters	• Co-drafting workshops: SSF communities actively write fishing laws and regulations • Citizen assemblies: Deliberative forums with guaranteed SSF representation quotas • Participatory budgeting: Communities decide how fisheries funds are allocated • Traditional knowledge protocols: Legal frameworks recognizing indigenous practices • Digital participation platforms: Online tools for remote community input

(Continued)

Table 5.1 (Continued)

Legal Issue	Potential Legal Effects	Innovative Approaches
	• Mandate relevant public authorities to promptly inform SSF individuals, enabling active involvement in formulation, planning, and evaluation of fisheries management strategies	• Adaptive comanagement: Real-time data sharing between fishers and managers • Community science programs: Fisher-collected data integrated into management decisions
4. Partial/Exclusive Zones and/or Preferential Access	• Enhance stewardship of SSF communities over their fishery resources, ecosystems, and biodiversity • Safeguard SSF from inequitable competition with other sectors, particularly large-scale industrial fishing fleets • Mitigate conflicts regarding marine spatial utilization	• Dynamic seasonal zones: Mobile exclusive areas following fish migration patterns • Global Positioning System (GPS)/satellite enforcement: Technology-enabled monitoring of SSF-only zones • GIS community mapping: Tools allowing communities to document customary fishing grounds • Integrated coastal zone management: Legal frameworks prioritizing SSF in spatial planning • Multiuse zoning: Compatible activities (tourism, research) allowed with SSF priority • Climate migration corridors: Legal provisions for displaced fishing communities
5. Public Financial Mechanisms	• Guarantee requisite funds are allocated to the SSF sector for designated purposes (e.g., capacity-building, education, data collection and research, infrastructure enhancement) • Provide a safety net for SSF during crises or unforeseen disasters	• Blue bonds: Dedicated financing instruments for SSF infrastructure • Microinsurance schemes: Weather and disaster loss coverage for fishers • Payments for ecosystem services (PES): Rewarding SSF for conservation practices • Crowdfunding platforms: Community-led fundraising for SSF projects • Climate adaptation grants: Simplified application processes for SSF • Carbon credit mechanisms: Revenue from mangrove restoration and sustainable practices • Emergency response funds: Pre-positioned resources with rapid disbursement protocols • Open-data platforms: Transparent tracking of how SSF funds are spent

6. Gender Laws and Policies	• Secure women's representation within relevant public institutions (e.g., fisheries departments, councils, and advisory bodies) • Enhance acknowledgement of women as vital contributors in SSF • Formalize women's preharvesting and postharvesting activities as recognized professional work within the SSF sector	• Gender-disaggregated licensing: Separate recognition of women's specific fishing activities • Women's fisher cooperatives: Legal recognition with preferential credit access • Childcare provisions: Laws enabling women's participation in meetings and training • Gender quotas with support: Mandated representation plus capacity-building programs • Mobile banking/digital payments: Direct financial access for women in SSF • Gender-responsive value chains: Legal frameworks supporting women in postharvest activities • Female extension workers: Dedicated support staff for women fishers
7. Climate Change Laws and Policies	• Guarantee allocation of climate change funds to facilitate SSF preparedness and adaptation strategies • Offer assistance from relevant public authorities overseeing fisheries and/or climate change to enhance awareness and deliver training to SSF communities regarding climate change matters • Integrate SSF considerations into national climate change adaptation and mitigation plans	• Climate risk insurance pools: Collective insurance funded through climate allocations • Early warning systems: Legal mandates for timely information sharing with SSF • Climate-smart infrastructure: Cyclone shelters, cold storage designed for climate resilience • Livelihood diversification frameworks: Legal support for aquaculture, ecotourism alternatives • Community-based adaptation planning: SSF-led climate response strategies with legal backing • Climate data apps: Mobile platforms delivering weather, ocean temperature, and forecast info • Mangrove restoration programs: Legal incentives for SSF involvement in coastal protection • Climate migration policies: Legal pathways for displaced fishing communities

(Continued)

Table 5.1 (Continued)

Legal Issue	Potential Legal Effects	Innovative Approaches
8. Representation of SSF People in Institutional Arrangements	• Ascertain SSF people's representation in the composition of fisheries institutions, departments, and agencies • Create by law a public institution entirely devoted to SSF issues • Provide the duty of the public authority in charge of fisheries to include SSF issues in the fisheries development policy, empowering the SSF sector and the people involved in it • Provide the duties of the authorities in charge of fisheries to coordinate and integrate SSF issues in relevant sectoral meetings and discussions	• Mandatory representation quotas: Legal requirements for SSF seats on all fisheries boards and councils • Dedicated SSF authority/agency: Standalone government body with budget and decision-making power • SSF advisory councils: Legally mandated bodies with binding consultation requirements • Rotating community representatives: Time-limited positions ensuring diverse SSF voices • Capacity-building programs: Training for SSF representatives on policy, budgets, and advocacy • Per diem and transport allowances: Compensation enabling participation without financial burden • Virtual participation options: Digital platforms for remote community engagement in meetings • SSF focal points: Designated officers in every relevant government department • Cross-sectoral coordination mechanisms: Legal mandates for inter-ministerial SSF working groups • Annual SSF policy reviews: Mandatory assessments of SSF integration in national development plans • SSF impact assessments: Required analysis of how policies affect SSF before adoption • Community nomination processes: Bottom-up selection of representatives rather than top-down appointments

Source: Modified after Nakamura et al. (2021).
SSF = small-scale fisheries.

transboundary political and geographical constraints for effective resource management, considering increasing climatic variability (Badjeck et al., 2010). The establishment of a transnational network of women's fishing organizations would facilitate the exchange of information and assist initiatives to safeguard illegal migrants or imprisoned fishermen from nearby countries (FAO, 2019). The desk-based rapid appraisal underscored the need for more country-specific legal research to investigate and confirm current legislation. Comprehending and distributing pertinent legislation to SSF communities may enhance access to justice and enable them to adopt sustainable fishing methods.

5.2.3.3 Social Inclusion and Security

To reconcile the sustainability of fisheries resources with the resilience of fisherfolk, fisheries policy and governance must adopt a more human-centric approach that acknowledges the interconnectedness of ecological and social dimensions (Sannigrahi et al., 2020; Sarkar & Padaria, 2010). Worldwide, there is growing appreciation for women's contributions to the feminization of fisheries (Shankar et al., 2021). The government, nongovernmental stakeholders, and fisheries experts must promote small-scale fisherwomen as "professionals" and outlaw part-time fishing (Allison & Ellis, 2001; FAO, 2019). Governmental and NGOs must offer technical and financial support, along with initiatives to establish the Association of Women's Fisherfolk Groups, which emphasizes credit and savings to enhance women's skill development and decision-making capacity within the local fishing community by augmenting their skills and knowledge. Diversification via occupational multiplicity mitigates risk and addresses uncertainty, providing a safety net for poor and vulnerable communities while offering targeted support to at-risk women in the fishing industry. Fishing industries have historically been perceived as male-dominated, and there is a notable absence of voluntary reporting of regional or national sex-disaggregated official catch statistics for nearshore fisheries that recognize women fisherfolk (Harper et al., 2020; Rahman, 2020, 2022). Nonetheless, these invisible catch statistics obtained through household and community fishing surveys are essential for the comprehensive evaluation of natural resource management frameworks and methodologies pertaining to ecosystems reliant on artisanal SSF management (de la Torre-Castro et al., 2017; Gee & Bacher, 2017; Frangoudes et al., 2019; Lam et al., 2016), particularly in the context of climate change.

In Sundarbans small-scale fisheries, like other developing nations, the augmented recognition of various contributions made by women fisherfolk communities is in sharp contrast to their restricted involvement across the value chain (Sánchez-Triana et al., 2018; Scheelbeek et al., 2016). Specific attention should be paid to the need to work toward the equitable participation of women in such structures. Where appropriate and necessary, separate spaces and mechanisms should be provided to enable women to organize autonomously at various levels on issues of relevance to them. Efforts to mainstream gender should include the use of gender analysis in the design phase of policies, programs, and projects for small-scale fisheries to design gender-sensitive interventions (Kirchhof & Perreault, 2019). Gender-sensitive indicators should be used to monitor and address gender inequalities and to capture how interventions have contributed toward social change (Sarker et al., 2020: Toimil et al., 2019). Case studies at the regional or local level should be undertaken to evaluate women's information and understanding of the full value chain (Szabo et al., 2020). Apart from the increasing focus on women and gender in fisheries, gender factors remain insufficiently emphasized in global fisheries policy and management, mostly due to the lack of voluntary national data reporting (Frangoudes et al., 2019; Gee & Bacher, 2017; Harper et al., 2017).

Financial resources used for fisheries conservation should be directed toward initiatives that provide the greatest sustainable returns on investment. It is imperative to contemplate financing of this scale for a more fair and foundational market-based strategy that will produce outcomes in aquatic environments: the eradication of detrimental fisheries subsidies. Small-scale boat manufacturing often promotes local craftsmanship and receives little government funding. Subsidies further marginalize small-scale fishers, and marine biodiversity suffers more than if the market were truly competitive. If the government is following progressive taxation in the case of direct taxation, it should adopt the opposite of the same policy for a subsidy, which should lead to giving out more subsidies to the poor and fewer to the rich. Small-scale fishers, who represent a significant portion of employment and are more directly reliant on fisheries for their livelihoods, get fewer subsidies due to their lower fishing intensity and fuel consumption compared to bigger vessels. However, selecting discount rates is a crucial factor in the formulation and assessment of conservation and management policies aimed at addressing climate change (Dasgupta et al., 2008; Sumaila & Walters, 2005). Rising fuel costs will diminish overcapacity and overfishing by decreasing potential earnings, thereby forcing marginal fishermen out of the industry (Sumaila et al., 2006). There is a great need for indicators in setting and moving gender objectives as they serve not only for evaluating but also for quantifying present gaps and measuring further progress. Rescuing and rehabilitating children involved in child labor necessitates active community participation and the enforcement of child labor regulations in fisheries that comply with international standards. Rehabilitation and educational institutions must be created for children liberated from exploitative labor and trafficking in fisheries, together with protocols to assist their reintegration with families and communities (FAO/ILO, 2011). Social security provisions must be extended to aged working women, widows, and abandoned individuals in this sector to enhance their livelihoods and empower them in domestic responsibilities via alternative income opportunities (FAO, 2019). To rescue and rehabilitate female children engaged in employment, comprehensive community involvement and the enforcement of child labor regulations in the fishing industry must be fully integrated with worldwide initiatives. Implementing a national social security system and providing low-interest financing to SSF communities will improve their quality of life (Béné and Friend, 2011; FAO, 2019).

5.2.3.4 Capacity Building

Community capacity building can be broadly defined as the "increase in community groups' abilities to define, assess, analyze and act on health (or any other) concerns of importance to their members" (Laverack, 2024). In a stressed and ecologically sensitive natural resource context, community-based fishing management is crucial.

This management requires support from fisheries professionals and enforcement authorities. Community-based management of experimentation and learning emphasizes local and indigenous knowledge to sustain economically significant fisheries stocks, enhance targeted fish production, and improve harvesting proficiency (Carvalho et al., 2019) as adaptive strategies amid uncertainty. Regional or local studies emphasizing the indigenous knowledge of fisher folk throughout the entire value chain, highlighting the stochasticity of the environment is thus necessary for sustainable management of Sundarbans SSFs. Living in poverty and extreme environmental conditions, ecopsychiatry can tackle the "triple jeopardy" of social, health, and environmental inequities (Pearce et al., 2010) impacting this fishing community (Chowdhury & Jadhav, 2012), as human rights to well-being are intrinsically connected to a secure environment and ecological rights. It should be ensured that the range and diversity of the small-scale fisheries subsector along the entire value chain are appropriately represented through the creation of legitimate, democratic, and representative structures. Establishing an autonomous monitoring and compliance organization can assist communities and laborers in addressing remediation challenges. Mechanisms enabling the integration of monitoring outcomes into policy formulation and implementation should be incorporated. Expert knowledge, in addition to the information accessible to fishermen, is crucial for establishing the foundation necessary for effective management to attain sustainable production. The enforcement of fry collection restrictions through community-based initiatives and the establishment of community-based natural resource management (CBNRM) can enhance community support, mitigate corruption, and improve the likelihood of the bans' success. A more resilient local society possesses a greater need and capacity to safeguard and manage its environment, hence enhancing the ecosystem's resilience. The indigenous possess unique social institutions and ecological knowledge systems, which they have utilized for collective and systematic environmental protection efforts, demonstrating efficacy in conservation. Ongoing awareness initiatives must be implemented to inform fishermen about the current state of the fisheries and how they can optimize their catch while minimizing adverse impacts on this natural resource. Incentives should be established to encourage local hotels and ecotourism providers to adopt sustainable management methods.

Women fishers may be empowered by the development of the ornamental fish industry through the formation of women fisher cooperatives, women self-help groups, and other suitable activities (Mukherjee, 2018). Government initiatives have supported 35,000 beneficiaries through fingerling distribution and training programs, demonstrating efforts to improve resource accessibility (Rahman et al., 2018). Vocational training may be provided for several emerging industries, including poultry, cattle, pigeon farming, quail farming, beef fattening, platform shed goat husbandry, coastal aquaculture, and the cultivation of entrepreneurship for small trade firms (Roy et al., 2017). As facilitators of community cohesiveness, women may enhance communication, dispute resolution, collaboration, and unity within and across communities and institutions. Education and awareness Programs aimed at educating and raising awareness about ecological integrity, the significance of nursery and recruitment grounds, migratory pathways, and the effects of riverbank erosion are essential. The implementation of laws and rules to ensure the environmentally friendly use of resources. Increasing the number of educational institutions specializing in marine science and management, along with institutional interactions and joint research between the two countries, would uplift the capacity for the sustainable development of Sundarbans. Periodic studies and long-term data collection will evaluate the synergistic effects of climate change and pollution on fisheries and human health; assess the conservation status of currently prohibited fishing zones and the types and quantities of bycatch; analyze trends in production and fisheries catches, as well as the socioeconomic impacts on coastal fishing communities across the borders. Additionally, the use of stationary nets and traps for capture methods will promote ecologically sustainable practices, while catch monitoring programs will offer a comprehensive brief overview of the Sundarbans' socioeconomic and ecological aspects. Successfully managed fisheries that attain resource sustainability are likely to possess more resilience and adaptability to the uncertainties of climate change compared to those governed by a more laissez-faire approach. The resistance to climate change among the fisherfolk of Sundarbans may be attained by effective adaptive management via participatory and integrated programs focused on pro-poor development (Cinar et al., 2009) from the early stages of their life. An urgent necessity exists for the restructuring, synchronization, and dissemination of knowledge via certification programs among shrimp hatchery proprietors to enhance broodstock collection techniques and to establish independent breeding production units, thereby eliminating current detrimental harvesting practices. The promotion of alternative livelihoods will also limit new entrants into the prawn fry collecting sector. Shrimp fry collectors must be provided with knowledge and equipment to promote improved practices, as well as training to enhance the survival rates of shrimp fry that are harvested and sold (Environmental Justice Foundation, 2004).

5.2.3.5 International Cooperation

An integrated strategy for marine resource conservation between India and Bangladesh was established with the signing of a Memorandum of Understanding on Blue Economy and Maritime Cooperation in the Bay of Bengal and the Indian Ocean in 2015. In 2016, the Swatch of No Ground, Bangladesh's first MPA, was designated by the Ministry of Environment and Forests, located 45 km from the Sundarbans (Duke, 2017: Béné, 2003). Small-scale fisheries must be prioritized before establishing resource access agreements with foreign nations and third parties. In international competition for both catches and markets, small-scale fisheries may have a considerable international influence akin to large-scale fisheries; specifically, small-scale fishers may share stocks (transboundary, straddling, or highly migratory) inside their EEZs. Due to localized depletion, sometimes resulting from overfishing, small-scale boats extend their operations outside territorial seas or EEZs to exploit deep-sea or remote fisheries. The harvests from small-scale fishermen may substantially contribute to the global fish trade. The Sundarbans region, which delineates the border between India and Bangladesh, may experience heightened conflicts over shared fish stocks due to changes in the distribution of exploited species. Consequently, a treaty should be established to facilitate bilateral and multilateral agreements with neighboring nations, ensuring the peaceful and regulated sharing of common marine resources while considering customary practices (Das & Thatoi, 2020; Zaman et al., 2014). An active collaboration between Bangladesh and India, facilitated by the National Institute of Oceanography (NIO) under the Council of Scientific and Industrial Research (CSIR), India, and Dhaka University, aims to enhance the understanding of breeding grounds and nurseries for marine species (Arunrat et al., 2022; Chaluvadi et al., 2021). A joint fisheries stock assessment may be conducted at five-year intervals to address critical issues (Chanda et al., 2022; Chowdhury et al., 2018; Dasgupta et al., 2019), including total production of fish and other aquatic organisms, production by fishery, production by gear, production by species, and fish population dynamics, which are currently deficient. This information will assist in decision-making on the allocation of permits, regulation of gear use, and species capture, among other factors. The establishment of transnational networks of male and female fisher organizations will facilitate information exchange and bolster initiatives aimed at ensuring the safety of migrant fishermen or those jailed abroad. The implementation of explicit guidelines will assist the families of fishermen who have either perished at sea or been detained in other nations.

5.3 CONCLUSION

Fisheries management in the Sundarbans has traditionally focused on survival. It is often overshadowed by agricultural and forestry policies that prioritize resource extraction while neglecting the distinct requirements of small-scale fishers. The fisheries of the Sundarbans demand a shift from subsistence and small markets to more marketed and nationally acknowledged contributions to food security. Strong regulations and a national fisheries policy are crucial to unlocking the full potential of the Sundarbans fisheries. The abundant aquatic resources of the Sundarbans present significant prospects for a thriving SSF sector; however, the lack of a cohesive policy has hindered its advancement. Sustainability initiatives in fisheries have mainly functioned within a specific economic framework that prioritizes profit maximization and property rights as key motivators of rational behavior. Altering economic and political dialogues to emphasize human and planetary well-being, guided by five core principles of a well-being economy—purpose, nature, fairness, participation, and dignity—can aid in redirecting the fisheries sector. The goal of such a transformation is to ensure that fishing practices are only allowed when they provide benefits to the environment, society, fishers, and fishing communities, while also making certain that all costs and effects are considered and internalized (Kelling et al., 2026). Analyzing small-scale fisheries through the lens of food systems highlights the interconnectedness of food production and consumption, and recognizes that processes of change and transformation are inherently complex and fraught with uncertainty. This necessitates not only partnerships with other sectors, disciplines, and experts (Fanzo, 2021) but also diverse knowledge and viewpoints. Consequently, it is impossible to pinpoint any universal pathways or sets of win-win solutions, as these processes are highly context-dependent (Arthur et al., 2022; Béné et al., 2019; McClanahan et al., 2015).

BIBLIOGRAPHY

Abell, R., Allan, J. D., & Lehner, B. (2007). Unlocking the potential of protected areas for freshwaters. *Biological Conservation*, *134*(1), 48–63.

Ahamed, F., Hossain, M. Y., Fulanda, B., Ahmed, Z. F., & Ohtomi, J. (2012). Indiscriminate exploitation of wild prawn postlarvae in the coastal region of Bangladesh: A threat to the fisheries resources, community livelihoods and biodiversity. *Ocean & Coastal Management*, *66*, 56–62.

Ahmed, & Haq, S. M. A. (2017). Indigenous people's perceptions about climate change, forest resource management, and coping strategies: A comparative study in Bangladesh. *Environment, Development and Sustainability*, *21*, 679–708. https://doi.org/10.1007/s10668-017-0055-1

Akhtar, R., Afroz, R., Masud, M. M., Rahman, M., Khalid, H., & Duasa, J. B. (2020). Farmers' perceptions, awareness, attitudes and adaption behaviour towards climate change. *Journal of Asia Pacific Economy*, *23*, 246–262. https://doi.org/10.1080/13547860.2018.1442149

Allison, E. H., & Ellis, F. (2001). The livelihoods approach and management of small-scale fisheries. *Marine Policy*, *25*(5), 377–388. 10.1016/S0308-597X(01)00023-9

Alongi, D. M. (2020). The impact of climate change on mangrove forests. *Current Climate Change Reports*, *1*, 30–39. https://doi.org/10.1007/s40641-015-0002-x

Arthur, R. I., Skerritt, D. J., Schuhbauer, A., Ebrahim, N., Friend, R. M., & Sumaila, U. R. (2022). Small-scale fisheries and local food systems: Transformations, threats and opportunities. *Fish and Fisheries*, *23*(1), 109–124.

Arunrat, N., Sereenonchai, S., Chaowiwat, W., & Wang, C. (2022). Climate change impact on major crop yield and water footprint under CMIP6 climate projections in repeated drought and flood areas in Thailand. *Science of the Total Environment*, *807*(Pt. 2), 150741. https://doi.org/10.1016/j.scitotenv.2021.150741

Ashrafuzzaman, M. (2022). Climate change driven natural disasters and influence on poverty in Bangladesh and its journey towards resilience. *Journal of the British Academy*, *9*(s8), 55–77. https://doi.org/10.5871/jba/009s8.055

Ashrafuzzaman, M., & Furini, G. L. (2019). Climate change and human health linkages in the context of globalization: An overview from global to southwestern coastal region of Bangladesh. *Environment International*, *127*, 402–411. https://doi.org/10.1016/j.envint.2019.03.020

Avilez, O. (2003). Experiencia en el manejo y explotación de las áreas de manejo de recursos bentónicos en las caletas artesanales de la IV Región. In G. A. San Martín, A. Pinto, M. Montecinos, A. González, & J. A. Guerra (Eds.), *Taller de áreas de manejo: Experiencias y proyecciones* (pp. [insert page numbers if known]). Unidad de Áreas de Manejo, Departamento de Pesquerías, Subsecretaría de Pesca, Valparaíso, Chile.

Azam, G., Huda, M. E., Bhuiyan, M. A. H., Mohinuzzaman, M., Bodrud-Doza, M., & Islam, S. D. U. (2021). Climate change and natural hazards vulnerability of char land (Bar land) communities of Bangladesh: A review. In *Bangladesh II: Climate change impacts, mitigation, and adaptation in developing countries* (pp. 1–39).

Badjeck, M., Allison, E., Halls, A., & Dulvy, N. (2010). Impacts of climate variability and change on fishery-based livelihoods. *Marine Policy*, *34*(3), 375–383.

Banerjee, K., Gatti, R. C., & Mitra, A. (2017). Climate change-induced salinity variation impacts on a stenoecious mangrove species in the Indian Sundarbans. *Ambio*, *46*, 492–499. https://doi.org/10.1007/s13280-016-0839-9

Banerjee, S. B. (2008). Corporate social responsibility: The good, the bad and the ugly. *Critical Sociology*, *34*(1), 51–79. https://doi.org/10.1177/0896920507084623

Barahona, N., Montecinos, M., & Elias, I. (2005). What are the key elements for the sustainability of "S-fisheries"? Insights from South America. *Bulletin of Marine Science*, *76*(2), 527–556.

Barría, P. M. (2003). *Investigación captura total permisible pez espada, 2003: Informe avance corregido*. Subsecretaría de Pesca, Chile: Instituto de Fomento Pesquero.

Barrios-Garrido, H., Marchena, M. D., Medina, A., Mejía, K., Nava, A., & Rincón, J. A. (2003). Descriptive analysis of the fishing arts that affect the marine turtles in the Gulf of Venezuela. In J. Seminoff (Ed.), *Proceedings of the twenty-second annual symposium on sea turtle biology and conservation* (p. 156). U.S. Department of Commerce, National Oceanic and Atmospheric Administration. NOAA Technical Memorandum NMFS-SEFSC-503.

Barua, P., & Mitra, A. (2024). Climate resilience aquaculture development for achieved sustainable development goals. *International Journal of Energy, Environment and Economics*, *31*(1), 24–44.

Basurto, X., Gelcich, S., & Ostrom, E. (2013). The social–ecological system framework as a knowledge classificatory system for benthic small-scale fisheries. *Global Environmental Change*, *23*(6), 1366–1380. https://doi.org/10.1016/j.gloenvcha.2013.08.001

Béné, C. (2003). When fishery rhymes with poverty: A first step beyond the old paradigm on poverty in small-scale fisheries. *World Development*, *31*(6), 949–975. https://doi.org/10.1016/S0305-750X(03)00045-7

Béné, C., Barange, M., Subasinghe, R., Pinstrup-Andersen, P., Merino, G., Hemre, G.-I., & Williams, M. (2016). Contribution of fisheries and aquaculture to food security and poverty reduction: Assessing the current evidence. *World Development*, *79*, 177–196. https://doi.org/10.1016/j.worlddev.2015.11.007

Béné, C., & Friend, R. M. (2011). Poverty in small-scale fisheries: Old issue, new analysis. *Progress in Development Studies*, *11*(2), 119–144. https://doi.org/10.1177/146499341001100203

Béné, C., Prager, S. D., Achicanoy, H. A. E., Alvarez Toro, P., Lamotte, L., Bonilla Cedrez, C., & Mapes, B. R. (2019). Understanding food systems drivers: A critical review of the literature. *Global Food Security*, *23*, 149–159. https://doi.org/10.1016/j.gfs.2019.04.009

Bera, B., Bhattacharjee, S., Sengupta, N., Shit, P. K., Adhikary, P. P., Sengupta, D., & Saha, S. (2022). Significant reduction of carbon stocks and changes of ecosystem service valuation of Indian Sundarban. *Scientific Reports*, *12*(1), 7809.

Berkes, F. (2007). Community-based conservation in a globalized world. *Proceedings of the National Academy of Sciences*, *104*(39), 15188–15193. https://doi.org/10.1073/pnas.0702098104

Bertram, D. F. (1995). The roles of introduced rats and commercial fishing in the decline of ancient Murrelets on Langara Island. *British Columbia. Conservation Biology*, *9*(4), 865–872. https://doi.org/10.1046/j.1523-1739.1995.09040865.x

Bird, K., Hulme, D., Moore, K., & Shepherd, A. (2002). *Chronic poverty and remote rural areas* (CPRC Working Paper No. 13). Chronic Poverty Research Centre. https://doi.org/10.2139/ssrn.1754490

Botsford, L. W., Castilla, J. C., & Peterson, C. H. (1997). The management of fisheries and marine ecosystems. *Science*, *277*(5325), 509–515.

Bridgespan Group. (2005). Building a viable civic and educational infrastructure: The essential role of nonprofit leadership in delivering results. *The Bridgespan Group*. https://www.bridgespan.org/insights/library/leadership/building-a-viable-civic-and-educational-infrastr

Brown, K., Cinner, J., & Adger, N. (2019). Moving climate change beyond tragedy of the commons. *Global Environmental Change*, *54*, 61–63. https://doi.org/10.1016/j.gloenvcha.2018.11.009

Buchanan, J. M. (1973). The institutional structure of externality. *Public Choice*, *14*, 69–82.

Bunn, S., & Arthington, A. (2002). Basic principles and ecological consequences of altered flow regimes for aquatic biodiversity. *Environmental Management*, *30*, 492–507. https://doi.org/10.1007/s00267-002-2737-0

Bush, S. R., & Oosterveer, P. J. M. (2019). *Governing sustainable seafood*. Routledge.

Caihong, F., Travers-Trolet, M., Velez, L., Grüss, A., Bundy, A., Shannon, L. J., & Elizabeth, A. (2019). Risky business: The combined effects of fishing and changes in primary productivity on fish communities. *Ecological Modelling*, *368*(2), 120–128. https://doi.org/10.1016/j.ecolmodel.2017.12.003

Carvalho, T. P., Soares, F. A. A. M. N., Vita, R., Francisco, R. da P., Basto, J. P., & Alcalá, S. G. S. (2019). A systematic literature review of machine learning methods applied to predictive maintenance. *Computers & Industrial Engineering*, *137*, 106024. https://doi.org/10.1016/j.cie.2019.106024

CBD (2009). *Connecting biodiversity and climate change mitigation and adaptation: Report of the second ad hoc technical expert group on biodiversity and climate change*. Montreal, Convention on Biological Diversity Technical Series No. 41, 126 pages.

Cespedes, R. (2012). A reference model for the electrical energy system based on Smart Grids. In *Proceedings of the 2012 Sixth IEEE/PES Transmission and Distribution: Latin America Conference and Exposition* (T&D-LA) (pp. 1–6). IEEE. http://dx.doi.org/10.1109/tdc-la.2012.6319068

Chakraborty, S., Zaman, S., & Mitra, A. (2020). Bioaccumulation of heavy metals by dominant oyster in Indian Sundarbans. *Journal of Science, Engineering and Health Management*, *1*(2), 34–39.

Chaluvadi, R., Varikoden, H., Mujumdar, M., Ingle, S. T., & Kuttippurath, J. (2021). Changes in large-scale circulation over the indo-Pacific region and its association with 2018 Kerala extreme rainfall event. *Atmospheric Research*, *263*, 1–36. https://doi.org/10.1016/j.atmosres.2021.105808

Chanda, A., Das, S., & Ghosh, T. (Eds.). (2022). *Blue carbon dynamics of the Indian Ocean: The present state of the art*. Springer Nature.

Chatterjee, N., Mukhopadhyay, R., & Mitra, D. (2015). Decadal changes in shoreline patterns in Sundarbans, India. *Journal of Coastal Sciences*, *2*(2), 54–64.

Chowdhury, A., Maiti, S. K., & Bhattacharyya, S. (2018). How to communicate climate change 'impact and solutions' to vulnerable population of Indian Sundarbans? From theory to practice. *Springerplus*, *5*, 1–17. https://doi.org/10.1186/s40064-016-2816-y

Chowdhury, A. N., & Jadhav, S. (2012). Eco-psychiatry: Culture, mental health and ecology with special reference to India. In B. S. Chavan, N. Gupta, P. Arun, A. Sidana, & S. Jadhav (Eds.), *Community mental health in India* (pp. 521–542). Jaypee Brothers Medical Publishers.

Chuenpagdee, R. (2011). Interactive governance for marine conservation: An illustration. *Bulletin of Marine Science*, *87*(1), 197–211. https://doi.org/10.5343/bms.2010.1061

Chuenpagdee, R., & Jentoft, S. (Eds.). (2019). *Transdisciplinarity for small-scale fisheries governance: Analysis and practice*. Springer International Publishing. https://doi.org/10.1007/978-3-319-94938-3

Çınar, M. E., Balkıs, H., Albayrak, S., Dağlı, E., & Karhan, S. Ü. (2009). Distribution of polychaete species (Annelida: Polychaeta) on the polluted soft substrate of the Golden Horn estuary (sea of Marmara), with special emphasis on alien species. *Cahiers de Biologie Marine*, *50*, 11–17.

Cochrane, K. L., Bahri, T., Dabbadie, L., Fernandez-Reguera, D., Kalioski, D. C., Ma, X., & Vannuccin, S. (2023). The impact of climate change on coastal fisheries and aquaculture. *Reference Module in Earth System and Environment*, *30*(3), 45–65.

Cochrane, K. L., & Garcia, S. M. (Eds.). (2009). *A fishery manager's guidebook*. John Wiley & Sons.

Colls, A., Ash, N., & Ikkala, N. (2009). *Ecosystem -based adaptation: A natural response to climate change*. Gland, Switzer-land, IUCN. available at: portals.iucn.org/ library/sites/library/files/documents/2009-049.pdf

Das, S. K., & Thatoi, H. (2020). Mangrove plant mediated green synthesis of nanoparticles and their pharmaceutical applications: An overview. In J. K. Patra, R. R. Mishra, & H. Thatoi (Eds.), *Biotechnological utilization of mangrove resources* (pp. 355–369). Academic Press. https://doi.org/10.1016/B978-0-12-819532-1.00016-0

Dasgupta, S., Hamilton, K., Pagiola, S., & Wheeler, D. (2008). Environmental economics at the World Bank. *Review of Environmental Economics and Policy*, *2*(1), 4–25. https://doi.org/10.1093/reep/rem025

Dasgupta, S., Sobhan, I., & Wheeler, D. (2019). The impact of climate change and aquatic salinization on mangrove species in the Bangladesh Sundarbans. *Ambio*, *46*, 680–694. https://doi.org/10.1007/s13280-017-0911-0

Dasgupta, S., Wheeler, D., Sobhan, M. I., Bandyopadhyay, S., Nishat, A., & Paul, T. (2020). *Coping with the vulnerability of the Sundarban in a changing climate: Lessons from multidisciplinary studies*. World Bank.

de la Torre-Castro, M., Fröcklin, S., Börjesson, S., Okupnik, J., & Jiddawi, N. S. (2017). Gender analysis for better coastal management – Increasing our understanding of social-ecological seascapes. *Marine Policy*, *83*, 62–74. https://doi.org/10.1016/j.marpol.2017.05.015

Donato, D. C., Kauffman, J. B., Murdiyarso, D., Kurnianto, S., Stidham, M., & Kanninen, M. (2011). Mangroves among the most carbon-rich forests in the tropics. *Nature Geoscience*, *4*, 293–297. https://doi.org/10.1038/ngeo1123

Dubey, S. K., Trivedi, R. K., Chand, B. K., Mandal, B., & Rout, S. K. (2017). Farmers' perceptions of climate change, impacts on freshwater aquaculture and adaptation strategies in climatic change hotspots: A case of the Indian Sundarban delta. *Environmental Development*, *21*, 38–51. https://doi.org/10.1016/j.envdev.2016.12.002

Dugan, P. (2005). Fish out of water: Competing for water for Africa's freshwater ecosystems. In M. L. Thieme, R. A. Abell, & D. M. Olson (Eds.), *Freshwater ecoregions of Africa: A conservation assessment* (pp. 128–132). Island Press.

Duke, N. C. (2017). Mangrove floristics and biogeography revisited: Further deductions from biodiversity hot spots, ancestral discontinuities, and common evolutionary processes. In V. H. Rivera-Monroy, S. Y. Lee, E. Kristensen, & R. R. Twilley (Eds.), *Mangrove ecosystems: A global biogeographic perspective* (pp. 17–53). Springer International Publishing. https://doi.org/10.1007/978-3-319-62206-4_2

Dutta, S., Maiti, S., Garai, S., Ghosh, S., Ghosh, T., & Hazra, S. (2020). Analyzing adaptation strategies to climate change followed by the farming community of the Indian Sunderbans using analytical hierarchy process. *Journal of Coastal Conservation*, *24*, 61. https://doi.org/10.1007/s11852-020-00776-2

Dynesius, M., & Nilsson, C. (1994). Fragmentation and flow regulation of river systems in the northern third of the world. *Science*, *266*, 753–762.

Ellison, J. C. (2019). Vulnerability assessment of mangroves to climate change and sea-level rise impacts. *Wetlands Ecology and Management*, *23*, 115–137. https://doi.org/10.1007/s11273-014-9397-8

Environmental Justice Foundation. (2004). *Farming the sea, costing the earth* (Report). Environmental Justice Foundation. https://ejfoundation.org/resources/downloads/Farming-Sea-Costing-Earth-ok.pdf

Fabinyi, M., & Barclay, K. (2022). Fishing livelihoods and wellbeing. In *Asia-Pacific fishing livelihoods* (pp. 91–109). Springer International Publishing. https://doi.org/10.1007/978-3-030-79591-7_5

Fagan, W. F. (2002). Connectivity, fragmentation, and extinction risk in dendritic metapopulations. *Ecology*, *83*(12), 3243–3249. https://doi.org/10.1890/0012-9658(2002)083[3243:CFAERI]2.0.CO;2

Fanzo, J. (2021). A path to sustainable food systems. *Current History*, *120*(829), 313–319.

FAO. (2019). *FAO yearbook. Fishery and aquaculture statistics 2016/FAO annuaire. Statistiques des pêches et de l'aquaculture 2017/ FAO anuario*. Estadísticas de Pesca y acuicultura 2017. Rome/Roma. https://www.fao.org/fishery/en/publication/256603

FAO. (2022). *The state of world fisheries and aquaculture 2022. Towards blue transformation*. Rome, Itali: Food and Agriculture 'organization. https://doi.org/10.4060/cc0461en

FAO and ILO. (2011). *Fao-Ilo good practice guide for addressing child labour in fisheries and aquaculture: Policy and practice* (Preliminary version, December 2011). Food and Agriculture Organization of the United Nations & International Labour Organization. https://www.fao.org/fishery/docs/DOCUMENT/child_labour_FAO-ILO/child_labour_FAO-ILO.pdf

Faridah-Hanum, I., Latiff, A., Hakeem, K. R., & Ozturk, M. (2014). *Mangrove ecosystems of Asia*. Springer. https://doi.org/10.1007/978-1-4614-8582-7

Finkbeiner, E. M., & Basurto, X. (2015). Re-defining co-management to facilitate small-scale fisheries reform: An illustration from Northwest Mexico. *Marine Policy*, *51*, 433–441. https://doi.org/10.1016/j.marpol.2014.10.010

Frangoudes, K., Gerrard, S., & Kleiber, D. (2019). Situated transformations of women and gender relations in small-scale fisheries and communities in a globalized world. *Maritime Studies*, *18*, 241–248. https://doi.org/10.1007/s40152-019-00159-w

Frawley, T. H., Finkbeiner, E. M., & Crowder, L. B. (2019). Environmental and institutional degradation in the globalized economy: Lessons from small-scale fisheries in the Gulf of California. *Ecology and Society*, *24*(1), 7. https://doi.org/10.5751/ES-10693-240107

Frawley, T. H., González-Mon, B., Nenadovic, M., Gladstone, F., Nomura, K., Zepeda-Domínguez, J. A., et al. (2024). Self-governance mediates small-scale fishing strategies, vulnerability and adaptive response. *Global Environmental Change*, *84*, 102805. https://doi.org/10.1016/j.gloenvcha.2024.102805

Funari, E., Manganelli, M., & Sinisi, L. (2019). Impact of climate change on waterborne diseases. *Annali dell'Istituto Superiore di Sanità*, *48*(4), 473–487. https://doi.org/10.4415/ANN_12_04_13

Füssel, H. M., & Klein, R. J. T. (2019). Climate change vulnerability assessments: An evolution of conceptual thinking. *Climatic Change*, *75*, 301–329. https://doi.org/10.1007/s10584-006-0329-3

Galappaththi, E. K., Susarla, V. B., Loutet, S. J., Ichien, S. T., Hyman, A. A., & Ford, J. D. (2022). Climate change adaptation in fisheries. *Fish and Fisheries*, *23*(1), 4–21.

Galappaththi, M., Collins, A. M., Armitage, D., & Nayak, P. K. (2021). Linking social wellbeing and intersectionality to understand gender relations in dried fish value chains. *Maritime Studies*, *20*(1), 355–370. https://doi.org/10.1007/s40152-021-00234-1

Galat, D. L., & Zweimuller, I. (2001). Conserving large-river fishes: Is the highway analogy an appropriate Para digm? *Journal of North American Benthological Society*, *20*, 266–279.

Gee, J., & Bacher, K. (2017). Engendering statistics for fisheries and aquaculture. *Asian Fisheries Science*, *30*(Suppl.), 277–290.

Ghassen, H., François, L., Yunne-Jai, S., Laure, V., Tarek, H., Mohamed, S., Romdhane, H., & Frida, B. (2021). An end-to-end model to evaluate the sensitivity of ecosystem indicators to track fishing impacts. *Ecological Indicators*, *98*(2), 121–130. https://doi.org/10.1016/j.ecolind.2018.10.061

Ghosh, A. (2021). Climate change impact and comprehensive disaster management approach in Bangladesh: A review. In *Bangladesh I: Climate change impacts, mitigation, and adaptation in developing countries* (pp. 1–25).

Girot, P., Ehrhart, C., & Oglethorpe, J. (2012). *Integrating community and ecosystem based approaches in climate change adaptation responses* (19). Gland Switzerland: Briefing Paper, Ecosystem & Livelihoods Adaptation Networks (ELAN), a partnership between International Union for the Conservation of Nature (IUCN), WorldWildlife Fund (WWF), CARE International, and the International Institute of Environment and Development (IIED), IUCN.

Grindle, M. S., & Thomas, J. W. (1991). *Public choices and policy change: The political economy of reform in developing countries*. Johns Hopkins University Press.

Groeneveld, J., Müller, B., Buchmann, C. M., Dressler, G., Guo, C., Hase, N., . . . Schwarz, N. (2017). Theoretical foundations of human decision-making in agent-based land use models–a review. *Environmental Modelling & Software*, *87*, 39–48. https://doi.org/10.1016/j.envsoft.2016.10.008

Hagenlocher, M., Renaud, F. G., Haas, S., & Sebesvari, Z. (2018). Vulnerability and risk of deltaic social-ecological systems exposed to multiple hazards. *Science of the Total Environment*, *631*, 71–80.

Haider, L. J., Boonstra, W. J., Peterson, G. D., & Schlüter, M. (2018). Traps and sustainable development in rural areas: A review. *World Development*, *101*(C), 311–321. https://doi.org/10.1016/j.worlddev.2017.05.038

Hale, R., Bain, R., Goodbred, S., Jr., & Best, J. (2019). Observations and scaling of tidal mass transport across the lower Ganges-Brahmaputra delta plain: Implications for delta management and sustainability. *Earth Surface Dynamics*, *7*(1), 231–245. https://doi.org/10.5194/esurf-7-231-2019

Haque, K. N. H., Chowdhury, F. A., & Khatun, K. R. (2022). Participatory environmental governance and climate change adaptation: Mainstreaming of tidal river management in south-West Bangladesh. In H. Ha (Ed.), *Land and disaster management strategies in Asia* (pp. 189–208). Springer. https://doi.org/10.1007/978-81-322-1976-7_13

Harper, S., Adshade, M., Lam, V. W. Y., Pauly, D., & Sumaila, U. R. (2020). Valuing invisible catches: Estimating the global contribution by women to small-scale marine capture fisheries production. *PLoS One*, *15*(3), 55–75. https://doi.org/10.1371/journal.pone.0228912

Harper, S., Grubb, C., Stiles, M., & Sumaila, U. R. (2017). Contributions by women to fisheries economies: Insights from five maritime countries. *Coastal Management*, *45*(2), 91–106. https://doi.org/10.1080/08920753.2017.1278143

Hazra, R., & Ghosh, T. (2019). Agricultural productivity, household poverty and migration in the Indian Sundarban Delta. *Elementa: Science of the Anthropocene*, *6*(3), 1–13. https://doi.org/10.1525/elementa.196

He, B. J., Sharifi, A., Feng, C., Yang, J., Prasad, D., Jupesta, J., & Pignatta, G. (2022). Climate emergency, actions and environmental sustainability. In B.-J. He, D. Prasad, G. Pignatta, & J. Jupesta (Eds.), *Climate change and environmental sustainability* (1st ed., pp. 127–144). Springer Nature. https://doi.org/10.1007/978-981-16-0902-2_6

Helgeson, J., & O'Fallon, C. (2021). Resilience dividends and resilience windfalls: Narratives that tie disaster resilience co-benefits to long-term sustainability. *Sustainability*, *13*(8), 4554. https://doi.org/10.3390/su13084554

Helgeson, J., van der Linden, S., & Chabay, I. (2012). *The role of knowledge, learning and mental models in public perception of climate change related risks*. Wageningen Academic Publishers. https://doi.org/10.3920/978-90-8686-762-2_16

Hogan, F., Cadrin, S., & Haygood, A. (2013). Fishery management complexes: An impediment or aid to sustainable harvest? A discussion based on the northeast skate complex. *North American Journal of Fisheries Management*, *33*(2), 406–421.

Hondula, D. M., Balling, R. C., Vanos, J. K., & Georgescu, M. (2020). Rising temperatures, human health, and the role of adaptation. *Current Climate Change Reports*, *1*(3), 144–154. https://doi.org/10.1007/s40641-015-0016-4

Hoq, M. E. (2007). An analysis of fisheries exploitation and management practices in Sundarbans mangrove ecosystem, Bangladesh. *Ocean & Coastal Management*, *50*(5–6), 411–427. https://doi.org/10.1016/j.ocecoaman.2006.11.001

Hulme, D., Moore, K., & Shepherd, A. (2001). *Chronic poverty: Meanings and analytical frameworks* (CPRC Working Paper No. 2). Chronic Poverty Research Centre. https://doi.org/10.2139/ssrn.1754546

Huq, S., Roberts, E., & Fenton, A. (2013). Loss and damage. *Nature. Climate Change*, *3*, 947–949. https://doi.org/10.1038/nclimate2026

Hutton, C., Adger, W., Hanson, S., Rahman, M., Salehin, M., & Eds.). (2021). *Ecosystem services for well-being in deltas: Integrated assessment for policy analysis*. Springer. https://doi.org/10.1007/978-3-319-71093-8

Islam, M. M., Begum, P., Begum, A., & Herbeck, J. (2021). When hazards become disasters: Coastal fishing communities in Bangladesh. *Environmental Hazards*, *20*(5), 533–549.

Islam, M. N., van Amstel, A., Islam, M. N., Tamanna, S., van Amstel, A., & Noman, M. (2018). Climate change impacts from the global scale to the regional scale: Bangladesh. *Bangladesh I: Climate change impacts, mitigation, and adaptation in developing countries*, 1–39. Springer International Publishing.

Jabir, A. G., Hasan, M. J., & Anam, M. M. (2021). Correlation between temperature, sea level rise and land loss: An assessment along the Sundarbans coast. *Journal of King Saud University – Engineering Sciences*, *50*(5), 65–75. https://doi.org/10.1016/j.jksues.2021.04.006

Jager, H. I., Chandler, J. A., Lepla, K. B., et al. (2001). A theoretical study of river fragmentation by dams and its effects on white sturgeon populations. *Environmental Biology of Fishes*, *60*, 347–361. https://doi.org/10.1023/A:1011036127663

Jalan, J., & Ravallion, M. (1999). Are the poor less well insured? Evidence on vulnerability to income risk in rural China. *Journal of Development Economics*, *58*(1), 61–82. https://doi.org/10.1016/S0304-3878(98)00103-5

Jentoft, S., & Chuenpagdee, R. (2015). *Interactive governance for small-scale fisheries: Global reflections*. Springer. https://doi.org/10.1007/978-3-319-17034-3

Joa, B., Winkel, G., & Primmer, E. (2018). The unknown known—A review of local ecological knowledge in relation to forest biodiversity conservation. *Land Use Policy*, *79*, 520–530. https://doi.org/10.1016/j.landusepol.2018.09.001

Johnson, D. S. (2010). Institutional adaptation as a governability problem in fisheries: Patron-client relations in the Junagadh fishery, India. *Fish and Fisheries*, *11*, 264–277. https://doi.org/10.1111/j.1467-2979.2010.00376.x

Jonah, F. E., Adams, O., Aheto, D. W., Jonah, R. E., & Mensah, E. A. (2017). Coastal zone management challenges in Ghana: Issues associated with coastal sediment mining. *Journal of Coastal Conservation*, *21*(3), 343–353.

Joydeb, G. (2016). Gender specific vulnerability in climate change and possible sustainable livelihoods of coastal people: A case from Bangladesh. *Journal of Integrated Coastal Zone Management*, *16*(1), 79–88. https://doi.org/10.5894/rgci656

Karl, H. A. (1999). The coastal environment: Toward integrated coastal and marine sanctuary management. *Eos, Transactions American Geophysical Union*, *80*. https://doi.org/10.1029/99EO00137

Karmakar, S. (2021). Gender and climate change: The condition of women in the Indian Sundarbans. *Scholars Journal of Arts, Humanities and Social Sciences*, *9*(1), 1–5. https://doi.org/10.36347/sjahss.2021.v09i01.001

Kaur, M., & Sood, S. K. (2020). Hydro-meteorological hazards and role of ICT during 2010-2019: A scientometric analysis. *Earth Science Informatics*, *13*(4), 1201–1223. https://doi.org/10.1007/s12145-020-00495-0

Kelling, I., Bennett, N., Barclay, K., Jeffs, A., Pita, C., Krogh-Poulsen, B., & Wiese, K. (2026). Beyond growth: Reshaping fisheries for a wellbeing economy. *Marine Policy, 183*, 106898.

Kirchhof, A., & Perreault, J. (2019). *Nature and the iron curtain: Environmental policy and social movements in communist and capitalist countries, 1945–1990*. University of Pittsburgh Press. https://doi.org/10.2307/j.ctvdtpjs6

Klinsky, S., Roberts, T., Huq, S., Okereke, C., Newell, P., Dauvergne, P., & Keck, M. (2017). Why equity is fundamental in climate change policy research. *Global Environmental Change, 44*, 170–173. https://doi.org/10.1016/j.gloenvcha.2016.08.002

Koch, N., & Perreault, T. (2019). Resource nationalism. *Progress in Human Geography, 43*(4), 611–631. https://doi.org/10.1177/0309132518781497

Krishnan, R., Shrestha, A. B., Ren, G., Rajbhandari, R., Saeed, S., Sanjay, S., Syed, M. A., Vellore, R., Xu, Y., You, Q., & Ren, Y. (2019). Unravelling climate change in the Hindu Kush Himalaya: Rapid warming in the mountains and increasing extremes. In P. Wester, A. Mishra, A. Mukherji, & A. B. Shrestha (Eds.), *The Hindu Kush Himalaya assessment—Mountains, climate change, sustainability and people* (pp. 57–97). Cham: Springer. https://doi.org/10.1007/978-3-319-92288-1_3

Lam, D. P., Martín-López, B., Wiek, A., Bennett, E. M., Frantzeskaki, N., Horcea-Milcu, A. I., & Lang, D. J. (2020). Scaling the impact of sustainability initiatives: a typology of amplification processes. *Urban Transformations, 2*(1), 3.

Lam, V. W. Y., Cheung, W. W. L., Reygondeau, G., & Sumaila, U. R. (2016). Projected change in global fisheries revenues under climate change. *Scientific Reports, 6*(2), 32–55. https://doi.org/10.1038/srep32607

Laverack, G. (2024). Community capacity building. In *Encyclopedia of quality of life and well-being research* (pp. 1155–1158). Cham: Springer International Publishing.

Leal Filho, W. (2018). *Climate change impacts and adaptation strategies for coastal communities*. Springer. https://doi.org/10.1007/978-3-319-70703-7

Lei, X., Yong, C., Robert, B., Kisei, R., Tanaka, N., & Yping, R. (2021). Evaluating the impacts of fishing and migratory species in a temperate bay of China using the ecosystem model OSMOSE-JZB. *Fisheries Research, 243*(2), 150–165. https://doi.org/10.1016/j.fishres.2021.106072

Levy, D. (2008). Political contestation in global production networks. *Academy of Management Review, 33*(4), 943–963. https://doi.org/10.5465/AMR.2008.34422006

Liberati, D., Sconocchia, P., Ricci, A., Gigliotti, G., Tacconi, C., Grenni, P., Tariciotti, S., Caracciolo, A. B., Massacci, A., Lagomarsino, A., & De Angelis, P. (2018). Phytodehydration of confined sludge: A sustainable approach for the management of polluted ponds. *Journal of Soils and Sediments, 18*, 2292–2304. https://doi.org/10.1007/s11368-017-1760-y

Mallalieu, K. I. (2015). *Rejuvenating ACP smallscale fisheries using ICTs*. CTA.

Mandal, S., & Choudhury, B. U. (2015). Estimation and prediction of maximum daily precipitation at Sagar Island using best fit probability models. *Theoretical and Applied Climatology, 121*, 87–97. https://doi.org/10.1007/s00704-014-1212-1

Manson, S. M., Jordan, N. R., Nelson, K. C., & Brummel, R. F. (2020). Modeling the effect of social networks on adoption of multifunctional agriculture. *Environmental Modelling & Software, 75*, 388–401. https://doi.org/10.1016/j.envsoft.2014.09.015

Matsubara H., Sakai, H., & Iwata, A. (2001). A river metapop-ulation structure of a Japanese freshwatergoby, Odontobutis obscura , deduced fromallozyme genetic indices. *Environmental Biology of Fishes*, 61, 285–294 *(PDF) River fisheries: ecological basis for management and conservation* https://www.researchgate.net/publication/29453557_River_fisheries_ecological_basis_for_management_and_conservation [accessed Feb 22 2026].

Mayden, R. L., & Wood, R. M. (1995). Systematics, species concepts, and the evolutionarily significant unit in biodiversity and conservation biology. *American Fisheries Society*, *17*, 58–113.

McClanahan, T., Allison, E. H., & Cinner, J. E. (2015). Managing fisheries for human and food security. *Fish and Fisheries*, *16*(1), 78–103. https://doi.org/10.1111/faf.12045

Mercer, J., Kelman, I., Alfthan, B., & Kurvits, T. (2012). Ecosystem-based adaptation to climate change in Caribbean Small Island developing states: Integrating local and external knowledge. *Sustainability*, *4*(8), 1908–1932. https://doi.org/10.3390/su4081908

Mitra, A., Abdel-Gawad, F. K., Bassem, S., Barua, P., Assisi, L., Parisi, C., Temraz, T. A., Vangone, R., Kajbaf, K., Kumar, V., et al. (2023). Climate change and reproductive biocomplexity in fishes: Innovative management approaches towards sustainability of fisheries and aquaculture. *Water*, *15*, 725. https://doi.org/10.3390/w15040725

Mitra, A., & Barua, P. (2022). Coping practices of women fisherfolk in responses to climate change at UNESCO declared world heritage site of sundarbans. In U. Chatterjee, A. O. Akanwa, S. Kumar, S. K. Singh, & A. Dutta Roy (Eds.), *Ecological footprints of climate change*. Cham: Springer Climate. Springer. https://doi.org/10.1007/978-3-031-15501-7_28

Mollick, A. S., Roy, M., Khan, N. I., Islam, W., Sadath, N., & Nath, T. K. (2022). Assessing good governance in protected areas (PA) co-management: A case study of the Sundarbans mangrove forests of Bangladesh. *Journal of Sustainable Forestry*, *41*(3–5), 277–301. https://doi.org/10.1080/10549811.2021.1923531

Mondal, M. A. I., Choudhury, A. H., Siddiqui, A. A. M., Rashed-Un-Nabi, M., & Kader, M. A. (2023). Bio-socio-economic analysis of ESBN fishery of Kattoli coastal area of Chattogram, Bangladesh. *Asian Journal of Medical and Biological Research*, *9*(4), 114–125.

Mondal, S., & Chakraborty, S. (2022). Impact of climate change on health, migration and child protection in the Sundarbans region in India: Making interventions in health, migration, and child protection in Sundarbans more climate responsive. *South Asia Consortium for Interdisciplinary Water Resources Studies*, *1*, 52–80.

Moosavi, S. (2017). Ecological coastal protection: Pathways to living shorelines. *Procedia Engineering*, *196*, 930–938. https://doi.org/10.1016/j.proeng.2017.08.027

Mortuza, A. (2018). Trends in climate change and some of its determinants in Bangladesh. *Research and Science Today*, *15*(1), 66–72.

Mosepele, K., Kolding, J., & Bokhutlo, T. (2017). Fish community dynamics in an inland floodplain system of the Okavango Delta. *Botswana. Ecohydrology & Hydrobiology*, *17*(2), 89–102. https://doi.org/10.1016/j.ecohyd.2017.01.005

Mukherjee, S. (2018). *Indian suffragettes: Female identities and transnational networks*. Oxford University Press.

Murali, K., & Sundar, V. (2021). Reassessment of tidal energy potential in India and a decision-making tool for tidal energy technology selection. *International Journal of Ocean and Climate System*, *8*(2), 85–97. https://doi.org/10.1177/1759313117694629

Mutombo, K., & Ölçer, A. (2016). Towards port infrastructure: A global port climate risk analysis. *WMU Journal of Maritime Affairs*, *16*, 161. https://doi.org/10.1007/s13437-016-0113-9

Mutombo, K., & Ölçer, A. (2017). Towards port infrastructure adaptation: A global port climate risk analysis. *WMU Journal of Maritime Affairs*, *16*(2), 161–173.

Myers, T. A., Maibach, E. W., Roser-Renouf, C., Akerlof, K., & Leiserowitz, A. A. (2018). The relationship between personal experience and belief in the reality of global warming. *Nature Climate Change*, 1170–1176. https://doi.org/10.1038/nclimate1754

Nakamura, J., Chuenpagdee, R., & El Halimi, M. (2021). Unpacking legal and policy frameworks: A step ahead for implementing the small-scale fisheries guidelines. *Marine Policy*, *129*, 104568.

Nalau, J., Becken, S., & Mackey, B. (2018). Ecosystem-based adaptation: A review of the constraints. *Environmental Science & Policy*, *89*, 357–364. https://doi.org/10.1016/j.envsci.2018.08.014

Nayak, S., & Takemi, T. (2022). Assessing the impact of climate change on temperature and precipitation over India. In T. Sumi, S. A. Kantoush, & M. Saber (Eds.), *Wadi flash floods* (1st ed., pp. 121–142). Springer Singapore. https://doi.org/10.1007/978-981-16-2904-4_8

Neogi, S. B., Dey, M., Kabir, S. L., Masum, S. J. H., Kopprio, G., Yamasaki, S., & Lara, R. (2016). Sundarban mangroves: Diversity, ecosystem services and climate change impacts. *Asian Journal of Medical and Biological Research*, *2*(4), 488–507. https://doi.org/10.3329/ajmbr.v2i4.30988

Newell, B., Crumley, C. L., Hassan, N., Lambin, E. F., Pahl-Wostl, C., Underdal, A., & Wasson, R. (2019). A conceptual template for integrative human–environment research. *Global Environmental Change*, *15*, 299–307. https://doi.org/10.1016/j.gloenvcha.2005.06.003

Noor Islam, S., Gnauck, A., Voigt, H.-J., & Eslamian, S. (2014). Hydrological changes in mangrove ecosystems, in handbook of engineering hydrology. In S. Eslamian & Taylor, & Francis (Eds.), *Modeling, climate changes and variability* (Vol. 2, pp. 353–373). USA: CRC Group.

O'Sullivan, D., Evans, T., Manson, S., Metcalf, S., Ligmann-Zielinska, A., & Bone, C. (2019). Strategic directions for agent-based modeling: Avoiding the YAAWN syndrome. *Journal of Land Use Science*, *11*(2), 177–187. https://doi.org/10.1080/1747423X.2015.1030463

Oberdorff, T., Guégan, J. F., & Hugueny, B. (1995). Global scale patterns in freshwater fish species diversity. *Ecography*, *18*, 345–352. https://doi.org/10.1111/j.1600-0587.1995.tb00137.x

Oliver, S. L., Corburn, J., & Ribeiro, H. (2019). Challenges regarding water quality of eutrophic reservoirs in urban landscapes: A mapping literature review. *International Journal of Environmental Research and Public Health*, *16*, 40. https://doi.org/10.3390/ijerph16010040

Omar, S., & Chhachhar, A. R. (2012). A review on the roles of ICT tools towards the development of fishermen. *Journal of Basic and Applied Scientific Research*, *2*, 9905–9911.

Orensanz, J. M. (Lobo), Parma, A. M., Jerez, G., Barahona, N., Montecinos, M., & Elias, I. (2005). What are the key elements for the sustainability of "S-fisheries"? Insights from South America. *Bulletin of Marine Science*, *76*(2), 527–556. https://doi.org/10.5343/bms.2005.0503

Paniagua, P., & Rayamajhee, V. (2024). A polycentric approach for pandemic governance: Nested externalities and co-production challenges. *Journal of Institutional Economics*, *18*, 537–552. https://doi.org/10.1017/S1744137421000795

Parsaulian, B., Irianto, A., & Aimon, H. (2024). Re-thinking Indonesian fisheries policy: Empowerment of a hidden asset for sustainable fisheries in west Pasaman District, West Sumatra, Indonesia. *International Journal of Environmental Impacts*, *7*(1), 1–16.

Paszkowski, A., Goodbred, S., Jr., Borgomeo, E., Khan, M. S. A., & Hall, J. W. (2021). Geomorphic change in the Ganges-Brahmaputra-Meghna delta. *Nature Reviews Earth & Environment*, *2*, 480–491. https://doi.org/10.1038/s43017-021-00213-4

Payne, H. E., & Castilla, J. C. (1994). Socio-biological assessment of common property resource management: Small-scale fishing unions in Central Chile. *Out of the Shell*, *4*, 10–14.

Pearce, C. R., Coe, A. L., & Cohen, A. S. (2010). Seawater redox variations during the deposition of the Kimmeridge clay formation, United Kingdom (upper Jurassic): Evidence from molybdenum isotopes and trace metal ratios. *Paleoceanography*, *25*, PA4213. http://doi.org/10.1029/2010PA001963

Pearce, D. (1989). Sustainable development: An economic perspective, International Institute for Environment and Development, UK. https://www.iied.org/8042iied

Perez, C., Nicklin, C., Dangles, O., Vanek, S., Sherwood, S., Halloy, S., Garrett, K. A., & Forbes, G. (2010). Climate change in the high Andes: Implications and adaptation strategies for small-scale farmers. *The International Journal of Environmental, Cultural, Economic and Social Sustainability*, *6*(5), 71–88. https://doi.org/10.18848/1832-2077/CGP/v06i05/54835

Pitchaikani, J. S. (2020). Importance of ICZM practices in Sundarbans estuary. *Indian Journal of Geo-Marine Sciences*, *49*(3), 352–356. http://nopr.niscpr.res.in/handle/123456789/54432

Pomeroy, R. S. (2012). Managing overcapacity in small-scale fisheries in Southeast Asia. *Marine Policy*, *36*(2), 520–527. https://doi.org/10.1016/j.marpol.2011.10.002

Purcell, S. W., Mercier, A., Conand, C., Hamel, J.-F., Toral-Granda, M. V., Lovatelli, A., & Uthicke, S. (2013). Sea cucumber fisheries: Global analysis of stocks, management measures and drivers of overfishing. *Fish and Fisheries*, *14*(1), 34–59. https://doi.org/10.1111/j.1467-2979.2011.00443.x

Raha, A. K., Zaman, S., Sengupta, K., Bhattacharyya, S. B., Raha, S., Banerjee, K., & Mitra, A. (2013). Climate change and sustainable livelihood programme: A case study from Indian Sundarbans. *Journal of Ecology*, *107*(335348), 64.

Rahman, M. M. (2020). Impact of increased salinity on the plant community of the Sundarbans mangrove of Bangladesh. *Community Ecology*, *21*(3), 273–284. https://doi.org/10.1007/s42974-020-00028-1

Rahman, M. M. (2022). Effectiveness of the coastal and marine conservation initiatives in Bangladesh: Analyzing the drawbacks of the legal, policy, and institutional framework. *Journal of Indian Ocean*, *18*, 149–172. https://doi.org/10.1080/19480881.2022.2111050

Rahman, M. M., Jiang, Y., & Irvine, K. (2018). Assessing wetland services for improved development decision-making: A case study of mangroves in coastal Bangladesh. *Wetlands Ecology and Management*, *26*, 563–580. https://doi.org/10.1007/s11273-018-9592-0

Rakib, M. R., Islam, M. N., Parvin, H., & van Amstel, A. (2018). Climate change impacts and other sustainable development goals. *Frontiers in Environmental Science*, *511*. https://doi.org/10.3389/fenvs.2018.00147

Rehman, S., Mohamed, R., & Ayoup, H. (2019). The mediating role of organizational capabilities between organizational performance and its determinants. *Journal of Global Entrepreneurship Research*, *9*, 30. https://doi.org/10.1186/s40497-019-0155-5

Roy, I., Raquib, T., & Sarker, A. K. (2017). Contribution of NGOs for socio-economic development in Bangladesh. *Science Journal of Business and Management*, *5*, 1–8. https://doi.org/10.11648/j.sjbm.20170501.11

Roy, S., Kumar, V., Manna, R. K., & Suresh, V. R. (2022). Sundarbans mangrove deltaic system—An overview of its biodiversity with special reference to fish diversity. *Journal of Applied Natural Sciences*, *8*(2), 1090–1099.

Ruben, R. (2005, June 26–29). *Smallholder farming in less-favoured areas: Options for pro-poor and sustainable livelihoods*. Paper presented at the ODI–IFPRI–Imperial College Research Workshop, London. http://www.ifpri.org/events/seminars/2005/smallfarms/ruben.pdf

Salehin, M., Chowdhury, M. M. A., Clarke, D., Mondal, S., Nowreen, S., Jahiruddin, M., & Haque, A. (2018). Mechanisms and drivers of soil salinity in coastal Bangladesh. In R. J. Nicholls, C. W. Hutton, W. N. Adger, S. E. Hanson, M. M. Rahman, & M. Salehin (Eds.), *Ecosystem services for well-being in deltas* (pp. 333–347). Palgrave Macmillan. https://doi.org/10.1007/978-3-319-71093-8_18

Sammonds, P., Shamsudduha, M., & Ahmed, B. (2021). Climate change driven disaster risks in the Southwestern coastal region of Bangladesh (SWCRB). *SN Social Sciences*, *2*(7), 102–122. https://doi.org/10.1007/s43545-022-00404-w

Sánchez-Triana, E., Ortolano, L., & Paul, T. (2018). Managing water-related risks in the West Bengal Sundarbans: Policy alternatives and institutions. *International Journal of Water Resources Development*, *34*(1), 78–96. https://doi.org/10.1080/07900627.2016.1202099

Sannigrahi, S., Zhang, Q., Joshi, P. K., Sutton, P. C., Keesstra, S., Roy, P. S., Pilla, F., Basu, B., Wang, Y., Jha, S., Paul, S. K., & Somnath, S. (2020). Examining effects of climate change and land use dynamic on biophysical and economic values of ecosystem services of a natural reserve region. *Journal of Cleaner Production*, *257*(92), 55–65. https://doi.org/10.1016/j.jclepro.2020.120424

Sarkar, R., Alam, A., & Siddiqui, A. R. (Eds.). (2023). *Agriculture and climatic issues in South Asia: Geospatial applications*. CRC Press.

Sarkar, S., & Padaria, R. N. (2010). Farmers' awareness and risk perception about climate change in coastal ecosystem of West Bengal. *Indian Research Journal of Extension Education*, *10*(2), 1–7.

Sarker, S. K., Matthiopoulos, J., Mitchell, S. N., Ahmed, Z. U., Mamun, M. B., & Reeve, R. (2020). Salinity reduces site quality and mangrove forest functions. From monitoring to understanding. *Biology and Conservation*, *236*(1), 79–91. https://doi.org/10.1016/j.biocon.2019.05.010

Scheelbeek, P. F. D., Khan, A. E., Mojumder, S., Elliott, P., & Vineis, P. (2016). Drinking water sodium and elevated blood pressure of healthy pregnant women in salinity-affected coastal areas. *Hypertension*, *68*(2), 464–470. https://doi.org/10.1161/HYPERTENSIONAHA.116.07743

Schlueter, M., McAllister, R., Arlinghaus, R., Bunnefeld, N., Eisenack, K., Hoelker, F., Milner-Gulland, E., Müller, B., Nicholson, E., Quaas, M., & Stöven, M. (2018). New horizons for managing the environment: A review of coupled social-ecological systems modeling. *Natural Resource Modeling*, *25*(1), 219–272. https://doi.org/10.1111/j.1939-7445.2011.00108.x

Schneider, B. R., & Heredia, B. (Eds.). (2003). *Reinventing leviathan: The politics of administrative reform in developing countries*. North-South Center Press.

Shankar, Y. S., Kumar, A., & Mohan, D. (2021). Climate change and water resources: Emerging challenges, vulnerability and adaptation in Indian scenario. In R. Jha, V. P. Singh, V. Singh, L. B. Roy, & R. Thendiyath (Eds.), *Climate change impacts on water resources* (1st ed., pp. 365–376). Springer Nature. https://doi.org/10.1007/978-981-16-5739-9_16

Shi, J., Visschers, V. H. M., & Siegrist, M. (2019). Public perception of climate change: The importance of knowledge and cultural worldviews. *Risk Analysis*, *35*(12), 2183–2201. https://doi.org/10.1111/risa.12406

Siam, H. S., Hasan, M. R., & Sultana, T. (2020). Socio-economic status of fisher community at NijhumDwip. *Asian Journal of Medical and Biological Research*, *6*(1), 50–70. https://doi.org/10.3329/ajmbr.v6i1.46479

Smith, A. C., Tasnim, T., Irfanullah, H. M., Turner, B., Chausson, A., & Seddon, N. (2021). Nature-based solutions in Bangladesh: Evidence of effectiveness for addressing climate change and social welfare. *Global Social Welfare*, *8*, 93–105. https://doi.org/10.1007/s40609-020-00192-2

So, N., Van Houdt, J. K., & Volckaert, F. A. (2006). Genetic diversity and population history of the migratory catfishes Pangasianodon hypophthalmus and Pangasius bocourti in the Cambodian Mekong River. *Fisheries Science*, *72*(3), 469–476.

Soga, M., & Gaston, K. J. (2020). The ecology of human–nature interactions. *Proceedings of the Royal Society B: Biological Sciences*, *28*(7), 300–320. https://doi.org/10.1098/rspb.2019.1882

Spijker, H. M., Stackpole, E. E., Almeida, S., Katsara, O., Liu, B., Shen, K., Schneider, R. J., Gao, F. B., & Richter, J. D. (2022). Ribosome profiling reveals novel regulation of C9ORF72 GGGGCC repeat-containing RNA translation. *RNA*, *28*, 123–138.

Stotz, W. (1997). Las áreas de manejo en la ley de Pesca y Acuicultura: Primeras experiencias evaluación de la utilidad de esta herramienta Para el recurso loco. *Estudios Oceanológicos*, *16*, 67–86.

Sugunan, V., Welcomme, R., Béné, C., Brummett, R., & Beveridge, M. (2007). Inland fisheries, aquaculture and water productivity. In D. Molden (Ed.), *Water for food, water for life—Comprehensive assessment for agricultural water* (pp. 459–484). Earthscan.

Sumaila, S. G., Alder, J., & Chuenpagdee, R. (2000). Addressing ecosystem effects of fishing using marine protected areas. *ICES Journal of Marine Science*, *57*, 752–760.

Sumaila, U. R., Alder, J., & Keith, H. (2006). Global scope and economics of illegal fishing. *Marine Policy*, *30*(6), 696–703. https://doi.org/10.1016/j.marpol.2005.11.001

Sumaila, U. R., & Walters, C. (2005). Intergenerational discounting: A new intuitive approach. *Ecological Economics*, *52*(2), 135–142. https://doi.org/10.1016/j.ecolecon.2003.11.012

Sunny, A. R., Sazzad, S. A., Prodhan, S. H., Ashrafuzzaman, M., Datta, G. C., Sarker, A. K., Rahman, M., & Mithun, M. H. (2021). Assessing impacts of COVID-19 on aquatic food system and small-scale fisheries in Bangladesh. *Marine Policy*, *126*, 104422.

Sunny, B. K., & Prapti, D. R. (2022, July). A systematic study of climate change impact on fish and fisheries sector in Bangladesh and role of geospatial science in mitigation. In *IOP conference series: Earth and environmental science* (Vol. 1064, No. 1, p. 012036). IOP Publishing.

Szabo, S., Hossain, M. S., & Adger, W. N. (2020). Soil salinity, household wealth and food insecurity in tropical deltas: Evidence from south-west coast of Bangladesh. *Sustainability Science*, *11*, 411–421.

Talarico, L., Caniglia, R., Carosi, A., Lorenzoni, M., Greco, C., Padula, A., . . . Mucci, N. (2023). Population structure, genetic diversity and demographic patterns unveil massive Mediterranean brown trout manipulations in a protected area of the northern Apennines (Italy). *The European Zoological Journal*, *90*(1), 470–486. https://doi.org/10.1080/24750263.2023.2223222

Tesfahunegn, F., Gebreyesus, B., Kirubel, M., & Abadi Tekle, A. T. (2019). Farmers perception on causes, indicators and determinants of climate change in northern Ethiopia: Implication for developing adaptation strategies. *Applied Geography*, *73*(2), 45–65.

Tesfahunegn, G. B., Ayuk, E. T., & Adiku, S. G. K. (2021). Farmers' perception on soil erosion in Ghana: Implication for developing sustainable soil management strategy. *PLoS One*, *16*(3), e0242444.

Tesfahunegn, G. B., & Gebru, T. A. (2021). Climate change effects on agricultural production: insights for adaptation strategy from the context of smallholder farmers in Dura catchment, northern Ethiopia. *GeoJournal*, *86*(1), 417–430.

Thorpe, A. (2004). *Mainstreaming fisheries into national development and poverty reduction strategies: Current situation and opportunities* (FAO fisheries circular no. 997). Food and Agriculture Organization of the United Nations. https://www.fao.org/3/y5734e/y5734e00.html

Titumir, R. A. M., Afrin, T., & Islam, M. S. (2020). Traditional knowledge, institutions and human sociality in sustainable use and conservation of biodiversity of the Sundarbans of Bangladesh. In O. Saito, S. Subramanian, S. Hashimoto, & K. Takeuchi (Eds.), *Managing socio-ecological production landscapes and seascapes for sustainable communities in Asia*. Springer Singapore: Science for sustainable societies.

Titumir, R. A. M., & Shah Paran, M. (2022). Ecosystem services and well-being in the Sundarbans of Bangladesh: A multiple evidence base trajectory. In *Assessing, Mapping and Modelling of Mangrove Ecosystem Services in the Asia-Pacific Region* (pp. 263–281). Singapore: Springer Nature Singapore.

Toimil, A., Losada, I. J., Nicholls, R. J., Dalrymple, R. A., & Stive, J. F. (2019). Addressing the challenges of climate change risks and adaptation in coastal areas: A review. *Coastal Engineering*, *56*(4), 60–80.

Toufique, K. A. (1997). Institutions and externalities in the inland fisheries of Bangladesh. *Land Economics*, *73*(3), 408–421.

Uddin, N., Enoch, S., Harihar, A., Pickles, R. S. A., & Hughes, A. C. (2023). Tigers at a crossroads: Shedding light on the role of Bangladesh in the illegal trade of this iconic big cat. *Conservation Science and Practice*, *5*(7), 90–110.

UNODRR. (2021). *UNODC approach to crimes in the fisheries sector*. United Nations Office on Drugs and Crime (UNODC). https://www.unodc.org/documents/Wildlife/UNODC_Approach_to_Crimes_in_the_Fisheries_Sector.pdf

Vachette, A. (2017). Integrating disaster risk reduction and climate change adaptation in Vanuatu: The art and practice of building resilience to hazards. In *Climate change adaptation in pacific countries: Fostering resilience and improving the quality of life* (pp. 119–136). Cham: Springer International Publishing.

Vignola, R., Locatelli, B., Martinez, C., & Imbach, P. (2009). Ecosystem-based adaptation to climate change: What role for policy-makers, society and scientists? *Mitigation and Adaptation Strategies for Global Change*, *14*(8), 691–696. https://doi.org/10.1007/s11027-009-9193-6

Walters, D. (1997). Developing and implementing value-based strategy. *Management Decision*, *35*(10), 709–720. https://doi.org/10.1108/00251749710192020

Ward, P., Myers, R. A., & Blanchard, W. (2004). Fish lost at sea: The effect of soak time and timing on pelagic longline catches. *Fishery Bulletin*, *103*(1), 179–195.

Welcomme, R.L. (1985) River fisheries. Food and Agriculture Organization of the United Nations, FAO Fisheries Technical Paper 262.

Widmalm, S. (Ed.). (2021). *Routledge handbook of autocratization in South Asia* (1st ed.). Routledge. https://doi.org/10.4324/9781003042211

Winemiller, K. O. (2004). Floodplain river food webs: Generalizations and implications for fisheries management. In *Proceedings of the second international symposium on the management of large rivers for fisheries* (Vol. 2, pp. 285–309). FAO.

Wingard, J. D. (2000). Community transferable quotas: Internalizing externalities and minimizing social impacts of fisheries management. *Human Organization*, *59*(1), 48–57.

Yao, M., Zhang, S., Lu, Q., Chen, X., Zhang, Y., Kong, Y., & Zhao, J. (2022). Fishing for fish environmental DNA: Ecological applications, methodological considerations, surveying designs, and ways forward. *Molecular Ecology*, *31*(20), 5132–5164. https://doi.org/10.1111/mec.16659

Yawson, D. O., Adu, M. O., & Armah, F. A. (2020). Impacts of climate change and mitigation policies on malt barley supplies and associated virtual water flows in the UK. *Scientific Reports*, *10*, 376–386.

Younus, M. A. F., & Kabir, M. A. (2018). Climate change vulnerability assessment and adaptation of Bangladesh: Mechanisms, notions and solutions. *Sustainability*, *10*(11), 4286.

Zaman, S., De, U. K., Pramanick, P., Thakur, S., Amin, G., Fazli, P., & Mitra, A. (2014). Forecasting heavy metal level on the basis of acidification trend in the major estuarine system of Indian Sundarbans. *Journal of Energy, Environment and Carbon Credits*, *4*(2), 1–9.

Zimmermann, A., Benda, J., Webber, H., & Jafari, Y. (2018). *Trade, food security and climate change: Conceptual linkages and policy implications*. Rome: FAO.

6 The Future of Small-Scale Mangrove Fisheries: *A Global Discussion on Sustainability and Community Resilience*

6.1 INTRODUCTION

The magnitude of the global contribution of small-scale fisheries (SSFs) with respect to the total fisheries sector highlights the critical importance of this subsector to sustainable development. Asia benefits from the highest absolute contributions in most dimensions (catch volume and value, people employed, livelihoods supported, and women employed), producing four times as much SSF catch as Africa, which is the region with the second-highest SSF catch. In Africa, SSF is the most important relative to total fisheries, contributing 67% of the total catch for the area and 75% of the total contribution to nutrition. In contrast, Europe's SSF supports the lowest absolute number of livelihoods of all the regions (Basurto et al., 2025). While differences in SSF around the globe are acknowledged, some common characteristics can be summarized as (1) targeting a mix of species using multiple types of boats and fishing gear, making it difficult to evaluate the state of the resources and the fishing intensity exerted; (2) low levels of capital investment by fishers and the use of labor-intensive harvesting, processing, and distribution methods to exploit the fishery resources; (3) a wide range of landing sites used by fishers along the coast (often in small communities), making it difficult to effectively record catches and fishing effort; (4) seasonal use of fishing resources, and fishing income that is often complemented by other economic activities (with the frequent presence of part-time fishers in the fishery); (5) significant provision of protein and jobs through fishing as for people in coastal areas; (6) attraction of migrant people to the coast in search of income, with typically less investment needed to enter the fishery in terms of capital and skills, or the option of becoming "workers" in private companies; (7) limited power of fishers to influence the fish market, given their small-scale capital commitment, and consequently a greater dependence on middlemen for marketing and loans; and (8) lack of social mechanisms for health. The complexity of SSF systems, due to their high heterogeneity and uncertainty, along with their limited capacity for data collection and analysis, makes assessing such fisheries in the region challenging. A number of issues were identified in the literature as: (1) sampling programs that are limited in scope and inconsistent; (2) research often focuses on select fisheries,

DOI: 10.1201/9781003729990-6

overlooking integrated ecosystem approaches; (3) independent data collection is expensive and logistically complex to maintain; (4) catch and effort statistics are used for assessments despite the data being unreliable; (5) a shortage of interdisciplinary skills; (6) insufficient funding and capacity to collect independent fisheries data are significant limitations; and (7) economic data is difficult to access based on mistrust in the fishing community.

The issues and characteristics described above reflect the importance of SSFs to livelihoods and food security, especially in regions such as Southeast Asia, where these activities still underpin the livelihoods of coastal communities. Small-scale fish capture and production represent an important means of securing livelihood and food security for millions of people in Southeast Asia, as well as a way of life (Chuenpagdee, 2011). In many SSFs, increased pressure on fish stocks has coincided with falling incomes and jobs, making already poor households even more vulnerable (Allison et al., 2011). In this context, Chinua Achebe's insight—"The world is like a mask dancing. If you want to see it well, you do not stand in one place"—offers a powerful interpretative lens. This highlights the need for flexible, multi-perspective approaches to SSF governance and assessment. Understanding SSFs requires moving beyond static, single-sector analyzes toward adaptive, interdisciplinary, and context-sensitive frameworks that capture their dynamic ecological, economic, and social dimensions.

6.2 SOUTHEAST ASIAN SMALL-SCALE MANGROVE FISHERIES

6.2.1 Indonesia: Archipelagic Complexity and Decentralization

Indonesia's small-scale mangrove fisheries are practiced across 17,000 islands with complex ecological and socioeconomic conditions. The country has roughly 23% of the total global mangrove coverage, and also has the largest single country estate of mangroves (Giri et al., 2011). However, Indonesia has struggled with a significant loss of mangroves, especially from the expansion of aquaculture and palm oil (Richards & Friess, 2016). Nurdin et al. (2021) conducted comprehensive research revealing that developing sustainable SSF livelihoods in Indonesia requires addressing various interconnected trends and factors. Their study demonstrated that ecosystem degradation significantly influences the sustainability of fishing operations, with climate change acting as a primary driver of environmental stress in marine ecosystems. The information regarding altered species compositions, reduced fish abundances due to rising sea temperatures and ocean acidification, technological gaps in Indonesian SSFs, inadequate access to modern equipment and facilities, limited cold storage infrastructure, and financial exclusion of Indonesian fishers is supported by multiple studies referenced in the surrounding context. The key references include Nurdin et al. (2021) on community-based management and technology, Thorburn (2000) and Siry (2011) on governance and infrastructure, and Fauzi and Anna (2013) on market access and postharvest losses among Indonesian fishers. These works collectively highlight the technological and financial constraints facing the sector and their effect on productivity, resilience, and livelihoods.

Fishing communities in Indonesia are utilizing complex community-based management systems that integrate traditional ecological knowledge with modern conservation practices. These systems include locally managed marine protected areas (Sasi) and seasonal fishing restrictions that allow for stock recovery (Nurdin et al., 2021). Coastal communities in Indonesia have adopted diversification strategies, including aquaculture development, tourism services, and small-scale trading, to reduce dependence on capture fisheries while increasing the stability of household income stability. Indonesia has developed village-level fisheries management (Sasi) systems where traditional management authorities are formalized with recognition from the government (Thorburn, 2000). Thus, Indonesia has developed a more formalized community-based management approach than the Sundarbans. The governance structure presents complexity, with mangrove management involving 22 different laws administered by 18 institutions, in contrast to the informal traditional governance systems in the Sundarbans (Siry, 2011). Indonesian SSFs have been more modernized than those of the Sundarbans, involving more advanced boats, gear, and communication systems, while traditional technologies are still dominant due to poverty and lack of infrastructure. Fishing gear and crafts in the Sundarbans are mostly traditional, reflecting limited capital investment capacity (Hoq, 2007). Indonesian fishers have better market access and higher prices for their catch than the Sundarbans fishers, who often face monopsonistic buyers and limited transportation options due to geographic isolation and poor infrastructure (Fauzi & Anna, 2013).

6.2.2 Philippines: Community-Based Fisheries Management

In the Philippines, SSFs make important contributions to food security, sustainable livelihoods, and poverty alleviation. There are two million active fishers, who operate an estimated 476,000 fishing vessels. Approximately 60% of these vessels do not have engines and use sails or oars (FAO, 2020). More than 80% of the fishers in the Philippines can be characterized as small-scale fishers. The Philippines has lost approximately 70% of its original mangrove cover, primarily due to aquaculture development and urban expansion (Long & Giri, 2011). The case of shrimp farming in the Philippines illustrates the benefits and risks associated with mangrove conversion for aquaculture. Intensive shrimp farming resulted in substantial export revenues in the 1980s and 1990s, but also raised environmental issues such as water pollution, disease issues, and habitat loss (Primavera, 2006). The Philippine government has developed a comprehensive policy framework for the conservation of mangroves, which includes projects like the Mangrove and Beach Forest Development Project and has incorporated mangrove management into the National Greening Program (DENR-FMB, 2018). Concerning community-based forest management programs, there has been some success in areas that demonstrate how local participation can inform and bring about effective conservation (Walters et al., 2008).

The prominent focus areas in Philippine mangrove are climate change adaptation, which has gained prominence policy-wise in the past decade, and increased occurrence of typhoons has highlighted the protective value of mangrove systems (Zhang et al., 2012). This country has been the pioneer in disaster risk reduction with ecosystem-based management, offering lessons to those regions affected by typhoons

(Sudmeier-Rieux et al., 2021). The Philippines has developed significant innovations in community-based fisheries management, which is also fundamentally different from the Sundarbans approach. There has been a gradual evolution in fisheries management over the past decades from a focus on the sustainability of single species to ecosystem-based approaches (Pomeroy et al., 2010). The Community-Based Coastal Resource Management (CB-CRM) model in the Philippines provides more formal recognition of community management rights than does the Sundarbans model. Compared with the ambiguous legal status of many Sundarbans fishing practices, the Philippine Fisheries Code provides clearer legal recognition of small-scale fishers' preferential access rights. Municipal waters (up to 15 km from shore) are reserved for small-scale fishers, affording them a greater level of tenure security than that encountered by Sundarbans fishers. Philippine barangay (village) councils play more formalized roles in fisheries management than do the informal traditional leadership systems in the Sundarbans, although both systems strive for community participation. The Philippines has developed ecosystem-based fisheries management approaches for small-scale tropical marine fisheries with emerging models of governance (Pomeroy et al., 2010). Both regions experience typhoon/cyclone hazards; however, the Philippines has also established more systematic early warning and disaster management systems for fishing communities, compared to the ad hoc approaches to disasters common to the Sundarbans (Cruz et al., 2007). Fabinyi and Dressler (2024) provide an extensive examination of policy debates in the Philippines, focusing on approaches to combat illegal, unreported, and unregulated (IUU) fishing. Their finding suggested that IUU fishing represents one of the most significant threats to sustainable fisheries management, significantly impacting conservation efforts and the livelihoods of legitimate small-scale fishers. This study also documents how unregulated fishing activities deplete fish stocks faster than they can naturally replenish, creating long-term sustainability challenges. They also noted that there are attempts to revise spatial regulation related to access to fishing activities. They show the governance difficulty in relation to rapidly expanding commercial fishing and designating marine protected areas, which leads to conflicting claims for fishing access between small-scale and large-scale fishers. Small-scale fishers frequently lose access to their traditional fishing grounds because of other fishers and are displaced systematically. Atillo et al. (2024) performed in-depth research on the socioeconomic situation of SSFs in a region of the central Philippines, documenting ongoing poverty and economic circumstances in fishing communities. Specifically, they noted how community resilience and adaptive capacity are influenced by economic marginalization. Armada et al. (2016) provide a comprehensive analysis of the challenges in implementing management initiatives aimed at promoting the recovery of depleted fisheries in the Philippines. Their research reveals how poverty among fishery-dependent communities creates barriers to effective conservation, as immediate subsistence needs often conflict with long-term sustainability objectives. The World Bank (2023) outlined the Philippine Fisheries and Coastal Resiliency Project (FISH-CORE) for US$176 million, with the objective of enhancing fisheries management, increasing fisheries production value, and sustaining incomes in selected coastal communities. This initiative represents one of the largest investments in SSF resilience in Southeast Asia, targeting over one million beneficiaries through integrated approaches to sustainable fisheries development. A report published by the Australian Centre for International Agricultural Research ANNUAL Report 2024-25 examines how to create governance solutions cognizant of local contexts and issues in Indonesia and the Philippines, with the purpose of producing lessons for marine environmental governance in the broader region.

6.2.3 Thailand: Market-Oriented Development

SSFs in Thailand are well established and have long coexisted with large-scale, industrialized fisheries. Marine SSF occurs in all 22 coastal provinces of Thailand. These provinces are divided into five main zones: Southern Region (Andaman Sea & Gulf of Thailand); Eastern Region (Gulf of Thailand); Central Region (Upper Gulf); Western Region and Northern & Northeastern Regions. SSFs are defined in terms of vessel size as vessels that are less than 10 Gross Register Tonnage (GRT). According to the latest SSF catch survey statistics (2019) from the Department of Fisheries (DoF), the sector was estimated to produce 161,463 tons of marine fishery catch, which is only approximately 11% of the country's total production (DoF, 2020). This proportion is lower than the global estimate of approximately 25%. This difference may be due to the recently introduced definition of SSF in Thailand and to possible survey data collection inaccuracies. Nevertheless, the survey is currently the only available comprehensive report and will thus be considered the most reliable source of information for this study. Note that while inland SSFs are also important in Thailand, they are not included in this analysis. The majority of the SSF marine catch (approximately 33%) is composed of various fish species. Crab contributes to another 11%, while shrimp and squid each represent approximately 4%, and bivalves represent approximately 1%. The remaining 46% of the catch is composed of jellyfish caught by dip nets. The gill net is the most common gear used to target fish, crabs, and shrimp. Data on the number of SSF households and fishing boats are derived from the 2014 national statistics published by the DoF in 2016. Based on that report, 168,140 fishers (during the peak season) and 57,800 fishing households were included in Thailand (BFAR, 2016). Approximately half of the people employed in the sector are household members, who we assume are all involved in SSF. Of the 58,119 fishing boats, 2,639 (approximately 5%) are nonpowered, while the majority (73%) are outboard motor boats. These figures suggest that SSFs are prominent in Thailand, which corresponds with their long-established and well-recognized cultural and traditional importance. Being a tropical country, Thailand is home to productive ecosystems, such as mangroves, coral reefs, and seagrass. In general, small-scale fishers are aware of the significance of these habitats for their fishing livelihoods and use conservation measures in their fishing practices (Chuenpagdee et al., 2006, 2022; Thiammueang et al., 2012). Compared with the subsistence-plus-market orientation of Sundarbans fisheries, Thailand's small-scale mangrove fisheries have undergone a significant transformation toward market-oriented production. Thailand lost approximately 84% of its mangrove cover between 1961 and 2009, primarily due to the expansion of shrimp aquaculture, which was initially encouraged by government policies (Barbier & Cox, 2003). Thai fishers typically have better integration into national and international markets than do Sundarbans fishers, who remain primarily dependent on local and regional markets due to poor transportation and communication infrastructure (Kongkeo, 2001). Thailand has promoted motorization and gear modernization more extensively than other regions, resulting in higher productivity but also greater capital requirements and potential for overcapacity. This contrasts with the labor-intensive, low-capital approaches that are dominant in the Sundarbans (Poggie et al., 1995). Government extension services and fisher organizations in Thailand provide more technical and financial support than does the limited institutional support available to Sundarbans fishers, who rely primarily on traditional knowledge and informal networks (Lebel et al., 2006). SSFs in Thailand are experiencing significant stress from resource depletion, especially in the Gulf of Thailand, where successive large-scale fishing has caused fishery resources to decline over the past several decades. This depletion has a direct impact on traditional fishing communities that have been relying on them for generations. Fabinyi et al. (2022) provide a comprehensive analysis of coastal transitions across Southeast Asia, including Thailand, documenting how SSFs and livelihoods are becoming more diverse and commodified as maritime zone developments intensify. Research in ocean and coastal management reveals how traditional fishing communities are forced to adapt to rapidly changing coastal environments driven by development pressures. The Asia–Pacific Rural and Agricultural Credit Association (APRACA) and FAO, in collaboration with the Bank for Agriculture and Agricultural Cooperatives (BAAC), implemented a project to analyze and improve the access of small-scale fishers to financial services in Thailand.

6.2.4 Vietnam: Fragmented Fisheries and Adaptive Challenges

The fisheries in Vietnam are not easily classified into small-scale and commercial/large-scale, as they are in other countries in Southeast Asia. Rather, they are more commonly classified as nearshore or offshore in government documents. Armitage and Marschke (2013) described (1) the multidimensional and linked social, ecological, and economic challenges confronting lagoon resource users and government officials, including the possibility that important features of the ecological system have been significantly altered, and (2) the spatial and temporal variation in the lived experience and conditions facing lagoon resource users even in the context of one relatively bound physical system. Major problems in the Vietnamese fisheries sector and their impacts have been listed in Table 6.1

Adoption of an integrated fishery (fishing and aquaculture) and coastal systems perspective, clarifying the security of access rights to aquatic resources, and building institutional conditions for greater collaboration and learning among resource users and decision makers. Policies should allow fishing households to diversify their income sources rather than relying on fishing in a risky environment. Apart from the sociodemographic factors that affect adaptive livelihood strategies, positive traits can be cultivated, and barriers can be reduced through awareness and skill building.

The local fishing households' adaptive livelihood strategies might have unintended implications, increasing their susceptibility to climate change. For example, the use of harmful fishing gear, like small mesh fishing nets, to catch fish can cause negative consequences for the ecosystem. To preserve fishery resources and the

TABLE 6.1
Major Problems in the Vietnamese Fisheries Sector and Their Impacts

Issue	Description	Impacts
Violation of mesh size regulations	Most fishing gear currently used violates regulations (mesh size, cod end, purse seine, trammel nets, etc.).	Illegal fishing practices, resource depletion
High percentage of "trash" fish	Small mesh size results in 60–80% trash fish in shrimp trawls, 40–80% in fish trawls, 90% in fixed nets, and 90–93% in push nets.	Overfishing, ecosystem imbalance, wasted resources
Coral reef and sediment damage	Coral reefs and sediment rocks are degraded/damaged by improper fishing activities.	Loss of marine biodiversity and habitats
Harmful fishing gear (stow/push nets)	Stow and push nets are still in use, depleting juvenile fish and resulting in low income for fishers due to small-sized catches.	Depletion of resources, economic hardship for fishers
Harmful fishing techniques (explosives, chemicals)	The use of explosives and chemicals in fishing is still uncontrolled.	Environmental damage, long-term stock depletion
Increase in small fishing boats	The number of small boats is growing (~22,500 new entrants per year), and poor people rely on them; this limits sustainable development.	Unsustainable growth, socioeconomic challenges
Temporary closed areas are not enforced	Temporary closed areas have not yet been enforced as a management measure.	Ineffective management, continued overfishing
Large fishing boats in shallow waters	Boats (200–450 hp) operate in shallow waters (15–25 m), damaging resources due to a lack of clear legal regulations.	Damage to coastal ecosystems, unregulated exploitation
Oil sludge and waste dumping	Fishing villages dump oil sludge/waste into the sea or on beaches, affecting the quality of fish and exports.	Pollution, food safety concerns, and export restrictions
Weak survey data and statistics	Data and statistics are too weak to support ecosystem-based fishery management.	Weak foundation for sustainable fishery management
Implementation Issues with the Fisheries Law	Law exists, but enforcement and control are weak, limiting effectiveness.	Low effectiveness of fisheries governance
High demand and price for trash fish	Trash fish landings provide significant income despite ecological issues.	Continued reliance on unsustainable catch for income
Overcapitalization (excess effort/horsepower)	Increased horsepower and reduced catch per unit effort are standard in coastal/inshore/offshore fisheries.	Resource depletion and economic inefficiency
Lack of regulation on large vessels in shallow waters	No rules prevent large vessels from entering shallow-water fishing areas.	Excessive competition, overfishing in shallow waters

(Continued)

Table 6.1 (Continued)

Issue	Description	Impacts
Lag in boat/engine repair and administration	Repair/administrative systems are not keeping pace with offshore fisheries expansion.	Operational inefficiency, reduced offshore growth
Use of destructive fishing methods	Explosives and chemicals are still used, destroying coral reefs.	Environmental destruction, habitat loss
Limited access to credit for poor fishing households	Poor households lack collateral/access to loans, and cannot benefit from sectoral support	Financial exclusion of poor fishers, poverty persistence
Overexploitation of coastal aquatic resources	Aquatic resources are overexploited, and several species are threatened with commercial extinction.	Biodiversity loss, extinction threats

ability of marine ecosystems to support coastal communities, a fisheries management plan is urgently needed (Musinguzi et al., 2016; Smit & Wandel, 2006). Local governments can promote and facilitate local-level adaptation, but they require diverse strategies to extend their resources. Climate change and other socioeconomic and environmental stressors are likely to increase in the future (Solomon et al., 2007), so proper adaptation planning and effective management strategies are needed to support threatened communities and households and sustain marine and coastal ecosystems. More research is needed to predict the trend line of climate-related risk at the micro level and to understand the complexity and challenges of local-level adaptation in different contexts. This research will guide policies and management efforts to increase the resilience of affected households, communities, and natural ecosystems.

Recommended actions include improved fisheries statistics, resources for provincial fisheries staff, and a coordinated and integrated approach involving a mixed strategy of resource management, resource restoration, economic and community development, and new governance arrangements (Pomeroy et al., 2010). The literature on the impacts of climate risks and changes on SSF livelihoods in estuaries in coastal Vietnam is limited. The few studies (Ha & Thang, 2017; Leslie et al., 2018; Truong et al., 2011) that have investigated the difficulties faced by fisheries communities are not based on a general analysis of fisheries livelihoods (Hoang et al., 2020).

In contrast to the SSFs found in the Sundarbans, the fisheries sector in Vietnam faces more severe regulatory and enforcement challenges—for example, mesh sizes, destructive fishing practices, and the extraction of resources. The SSFs in the Sundarbans are oriented around traditional governance and community-based management, whereas in Vietnam, the fragmented system and weak fisheries law enforcement have led to chronic overfishing and environmental degradation. Although both regions contend with poverty and lack of access to credit for fishers, in contrast with the Sundarbans, Vietnam's reliance on small-scale aquaculture and expansive, unregulated growth of fishing vessels exacerbate their resource situation. Unlike the Sundarbans, the absence of an explicit SSF classification in Vietnam presents challenges for the targeted policy interventions and sustainable management of SSFs.

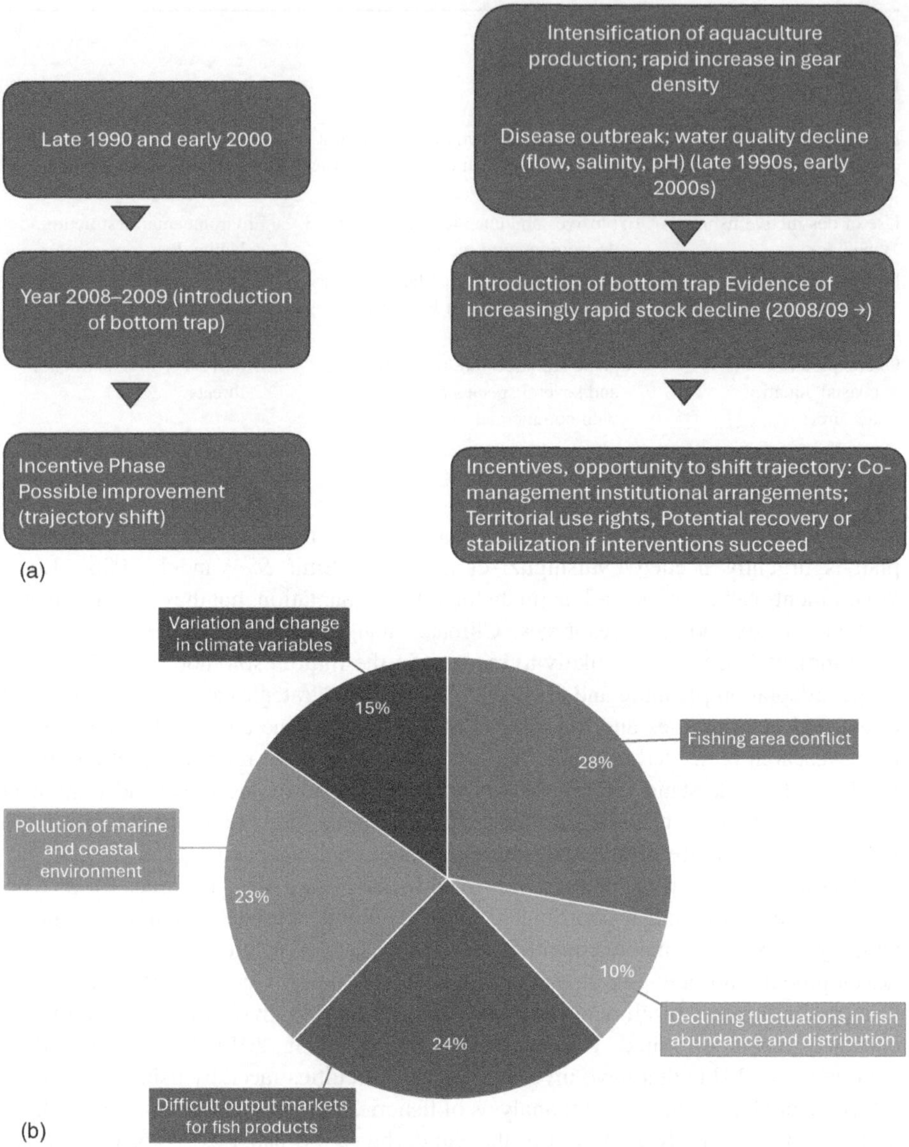

FIGURE 6.1 (a) Trajectories change in Vietnam Lagoon. (b) Livelihood stressors experienced by fishing households in recent years in Vietnam.
(Modified after Huynh et al. 2021a and 2021b)

6.2.5 Malaysia: Building Resilient Livelihoods through Social Capital

Small-scale fishing communities in the coastal areas of Malaysia face several challenges, including environmental change, encroachment of commercial vessels in nearshore fisheries zones, declining fish catch (Islam et al., 2014), and limited access to income and financial credit (Abu Samah et al., 2019). The fishing zone within five nautical miles from shore (zone A) has been reserved for artisanal fishers. The small-scale fishers along the coast usually have a few capital assets and resources, and their local economies are heavily reliant on fishing. They operate small fishing vessels which are powered by outboard engines and carry various kinds of fishing gear to target diverse fish species. The period of the monsoon, between November and February, is a time when fishing is prohibited and thus fishers are reckoning with an economic crisis (Islam, Noh, Sidique, et al., 2014; Islam et al., 2017). Fishers' income from fishing has been declining because of the overexploitation of resources and pollution (Islam et al., 2017). Islam et al. (2022) examined the contributions of social capital and various livelihood assets to the well-being of small-scale fisher households in Langkawi.

The study used structural equation modeling (SEM) and partial least squares (PLS) to examine the essential factors contributing to household well-being. Social capital, trust, human capital, and financial assets contribute significantly to household livelihoods and well-being. Natural capital and governance factors contributed moderately to household well-being. The key findings of the study suggest that social capital and trust factors play important roles in contributing to the livelihoods of small-scale fisher households in Langkawi. Although fishing is a crucial earning activity, income from tourism has increased. The government and NGOs should promote community engagement in social activities through targeted fisheries and tourism development programs (Rouf and Jensen 2001). Fishers' access to financial credit and physical assets is critically important for their livelihood security during crises. The establishment of a social safety net for the livelihood security of poor fishers should be prioritized. Creating alternative employment during the crisis would be a good option to secure their livelihoods. The critical policy implication of this study is that the key agencies of fisheries management, government, NGOs, and credit providers should encourage community participation in fisheries and tourism. This will enable local fishers to secure access to various assets and fisheries resources, thereby improving their well-being in Malaysia. Several local studies have identified several climate change symptoms. These symptoms are expected to worsen in the future and have been identified as affecting the natural environment. They negatively impact those who rely heavily on weather stability to conduct social and economic activities, including small-scale fishing.

Climate change has consistently been proven to reduce fishermen's productivity, endanger their lives, damage their physical infrastructure, expose them to health problems, and decrease their food supply. Although the situation is expected to worsen, several efforts can be made at the individual level to decrease and delay impacts, focusing on strengthening fishermen's adaptive ability. Social adaptation should be intensified by focusing on six main issues: (1) reducing the risks associated with fishing routines, (2) strengthening social relationships, (3) managing fishermen's climate change knowledge, (4) facilitating the community's learning of alternative skills, (5) involving fishermen in climate change adaptation planning, and (6) enhancing fishermen's access to credit (Shafril et al., 2017). A study by Kamsi et al. (2025) offers valuable insights into the adaptive capacity of SSF in Malaysia via the adaptive capacity index (ACI), which defines key indicator categories such as human, financial, social, physical, and natural capital. Each indicator within these categories is collected and normalized. Weights are assigned based on relative importance using a method such as the analytical hierarchy process (AHP) (paired comparisons). The weighted, normalized indicators are finally summed to obtain the ACI score. The findings reveal the important role of financial capital and physical capital in determining economic resilience, while facilitating better management of natural resources. The variations in adaptive capacity across regions underscore the importance of crafting location-specific adaptation strategies. In addition, the outcomes suggest that promoting aquaculture and sustainable fishing practices can increase economic stability and alleviate pressure on marine resources. This evidence may help policymakers to plan their measures in the way that they have a maximum effect in enhancing the resilience of SSFs. Considering the limitations of the study, a future study with a larger sample size (more than 750 respondents) is highly desirable to validate the ACI. Since the ACI validation encompassed only two states in Malaysia, it is necessary to revalidate the indicators for other states. While this study is of immense value, specific issues require consideration. First, using cross-sectional data can increase the difficulty of understanding how adaptive capacity changes over time and may obscure any associated resilience strategies or external stressors. While the ACI provides an inclusive measure of adaptive capacity, its weighting may not accurately represent each fishing community's specific circumstances; as a result, adjustments must be made according to context.

Future research should address its limitations and deepen our understanding of SSF adaptive capabilities. Studies conducted over time could clarify how adaptive capacity evolves in response to environmental circumstances, policy interventions, and market forces. Enhancing and validating the ACI across different contexts could increase its usefulness to practitioners and policymakers looking to strengthen fisher communities' resilience. Participation in research that directly engages fishermen could ensure that future strategies are practical and suitable for their culture. This can be achieved by studying the effects of globalization, digitalization, and climate change on SSF adaptation strategies (Kamsi et al., 2025). Teh et al. (2011) proposed the need for more encompassing fisheries monitoring and data collection methods that target the large population of small-scale fishers whose catch is unreported.

Malaysian and Sundarbans SSFs share core challenges related to environmental degradation, limited financial access, and socioeconomic vulnerability. Still, they have differed in governance, adaptation, and income diversification. Cross-contextual lessons—such as Malaysia's focus on social capital and alternative livelihoods, and the Sundarbans' traditional management—can inform context-specific, resilient SSF interventions. Malaysian SSFs struggle with the challenges from commercial vessel encroachment and monsoon-induced disruptions, leading to pronounced economic vulnerability. Similarly, Sundarbans SSFs face the consequences of overexploitation and biodiversity loss, intensified by fragmented enforcement and pressures on traditional governance systems. Socioeconomic vulnerability is pronounced in both contexts, with limited capital assets, restricted access to financial credit, and increased livelihood risk during environmental distress. Besides, Malaysian SSFs show a certain degree of income diversification through tourism, while in the Sundarbans, fishers are still heavily reliant on fishing. Social capital and community engagement are identified as critical determinants of household resilience in both regions, underscoring the value of social safety nets, alternative employment, and participatory development initiatives. Climate change exacerbates existing vulnerabilities, reducing productivity, and threatening infrastructure, health, and food security. In Malaysia, adaptive capacity is influenced by five capitals (human, financial, social, physical, and natural capital), where research supports locally tailored adaptation strategies and sustainable fisheries practices. The Sundarbans SSFs need a comprehensive adaptation plan that involves resource management, restoration, and governance reforms. Malaysia benefits from designated artisanal fishing zones and a rise in participatory governance, although enforcement and data collection remain inadequate. The Sundarbans relies on traditional community-based management (Iftekhar & Islam 2004), but these systems are increasingly strained by regulatory weakness and external pressures. In essence, both situations emphasize the necessity for enhanced monitoring, targeted policy interventions, and institutional support for the betterment of SSF sustainability.

6.3 AFRICAN SMALL-SCALE MANGROVE FISHERIES

Africa is home to roughly 3.2 million hectares of mangroves, among which West Africa accounts for 1.53 million hectares that support more than 55,000 shellfishers and 570,000 household beneficiaries (Chuku et al., 2022). The mangroves in East Africa, located in Kenya, Tanzania, and Mozambique, provide support to thousands of fishers. Women are engaged in shellfish harvesting, which is a female-dominated activity (79% in West Africa), while men concentrate on finfish and crustaceans. The key species are mangrove oysters (*Crassostrea tulipa*), periwinkles, cockles, and various crabs across biogeographical regions. African small-scale mangrove fisheries are the centuries-old traditional livelihood systems across the continent's vast coastlines. These fisheries evolved from indigenous knowledge systems, integrating seasonal patterns, traditional governance, and community-based resource management. Colonial disruptions and postindependence development policies often marginalize these systems, although they remain crucial for coastal food security.

Climate change impacts sea-level rise, temperature changes, and altered precipitation patterns, affecting mangrove ecosystems. Coastal development, oil exploration (particularly in Nigeria and Angola), and agricultural expansion drive habitat loss. Most African countries do not have specific mangrove fishery policies, which leads to the absence of governance that creates open-access scenarios that allow overexploitation. In addition to overexploitation, postharvest losses, inadequate infrastructure, and limited market access also exacerbate livelihood challenges. Alternatively, gender inequality in resource access and decision-making roles still exists while women oversee only harvesting activities. Africa merits a comprehensive analysis because of its extensive mangrove coverage (second globally after Asia), high dependency rates among coastal communities, successful comanagement examples (Ghana, Gambia), diverse ecological and social contexts spanning multiple biogeographical zones, and significant potential for scaling sustainable management approaches. The continent's mangrove fisheries represent critical climate adaptation opportunities while supporting millions of livelihoods across diverse socioeconomic contexts Rouf & Jensen 2001).

6.3.1 Nigeria: Environmental Justice and Resilience

Small-scale mangrove fisheries in Nigeria are closely linked to the traditional culture of the country and have been in existence long before the colonial administration in the coastal communities (FAO, 2022). Nigeria is the most populous country in sub-Saharan Africa, with a population of over 200 million people and a total land area of 923,773 km^2. The brackish water subsector, comprising estuaries, mangrove swamps, and saline creeks, has been the backbone of artisanal fisheries for centuries. After the 1960s, oil discovery in the Niger Delta significantly transformed the sector, introducing both opportunities and environmental challenges. Nigeria dominates West African mangrove fisheries, with 39,340 shellfishers across approximately 500 coastal settlements, equivalent to more than 70% of the regional total (Chuku et al., 2022). The Niger Delta is home to Africa's largest mangrove ecosystem, covering 689,417 hectares and dominated by Rhizophora species, which constitute more than 90% of the mangrove coverage (Spalding, 2023). Women mostly harvest shellfish such as oysters and periwinkles during low tide, whereas men engage in creek fishing (Cliffe, & Akinrotimi 2015). The mangrove oyster (*Crassostrea gasar*) is one of the promising fishery resources, with regional price variations from US $0.10–0.90/kg, indicating a lack of market efficiency. The country is naturally endowed with a vast network of rivers, estuaries/flood plains, natural and artificial lakes and reservoirs, as well as brackish waters, lagoons, and creeks that provide rich biodiversity to aquatic fisheries (Olaoye & Ojebiyi, 2018). In addition, it has over 800 km of coastline, often surf-beaten with sand and mud beaches, and the sea is five nautical miles from the coast. Coastal waters exclusive to artisanal fisheries cover nine coastal states in Nigeria, with a total area of 39,644 km^2.

Small-scale or artisanal fisheries in this area are described as capture fisheries from natural water bodies, including coastal waters, rivers, streams, lakes, and ponds, by small-scale fisherfolk using both traditional and modern fishing gear (Olaoye & Ojebiyi, 2018). Artisanal fisheries include local fishermen and women who engage in fishing activities on a full-time or part-time basis at the subsistence level, using different gear and methods, which may be destructive, inexpensive, and locally sourced (Olaoye & Ojebiyi, 2018). It is a heterogeneous, multispecies, multi-gear, multicraft, and multitechnique, nonindustrial fishery that covers the activities of small-scale canoes operating in coastal areas, creeks, lagoons, inshore water, and inland rivers that target exclusively freshwater, brackish water, and inshore pelagic fish stocks (Akintola & Fakoya, 2016; Akintola et al., 2022; Olatunji & Olah, 2012). Compared with the Sundarbans model, Nigeria's Niger Delta SSFs face unique challenges, particularly those related to the impacts of the oil industry and different ecological contexts. Nigeria's Niger Delta SSFs face pollution pressures from oil extraction that are absent in the Sundarbans, creating distinctive challenges for resource management and community health. The region has West Africa's most extensive mangrove system, which covers about 1.9 million hectares (Zabbey, 2009). Oil spills and gas flaring have degraded fish habitats and reduced catches, forcing fishers to travel greater distances and invest in new gears (Zabbey et al., 2010). Nigerian fishing communities have organized more visible resistance movements against environmental degradation than those in the Sundarbans, where protests are mostly suppressed or coopted by political interests (Obi, 2010). This difference reflects varying political spaces and civil society capacities. Oil companies provide some compensation and livelihood programs for affected communities (Idemudia, 2012), representing a different approach to addressing environmental impacts compared to the Sundarbans, where fishers bear most costs of environmental degradation individually.

Nigerian SSF governance is characterized by weaker traditional institutions than in the Sundarbans, where traditional leadership systems retain significant legitimacy and effectiveness despite facing external pressures. Environmental degradation poses the primary threat. Oil industry operations cause mangrove fragmentation and pollution, while increasing fuelwood demand depletes red mangroves. The SSF sector in Nigeria is confronted with problems of food insecurity caused by a lack of postharvest infrastructure and market access restrictions (Adeleke & Wolff, 2016). There are no officially recognized comanagement arrangements or women's shellfish cooperative groups in Nigeria. The lack of formal international recognition of Niger Delta mangroves as biodiversity hotspots has resulted in limited conservation funding and attention compared with other global mangrove systems (James et al., 2007). International development organizations have invested more heavily in capacity building for Nigerian fishing communities than for the Sundarbans. However, the effectiveness of implementation varies significantly and often fails to address structural inequalities (Mmom & Arokoyu, 2010). Women in Nigeria have the same prominent roles in fish processing and marketing as those in the Sundarbans. Nevertheless, they are less formalized in decision-making processes and have fewer emerging women's organizations for support, compared to the Sundarbans of Bangladesh (Tindall & Holvoet, 2008). The Niger Delta region of Nigeria is under environmental challenges due to frequent oil spills and industrial pollution, with excessive contamination weakening mangrove resilience and leading to ecosystem degradation. These impacts directly affect fisheries that depend on healthy mangrove habitats.

Ogbonna et al. (2023)referred that mangroves are the significant sinks for organic carbon, with growing interest in restoring and conserving organic carbon for greenhouse gas emission mitigation. Their research provides a framework for mangrove restoration and conservation in Nigeria, highlighting its importance in climate regulation and fisheries support. Rapid coastal development and infrastructure expansion have led to significant mangrove habitat loss, reducing the nursery areas essential for fish reproduction and juvenile development (Zu Ermgassen et al. 2020). Limited regulatory enforcement in mangrove protection zones allows continued degradation despite existing environmental laws, creating ongoing pressure on these critical ecosystems. Economic pressure on coastal communities often leads to overexploitation of remaining mangrove resources, including overfishing and destructive harvesting practices that further degrade ecosystem health. Fakoya (2024) outlines the importance of mangroves, with a specific focus on SSFs in Africa. This research emphasizes the interconnected nature of mangrove conservation and fisheries sustainability, demonstrating how community stewardship can provide adequate protection while maintaining livelihood opportunities. Nigerian communities are developing restoration programs that combine mangrove replanting with alternative livelihood development, including beekeeping and ecotourism initiatives that provide income while supporting conservation goals. Efforts to integrate traditional governance systems with formal management structures are promising in Nigeria, as they combine customary law and practices with contemporary conservation science.

The small-scale mangrove fisheries of Nigeria, rooted in traditional practices, face unique challenges compared with the Sundarbans. While both regions rely on artisanal fishing and are crucial for local livelihoods, Nigeria's sector contends with severe oil-related pollution and mangrove degradation, issues largely absent in the Sundarbans. Unlike the Sundarbans' traditional governance, Nigerian fisheries have weaker local institutions but more visible resistance movements to environmental harm. Women play prominent roles in both regions, though with less formal recognition in Nigeria. Additionally, international development interventions are more pronounced in Nigeria, but both regions struggle with market access, resource depletion, and socioeconomic vulnerability.

6.4 LATIN AMERICAN SMALL-SCALE MANGROVE FISHERIES

SSFs in Latin America and the Caribbean region can be traced back to precolonization periods (i.e., before the 15th century) (Begossi 2008; Salas et al. 2007; Leslie et al. 2018. At the time, the inhabitants of the Americas relied heavily on fishing resources for subsistence. Historically, the development of fishing practices and knowledge about resources and fishing grounds was passed on from elders to younger generations through various strategies, some of which are still present in specific communities (Salas et al. 2007). After colonization, the monetary value attributed to fish and their trade triggered a shift from a subsistence-based fishery to a commercial fishery. This change created opportunities for fishing enterprises that actively sought to expand their fishing efforts in the region, particularly during the 1980s (Allsopp, 1975; Salas et al., 2007). This expansion coincided with liberal political agendas combined with neoliberal practices such as financialization, deregulation, and privatization, among other drivers of change (Pinkerton, 2017). In the same vein, nation-states made a significant push to develop industrial fisheries and increase fishing fleets, arguing that such development would generate jobs and help secure food availability for coastal communities (Chuenpagdee, 2011). According to the FAO (2016), almost 90% of all motorized fishing vessels in Latin America and the Caribbean region are considered small-scale or less than 12 m in length. Despite the recognition that there is no single agreed-upon definition of SSFs (FAO, 2015) and that the criteria used in Latin America and the Caribbean to define SSFs vary remarkably across countries, SSFs are generally characterized by a small number of fishing crews (3–5 fishers), are primarily community-based, operate in nearshore areas, use low levels of fishing technology, and have limited capital investment. They typically target multiple species and employ a wide range of gear and fishing techniques that vary spatially and temporally, as determined by the dynamics of resource availability (Salas et al., 2007). SSFs in the region are closely linked to the history and culture of local fishing communities and strongly influence the regional economy through the generation of employment, income, and livelihoods (de Oliveira Leis et al., 2018).

6.4.1 Brazil: Legal Frameworks and Social Protection

Brazil has the world's second-largest mangrove area (after Indonesia), which stretches ~1.3 million hectares. Brazil's approach to small-scale mangrove fisheries management demonstrates more formalized conservation and social protection frameworks than the Sundarbans model does. The legal recognition of the customary rights of fishermen in the Sundarbans is limited, whereas Brazil has created more extensive legal frameworks that acknowledge the rights of traditional fishing communities (*Pescadores artesanais*). Since the 1980s, the expansion of shrimp aquaculture has put a great deal of strain on Brazil's mangrove systems, which span 1.0–1.4 million hectares (Lacerda et al., 2019). Compared to Bangladesh and India, Brazil's constitution offers more robust legal underpinnings for community-based management by acknowledging the rights of traditional peoples to their lands and resources (Little, 2002). Traditional communities can continue to have access to resources within protected areas through the National System of Conservation Units (SNUCs) (Diegues, 2008). Brazilian fishers have access to unemployment insurance during closed seasons and other social protection measures that are absent for Sundarbans fishers, reflecting different state capacities and social policy frameworks (Abdallah & Sumaila, 2007). Brazil has adopted more thorough integrated coastal zone management as opposed to the sectoral approaches typical in South Asia. Similar to other areas, the growth of intensive prawn farming has been characterized by initial economic gains followed by environmental deterioration and social unrest (Nunes et al., 2011). In contrast to Sundarbans communities, which have few formal participation opportunities despite robust traditional governance systems, Brazilian coastal management incorporates formal mechanisms for fishing communities to participate in planning processes (Berkes & Seixas, 2005). Brazil has developed stronger ties between fishing communities and research institutions than Sundarbans, where academic research often remains disconnected from community needs and priorities (Begossi, 2008). Silva et al. (2024) thoroughly examined the challenges facing mangroves in the semiarid equatorial coast of Brazil in the Anthropocene. They documented how human activities in the current geological epoch fundamentally altered these ecosystems, affecting their capacity to support traditional fisheries.

Brondizio et al. (2024) examined the flow of mangrove ecosystem services to coastal communities in the Brazilian Amazon. Their research demonstrated the complex relationships between mangrove health and community well-being, revealing how ecosystem degradation directly impacts fishing livelihoods and food security. Brazilian coastal areas face intense urbanization pressure that fragments and destroys mangrove habitats. In addition to decreasing fishing grounds, this urbanization also raises pollution levels and modifies water flow patterns that are vital to the health of mangroves. Significant mangrove areas have been converted as a result of the expansion of industrial shrimp farming and other aquaculture operations, which have changed ecosystem dynamics and reduced traditional fishing grounds. Mangrove salinity levels and nutrient distributions are impacted by changes in rainfall patterns and freshwater flows, which also have an effect on the species composition and abundance of fish communities that rely on these habitats. Brazilian initiatives focus on integrated coastal zone management programs that balance development needs with ecosystem conservation, attempting to maintain fishing opportunities while accommodating growth pressures. Programs to strengthen fisher organizations and promote collective action have shown success in improving both conservation outcomes and fisher livelihoods through enhanced political voice and market access. The development of sustainable aquaculture practices within mangrove systems, which complement rather than replace traditional fishing, provides additional income while maintaining ecosystem services.

Brazilian small-scale mangrove fisheries differ from those in the Sundarbans by adopting more formalized legal frameworks and social protection measures. Brazil's constitution and conservation units actively recognize traditional fishing rights and support community-based management, whereas Sundarbans fishers rely on informal, customary systems with limited legal backing. Brazilian fishers benefit from unemployment insurance and participatory planning mechanisms, in contrast to the Sundarbans' lack of formal safety nets and restricted involvement in management. Both regions face pressures from aquaculture and habitat loss. However, Brazil's integrated coastal management and stronger research-community linkages enable more adaptive responses, while Sundarbans communities contend with greater marginalization and weaker institutional support.

6.5 FROM DELTA ROOTS TO GLOBAL CURRENTS: TRANSBOUNDARY GOVERNANCE, TRADITION, AND TRANSFORMATION THROUGH THE SUNDARBANS LENS

The Sundarbans' small-scale fishery operations show the relevance of multilevel governance regimes that transcend scale and jurisdiction; at the same time, they also illustrate the limits of informal governance systems in addressing external pressures. While the Sundarbans mainly relies on informal authority regimes that lack recognition in law, other countries like the Philippines and Brazil have built more formalized community-based systems of management. Indonesia represents a middle path with government recognition of traditional authority (Adat) systems (Thorburn, 2000). The Sundarbans are unique in their transborder governance context, but other regions experience similar coordination challenges, much further up or down the scale, for example, Indonesia across provinces, Nigeria across states, or Brazil across municipalities. The poor record of development assistance agencies in the Sundarbans illustrates the challenges of coordinating across different institutional frameworks (Thompson & Islam, 2010). Most regions struggle to provide an appropriate local authority for management decisions.

The local management in Sundarbans is informal and differs from the more formalized local management in the Philippines and Brazil, although both face the challenge of identifying and balancing local autonomy with the necessities of a wider regional coordination. The fisheries sector faces challenges related to food security, habitat degradation, urbanization, and industrial development. These challenges manifest differently across regions but show typical patterns of vulnerability among small-scale fishers. The Sundarbans demonstrates how impacts on fisheries tend to affect the poorest elements of society—the landless subsistence fishermen (Thompson & Islam, 2010). This pattern is common across regions, where small-scale fishers represent marginalized groups with limited political power. Limited options for alternative livelihoods impede Sundarbans communities from diversifying their livelihoods because of geographic isolation, limited skills, and limited access to alternative jobs. Thai and Philippine communities have greater access to alternative employment opportunities, whereas Brazilian communities benefit from tourism and aquaculture opportunities supported by government programs. Access to appropriate financial services is crucial for fisher welfare and resource management; however, most regions show inadequate financial inclusion for small-scale fishers. The Sundarbans' debt trap problems illustrate the consequences of inadequate financial access, whereas other regions show varying degrees of improvement through microfinance and government programs. The Sundarbans demonstrates complex interactions between conservation and livelihood needs, which are reflected differently across regions. All regions show tensions between short-term livelihood needs and long-term sustainability. Most regions have established marine protected areas, but community participation in planning and management varies significantly. Brazil shows the strongest community participation through formal consultation mechanisms, whereas Nigeria shows the weakest integration of community priorities due to limited institutional capacity. Sundarbans have rich traditional ecological knowledge that influences management practices, but this knowledge is not formally recognized in management systems. Fishing gear and crafts are mostly traditional, reflecting adaptive capacity developed over generations (Hoq, 2007).

Indonesia and the Philippines have better incorporated traditional knowledge through formal co-management arrangements, while Thailand has focused on modernizing its technical fisheries rather than following traditional practices. Growing threats to sustainable fisheries arise from excessive exploitation of resources across all regions. The Sundarbans illustrates how fragile development projects can impact ecology, whereas other regions show varying degrees of success in balancing development and conservation priorities (Islam & Wahab 2005). Comparative analysis identifies substantial differences in terms of market integration and value addition, with the Sundarbans being one of the most disadvantaged positions. Sundarbans fishers experience monopsonistic relationships with buyers and face barriers to accessing markets due to near-isolation and lack of infrastructure. Indonesian and Thai fishers operate in better integrated markets due to enhanced transportation and communication systems, while Philippine and Brazilian fishers have limited access to the market and some cooperative marketing. Most regions have retained traditional approaches to postharvest processing, apart from Thailand, which has undergone greater modernization. Women process the fish throughout the world, but receive limited institutional recognition and support for their work, other than in the case of Brazil, which offers social protection for women.

Sundarbans fishers get low prices for their catch because of limited market access and limited bargaining power to negotiate prices with fish traders. This inequitable relationship reflects broader patterns of exploitation of small-scale fishers. Other regions demonstrated different degrees of price improvement through cooperative marketing, better transportation infrastructure, and marketing certificates. As a result of the analysis of Sundarbans' experience and external comparisons, several institutional design principles have emerged for responsible and effective small-scale fishery management. Rather than replace traditional governance systems, policies should formalize and enhance traditional systems, while also identifying the limitations of those systems. In the Sundarbans case, traditional governance systems do an effective job of maintaining social cohesion and ecological knowledge, while their limitations in addressing structural inequalities and tensions related to external pressures limit their governance for fishers. Effective SSF management requires institutions operating at multiple levels, from local communities to national governments. The Sundarbans demonstrates both the strengths and the weaknesses of informal multilevel systems, whereas other regions show varying levels of success with formalized multilevel arrangements. Each system experiences considerable uncertainty and variability, therefore they need to have a greater capacity for Adaptive Management. The Sundarbans has a recognized strong traditional adaptive capacity but weaker formal adaptive management capacity, while Brazil has a more systematic Adaptive Management process in the context of Integrated Coastal Zone Management. Comparative analysis has shown the significance of having secure access rights for small-scale fishers, and the Sundarbans demonstrates the risk of insecure tenure. While the Sundarbans fishing practices have an ambiguous legal status, the Philippines and Brazil have greater legal recognition. The existence of a clear legal framework that provides some legal recognition of small-scale fishers' preferential rights to access waters where they live is critical for sustainable management and community contributions. Effective management also requires mechanisms that can exclude destructive fishing practices and inappropriate technologies. There are no formal exclusion mechanisms in the Sundarbans, and they rely on a traditional system that is increasingly inadequate in the face of increasing external pressures. The Sundarbans show both the opportunities and limitations of community participation in management. While traditional mechanisms of participation exist, there are very few formal opportunities for participation compared with either case study in the Philippines or Brazil. Brazil's formal social protection is a stark contrast to the absence of any safety nets in the Sundarbans, where fishers rely solely on traditional support networks and informal coping strategies with minimal state support. Poorly documented development assistance in the Sundarbans highlights the necessity to be better equipped. The geographic isolation and limited infrastructure in the Sundarbans constrain alternative livelihood development, whereas other regions show greater success through targeted programs and better market access. The example of the Sundarbans demonstrates how a capacity-building initiative did not address the structural differences and power inequities that affect the predicament of small-scale fishers. The identification of these broader socioeconomic and political disincentives will require a more effective approach. In the Sundarbans, an ecosystem-based management practice occurs informally through a traditional management system amidst the biophysical realities of the area, which includes a recognition of the interrelatedness of ecological and social systems, without formal structures. Brazil and the Philippines lean toward practices and formal approaches to ecosystem-based management that blend multiple users with stakeholders' needs. In the Sundarbans, although some traditional conservation practices are visible, there is limited user benefit-sharing from conservation activities. Systematic programs require user involvement and benefit-sharing with a range of effectiveness across sites. All indications create the need for climate adaptation programming, albeit regionally very different. The Sundarbans utilize traditional coping strategies, which may not be able to respond to projections of climate change, and the other regions are developing varying systems of systematic adaptation, utilizing user participation and accountability. Policies will not supplant traditional local governance systems; rather, they should reinforce and formally acknowledge traditional management systems while mitigating their shortcomings. Sundarbans demonstrated the effectiveness of traditional systems as well as their need for external support to address contemporary challenges. The Sundarbans illustrates how impacts on fisheries tend to affect the poorest elements of society, with development assistance often failing to address underlying structural inequalities. Other regions can learn from this experience to develop more effective poverty alleviation programs that focus on root causes instead of symptoms. The cross-boundary nature of the Sundarbans illustrates both challenges and opportunities inherent in coordinated management activities. Institutions must be able to adapt under changing ecological, social, and economic conditions (Joseph et al., 2019). Other regions with shared resources can learn from the successes and failures of coordinating mechanisms to meet local objectives while also meeting larger management and coordination requirements. A comparative analysis of small-scale mangrove fisheries management across different countries is presented in Table 6.2.

The Sundarbans illustrated both the adaptive capacity of traditional systems and the limits of those systems in relation to increasingly rapid external changes. Rigid regulations are not going to be effective as adaptive management approaches that can learn and adjust over time. Effective management will require a monitoring process that is active and involves the communities in the process. Fisher's knowledge and observations are critical in an adaptive management process, as traditional systems of the Sundarbans showed, but there are often no formal or regular mechanisms for engaging fishermen in the monitoring process. Even if establishing systematic mechanisms for learning from experience and sharing lessons (across states or regions) would improve the effectiveness of policy in general, the comparison of the Sundarbans with the other two cases demonstrates common patterns and situationally specific solutions and would require learning and adapting over time. Moreover, small-scale fisheries cannot be decoupled from coastal management, conservation, and development policies, and the Sundarbans experience shows how sectoral approaches often fail to resolve dimensional challenges systemically. At the same time, integrated approaches are more appropriate for more effectively addressing multiple and overlapping objectives. The pressure for effective management across multiple levels is greater and more urgent under the threat of climate change (Figure 6.2). The Sundarbans demonstrates both the importance and difficulty of scale integration in transboundary contexts. Combining traditional ecological knowledge with scientific research and management requires appropriate institutional mechanisms and respectful engagement processes. Sundarbans have rich traditional knowledge that can inform management, but formal mechanisms for knowledge integration are lacking.

TABLE 6.2
Comparative Analysis: Small-Scale Mangrove Fisheries Management of Different Countries Across the Globe

Dimension	Sundarbans (Reference Model)	Indonesia	Philippines	Thailand	Nigeria	Brazil
Ecosystem characteristics	• world's largest contiguous mangrove • 10,000 km^2 transboundary • complex tidal dynamics • multispecies habitat	• 23% global coverage • 17,000+ islands • diverse ecological conditions • **significant loss to aquaculture (X)**	• **70% original cover lost (X)** • island archipelago system • typhoon-prone • high biodiversity	• **84% loss (1961–2009) (X)** • both coastal waters (along the Andaman Sea and Gulf of Thailand) and inland freshwater systems, including flooded fields and canals • monsoon-influenced	• West Africa's largest Niger delta complex • **oil extraction impacts (X)**	• Northeast and the North regions, along the extensive coastline and vast river systems • high species diversity • climate change vulnerable
Fisheries characteristics	• 500,000+ dependents • multispecies, multi-gear • traditional techniques • seasonal patterns • tidal adaptations	• archipelagic diversity • mixed traditional-modern • variable market access • Low technology method • better technology adoption (+)	• community-based approaches • disaster preparedness • mixed subsistence-commercial • Labor intensive	• motorization & inshore vessels (under 10 GT), modernization (+) • family-based • market integration • higher productivity • export orientation	• **pollution impacts (X)** • longer fishing distances • traditional processing • limited formal support • Labor-intensive multispecies and multi-gear • nonmotorized	• pescadores artesanais recognition (+) • tourism integration • research linkages • certification schemes • low level of technological assistance • small vessels, though specific gear and vessel types vary by region.

Governance structure	• traditional authority systems • informal multilevel management • community "samitis" • **limited formal recognition (X)**	• **largely centralized and multilayered** • **simplified fishing logbook reporting (X)** • village sasi systems • government recognition of adat (*)	• decentralized and community-based management (*) • barangay council involvement • formal legal framework • ecosystem-based approaches	• multilevel structure that empowers local communities within a national legal framework (*) • government extension services • market-oriented policies • decentralized governance	• **weaker traditional institutions (X)** • limited formal frameworks • oil company involvement • international NGO presence • widespread conflicts between small-scale and industrial fisheries	• constitutional recognition (+) • a mix of decentralization efforts since 1988 and a persistent, often ineffective, centralized, and fragmented approach • integrated coastal management • formal participation mechanisms (+) • strong legal foundations
Livelihood and poverty	• **debt-trap (X)** • landless subsistence fishers • limited diversification • geographic isolation	• better market access (+) • microfinance availability • alternative livelihoods • variable across islands	• government credit programs (+) • community cooperatives • alternative employment • disaster support systems • limited access to markets	• higher incomes • market volatility exposure • technology investments • skills development	• oil company compensation (*) • high dependency • limited formal services • environmental health impacts • international aid programs • traditional method	• social protection measures (+) • unemployment insurance (+) • constitutional rights • diversified opportunities Traditional knowledge

(Continued)

Table 6.2 (Continued)

Dimension	Sundarbans (Reference Model)	Indonesia	Philippines	Thailand	Nigeria	Brazil
Unique challenges	• **tiger attacks (X)** • cyclone vulnerability • **poor development assistance (X)** • transboundary coordination	• institutional complexity • coordination across provinces • decentralization gaps • palm oil pressure	• typhoon impacts (+) • island connectivity • resource competition • population pressure	• massive habitat loss • overcapacity risks • market competition • coastal development	• **oil spills and gas flaring (X)** • environmental justice issues • security challenges • pollution health impacts	• shrimp farming impacts • climate adaptation needs • market access in remote areas • implementation gaps
Market access	• **monopsonistic buyers (X)** • limited transportation • local/regional markets • **low bargaining power (X)**	• better market integration (+) • improved transportation • national/ international reach • cooperative initiatives	• intermediate market access (*) • cooperative marketing • island connectivity issues • price volatility	• strong market integration (+) • export orientation • value chain participation • quality standards	• limited formal markets • traditional value chains • regional trade networks • informal processing	• certification program (+) • cooperative marketing • tourism linkages • government support
Conservation approach	• rich traditional knowledge • informal area closures • **limited formal protection (X)** • traditional adaptive strategies	• traditional-modern integration (+) • marine protected areas • restoration programs • research collaboration	• ecosystem-based management • community participation • climate adaptation focus • disaster risk reduction	• technical modernization • payment for ecosystem services (+) • sustainable aquaculture • policy reversals	• **limited formal conservation (X)** • international biodiversity focus • community resistance movements • weak enforcement	• strongest community participation • ecosystem services valuation (+) • integrated planning • climate adaptation

Women's roles	• crucial postharvest roles • limited formal recognition • traditional processing • emerging organizations	• processing and marketing • variable recognition • cooperative participation • skills development programs	• formal recognition improving (+) • cooperative leadership • processing enterprises • government support	• processing modernization • market participation • technology adoption • skills upgrading	• prominent processing roles (*) • **limited decision-making (X)** • traditional methods • health impact concerns	• formal recognition (+) • social protection inclusion (+) • cooperative leadership • rights-based approach
Climate adaptation	• traditional coping strategies • **high vulnerability (X)** • informal early warnings • limited systematic planning	• regional cooperation • restoration for protection (+) • disaster preparedness • research initiatives	• systematic disaster preparedness (+) • ecosystem-based adaptation (+) • early warning systems • government programs	• ecosystem restoration (+) • coastal protection focus • technology solutions • policy integration	• limited systematic approaches • traditional resilience • international support • vulnerability assessment	• comprehensive adaptation planning (+) • ecosystem-based approaches • community participation • policy integration
Policy lessons	• strengthen traditional systems (+) • address structural inequalities • improve coordination • enhance formal recognition	• institutional coordination (+) • traditional knowledge integration • decentralization balance • technology support	• legal framework development (+) • community empowerment • disaster risk integration • multistakeholder approaches	• policy learning capacity (+) • market mechanism development • environmental rehabilitation • stakeholder coordination	• environmental justice focus • institutional capacity building (+) • industry accountability • community empowerment	• rights-based approaches • integrated management (+) • participatory planning • social protection integration

Legend and Key Insights

Positive/Advanced Practice (+), **Challenge/Gap (X)**, Mixed/Neutral (*)

Key Comparative Insights

Brazil has the strongest formal recognition and social protection systems. The Philippines demonstrates effective community-based management integration. Thailand has a successful market orientation, but it has environmental costs. Indonesia has a complex governance situation, but integrates traditional knowledge and practices. Nigeria faces unique industrial motivations that require environmental justice approaches. The Sundarbans has a rich traditional system, but must gain formal recognition and support.

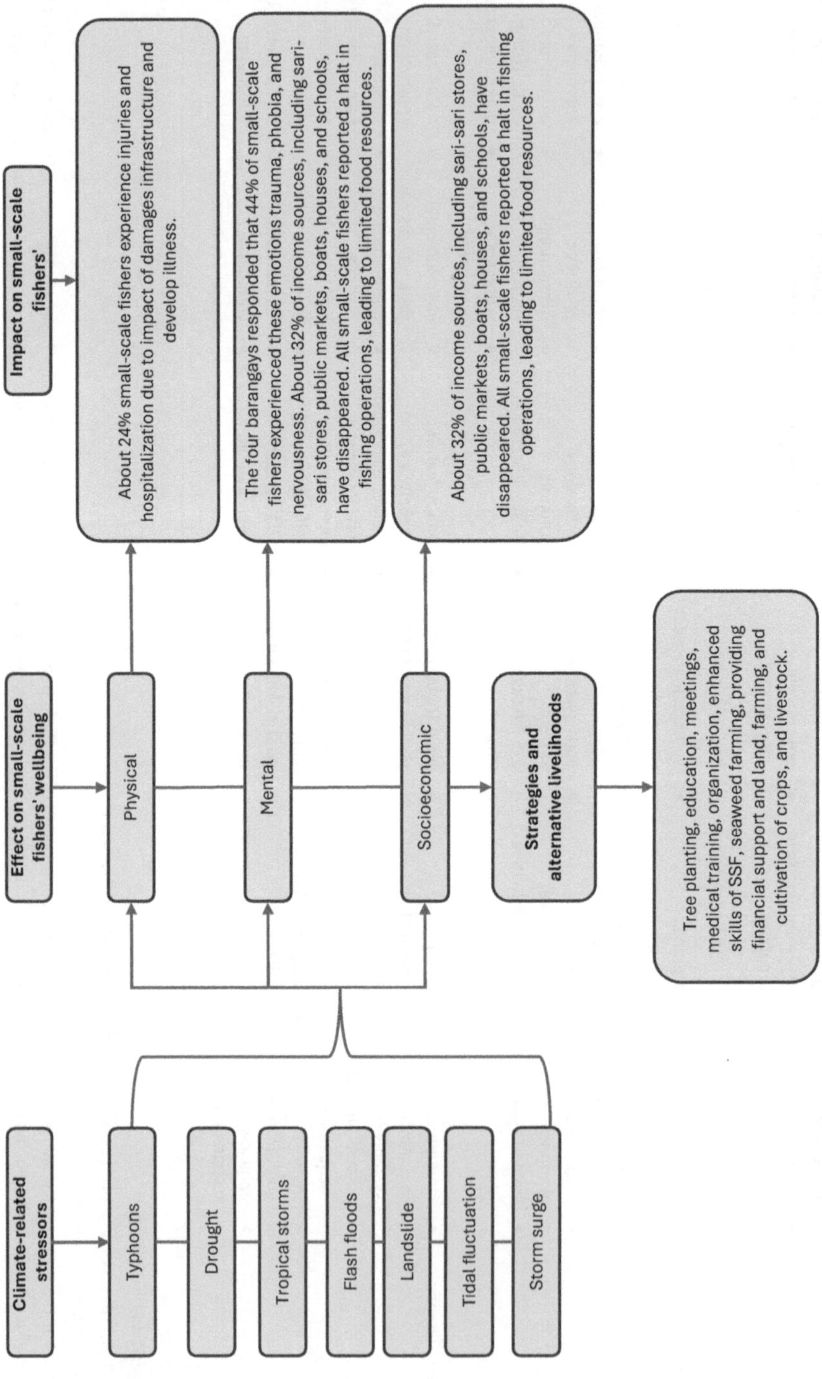

FIGURE 6.2 A framework illustrating the impact of climate-related stressors on the well-being of small-scale fishers. **(Adopted from Muringai et al. 2021.)**

6.6 CONCLUSION

The relationship between coastal communities and mangrove ecosystems represents one of the most complex and critical intersections of human welfare and environmental health in our contemporary world we live in. Through examining SSFs across diverse global settings—from the complex waterways of the Sundarbans to the complex archipelagos of Southeast Asia—a clear narrative has emerged about the urgent need for transformative fisheries management that transcends traditional boundaries between conservation and development. Perhaps the most remarkable idea that emerges from this global analysis is that social justice and environmental sustainability are not competing objectives but instead are mutually supportive. The communities that rely most profoundly on healthy ecosystems are generally the least able to advocate for their protection. Alternatively, the best conservation outcomes come from processes that promote community rights and enhance local livelihoods. This convergence suggests that the way forward is not to choose between protecting the welfare of communities or protecting ecosystems, but to work toward integrated approaches that incorporate and support both objectives. The example of Sundarbans serves as both inspiration and warning—it exemplifies the incredible resilience of traditional systems, while illustrating their vulnerability to external factors beyond their independent control. The window of opportunity for meaningful action is rapidly closing, but the elements for success through the integration of traditional fishing knowledge, scientific knowledge, innovative funding strategies, and increasing political recognition of the value of SSFs. Whether we seize this unprecedented opportunity will not only determine the fate of coastal communities but also the health of marine ecosystems for generations to come. The challenge lies in translating analysis into action, moving from knowledge to practice, and ensuring that the SSFs reform takes place in a proactive and equitable manner. The stakes could not be higher, and the urgency to act has never been more dramatic.

BIBLIOGRAPHY

Abdallah, P. R., & Sumaila, U. R. (2007). An historical account of Brazilian public policy on fisheries subsidies. *Marine Policy*, *31*(4), 444–450.

Abu Samah, A., Shaffril, H. A. M., Hamzah, A., & Abu Samah, B. (2019). Factors affecting small-scale fishermen's adaptation toward the impacts of climate change: Reflections from Malaysian fishers. *SAGE Open*, *9*(3), 2158244019864204.

Adeleke, M. L., & Wolff, M. (2016). Adaptation of the artisanal fisher folks to climate change in the coastal region of Ondo state, Nigeria. In *Innovation in climate change adaptation* (pp. 177–193). Cham: Springer International Publishing.

Akintola, S. L., & Fakoya, K. A. (2016). *Governance and social-institutional arrangement of small-scale fisheries and relationship with non-fishery users in Badagry Creek, Lagos state, Nigeria. Inter-sectoral governance of inland fisheries*. TBTI Publication Series, St John's, NL, Canada.

Akintola, S. L., Fakoya, K. A., Elegbede, I. O., Odunsi, E., & Jolaosho, T. (2022). Postharvest practices in small-scale fisheries. In *Sustainable fish production and processing* (pp. 79–110). Academic Press.

Allison, E. H., Béné, C., & Andrew, N. L. (2011). 12 poverty reduction as a means to enhance resilience in small-scale fisheries. *Small-Scale Fisheries Management: Frameworks and Approaches for the Developing World, 216.*

Allsopp, W. H. L. (1975, June). Management strategies in some problematic tropical fisheries. In *Unifying concepts in ecology: Report of the plenary sessions of the first international congress of ecology, the Hague, the Netherlands, September 8–14, 1974* (pp. 252–262). Dordrecht: Springer Netherlands.

Armada, N., White, A. T., &Christie, P. (2016). Status, trend and sustainability of small-scale fisheries in the Philippines. *Academia.edu*.https://www.academia.edu/28169239/Status_trend_and_sustainability_of_small_scale_fisheries_in_the_Philippines

Armitage, D., & Marschke, M. (2013). Assessing the future of small-scale fishery systems in coastal Vietnam and the implications for policy. *Environmental Science & Policy*, *27*, 184–194.

Atillo, R. C., et al. (2024). Socioeconomic status of small-scale fisheries in Central Philippines. *Ho Chi Minh City Open University Journal of Science -Social Sciences*, *14*(3). 10.46223/HCMCOUJS.soci.en.14.3.3094.2024

Barbier, E. B., & Cox, M. (2003). Does economic development lead to mangrove loss? A cross-country analysis. *Contemporary Economic Policy*, *21*(4), 418–432.

Barlow, A. C. D., McDougal, C., Smith, J. L. D., Gurung, B., Bhatta, S. R., Kumal, S., & Tamang, D. B. (2008). Temporal variation in tiger (Panthera tigris) populations and its implications for monitoring. *Journal of Mammalogy*, *89*(5), 1132–1139.

Basurto, X., Gutierrez, N. L., Franz, N., et al. (2025). Illuminating the multidimensional contributions of small-scale fisheries. *Nature*, *637*, 875–884. https://doi.org/10.1038/s41586-024-08448-z

Begossi, A. (2008). Local knowledge and training towards management. *Environment, Development and Sustainability*, *10*(5), 591–603.

Berkes, F., & Seixas, C. S. (2005). Building resilience in lagoon social–ecological systems: A local-level perspective. *Ecosystems*, *8*(8), 967–974.

Brondizio, E. S., et al. (2024). Flow of mangrove ecosystem services to coastal communities in the Brazilian Amazon. *Frontiers in Environmental Science*. https://doi.org/10.3389/fenvs.2024.1329006

Bureau of Fisheries and Aquatic Resources (BFAR). (2016). *Philippine fisheries profile 2016*. Diliman, Quezon City, Philippines: Department of Agriculture.

Castellanos-Galindo, G. A., et al. (2024). Impacts of climate change on mangrove subsistence fisheries: A global review. *Marine Life Science & Technology*. https://doi.org/10.1007/s42995-024-00231-3

Chandra, G., & Sagar, R. (2003). Fisheries in Sundarbans: Problems and prospects. *SSRN Electronic Journal*. 10.2139/ssrn.2084014

Chuenpagdee, R. (Ed.). (2011). *World small-scale fisheries: Contemporary visions*. Eburon Uitgeverij BV.

Chuenpagdee, R., Bugeja-Said, A., Isaacs, M., & Jentoft, S. (2022). Towards blue justice for small-scale fisheries. In *Blue justice: Small-scale fisheries in a sustainable ocean economy* (pp. 681-692). Cham: Springer International Publishing.

Chuenpagdee, R., Liguori, L., Palomares, M. L. D., & Pauly, D. (2006). Bottom-up, global estimates of small-scale marine fisheries catches. *Fisheries Centre Research Reports*, *14*(8), 1–110.

Chuku, E. O., et al. (2022). Women-led fisheries livelihoods in West Africa. *Frontiers in Marine Science*, *9*, 884715.

Cliffe, P. T., & Akinrotimi, O. A. (2015). *Role of women in fishery activities in some coastal communities of Rivers state*. Nigeria.

Cruz, R. V. O., Harasawa, H., Lal, M., Wu, S., Anokhin, Y., Punsalmaa, B., . . . Huu Ninh, N. (2007). Asia climate change 2007: Impacts, adaptation and vulnerability. In *Contribution of working group II to the fourth assessment report of the intergovernmental panel on climate change*. Cambridge University Press.

de Oliveira Leis, M., Chuenpagdee, R., & Medeiros, R. P. (2018). Where small-scale fisheries meet conservation boundaries: MPA governance challenges in southern Brazil. In *Viability and sustainability of small-scale fisheries in Latin America and the Caribbean* (pp. 453–472). Cham: Springer International Publishing.

DENR-FMB. (2018). *Philippine forestry statistics 2017–2018*. DENR-FMB.

Diegues, A. C. (2008). *Marine protected areas and artisanal fisheries in Brazil*. Chennai: Samudra Monograph. International Collective in Support of Fishworkers.

DoF. (2020). *DoF annual report, 2020*.

Fabinyi, M., & Dressler, W. (2024). Small-scale fisheries, large-scale fisheries and fisheries governance in the Philippines. *Journal of Environmental Policy & Planning*. https://doi.org/10.1080/1523908X.2024.2314451

Fabinyi, M., et al. (2022). Coastal transitions: Small-scale fisheries, livelihoods, and maritime zone developments in Southeast Asia. *Ocean & Coastal Management, 221*, 106086.

Fakoya, K. A. (2024). Shell fisheries as stewardship for mangroves. *Mundus Maris*. https://www.mundusmaris.org/small-scale-fisheries-academy/shell-fisheries-as-stewardship-for-mangroves/

FAO. (2015). *Voluntary guidelines for securing sustainable small-scale fisheries in the context of food security and poverty eradication*. Rome: Food and Agriculture Organization of the United Nations.

FAO. (2020). *The state of world fisheries and aquaculture 2020*. Rome.

FAO. (2022). *The state of world fisheries and aquaculture 2022. Towards blue transformation*. Rome, FAO. https://doi.org/10.4060/cc0461en

Fauzi, A., & Anna, Z. (2013). The complexity of the institution of payment for environmental services: A case study of two Indonesian PES schemes. *Ecosystem Services, 6*, 54–63.

Food and Agriculture Organization of the United Nations (FAO). (2016). 978-92-5-109374-0

Giri, C., Ochieng, E., Tieszen, L. L., Zhu, Z., Singh, A., Loveland, T., & Duke, N. (2011). Status and distribution of mangrove forests of the world using earth observation satellite data. *Global Ecology and Biogeography, 20*(1), 154–159.

Ha, H. D., & Thang, T. N. (2017). Fishery communities' perception of climate change effects on local livelihoods in tam Giang lagoon, Vietnam. In *Redefining diversity & dynamics of natural resources management in Asia, volume 3* (pp. 111–124). Elsevier.

Hoang, H. D., Momtaz, S., & Schreider, M. (2020). Assessing the vulnerability of small-scale fishery communities in the estuarine areas of Central Vietnam in the context of increasing climate risks. *Ocean & Coastal Management, 196*, 105302.

Hoq, M. E. (2007). An analysis of fisheries exploitation and management practices in Sundarbans mangrove ecosystem, Bangladesh. *Ocean & Coastal Management, 50*(5–6), 411–427.

Hoq, M. E., Islam, M. N., Kamal, M., & Wahab, M. A. (2008). Abundance and seasonal distribution of Penaeus monodon postlarvae in the Sundarbans mangrove, Bangladesh. *Estuarine, Coastal and Shelf Science, 81*(4), 949–954.

Hutchison, J., Philipp, D. P., Claussen, J. E., Aburto-Oropeza, O., Carrasquilla-Henao, M., Castellanos-Galindo, G. A., . . . Spalding, M. (2015). *Building an expert judgment-based model of mangrove fisheries*. American Fisheries Society.

Huynh, P. T., Le, N. D., Le, S. T., & Nguyen, H. X. (2021a). Vulnerability of fishery-based livelihoods to climate change in coastal communities in Central Vietnam. *Coastal Management*, *49*(3), 275–292.

Huynh, P. T., Le, N. D., Le, S. T., & Tran, T. N. (2021b). Adaptive livelihood strategies among small-scale fishing households to climate change-related stressors in central coast Vietnam. *International Journal of Climate Change Strategies and Management*, *13*(4-5), 492–510.

Idemudia, U. (2012). The resource curse and the decentralization of oil revenue: The case of Nigeria. *Journal of Cleaner Production*, *35*, 183–193.

Iftekhar, M. S., & Islam, M. R. (2004). Managing mangroves in Bangladesh: A strategy analysis. *Journal of Coastal Conservation*, *10*(1–2), 139–146.

Islam, G. M. N., Noh, K. M., Sidique, S. F., & Noh, A. F. M. (2014). Economic impact of artificial reefs: A case study of small scale fishers in Terengganu, peninsular Malaysia. *Fisheries Research*, *151*, 122–129.

Islam, M. M., Nahiduzzaman, M., Acosta, R., Mome, M. A., & Wahab, M. A. (2022). Status and potential of ecosystem approach to fisheries management (EAFM) in Bangladesh. *Ocean & Coastal Management*, *219*, 106068.

Islam, M. M., Shamsuzzaman, M. M., Mozumder, M. M. H., Xiangmin, X., Ming, Y., & Jewel, M. A. S. (2017). Exploitation and conservation of coastal and marine fisheries in Bangladesh: Do the fishery laws matter? *Marine Policy*, *76*, 143–151.

Islam, M. S., & Wahab, M. A. (2005). A review on the present status and management of mangrove wetland habitat resources in Bangladesh with emphasis on mangrove fisheries and aquaculture. *Hydrobiologia*, *542*(1), 165–190.

James, G. K., Adegoke, J. O., Saba, E., Nwilo, P., & Akinyede, J. (2007). Satellite-based assessment of the extent and changes in the mangrove ecosystem of the Niger Delta. *Marine Geodesy*, *30*(3), 249–267.

Joseph, K., Eslamian, S., Ostad-Ali-Askari, K., Nekooei, M., Talebmorad, H., & Hasantabar-Amiri, A. (2019). Environmental impact assessment as a tool for sustainable development. In W. Leal Filho (Ed.), *Encyclopedia of sustainability in higher education*. Springer Nature Switzerland.

Kamsi, N. S., Radin Firdaus, R. B., Sarker, M. N. I., & Gunaratne, M. S. (2025). Assessing the linkages of livelihood capitals of small-scale fishermen in Malaysia. *SAGE Open*, *15*(2), 21582440251338685.

Kongkeo, H. (2001). Current status and development trends of aquaculture in the Asian region. *FAO Aquaculture Newsletter*, *27*, 2–11.

Lacerda, L. D., Borges, R., & Ferreira, A. C. (2019). Neotropical mangroves: Conservation and sustainable use in a scenario of global climate change. *Aquatic Conservation: Marine and Freshwater Ecosystems*, *29*(8), 1347–1364.

Lebel, L., Anderies, J. M., Campbell, B., Folke, C., Hatfield-Dodds, S., Hughes, T. P., & Wilson, J. (2006). Governance and the capacity to manage resilience in regional social-ecological systems. *Ecology and Society*, *11*(1).

Leslie, M., Nguyen, S. T., Nguyen, T. K. D., Pham, T. T., Cao, T. T. N., Le, T. Q., … Chris, Y. (2018). Bringing social and cultural considerations into environmental management for vulnerable coastal communities: Responses to environmental change in Xuan Thuy National Park, Nam Dinh Province. *Vietnam. Ocean & Coastal Management*, *158*, 32–44.

Little, P. E. (2002). Territórios sociais e povos tradicionais no Brasil: Por uma antropologia da territorialidade. *Série Antropologia*, *322*, 1–32.

Long, J. B., & Giri, C. (2011). Mapping the Philippines' mangrove forests using landsat imagery. *Sensors*, *11*(3), 2972–2981. https://doi.org/10.3390/s110302972

Mmom, P. C., & Arokoyu, S. B. (2010). Mangrove forest depletion, biodiversity loss and traditional resources management practices in the Niger Delta, Nigeria. *Research Journal of Applied Sciences, Engineering and Technology*, *2*(1), 28–34.

Muringai, R. T., Mafongoya, P. L., & Lottering, R. (2021). Climate change and variability impacts on sub-Saharan African fisheries: A review. *Reviews in Fisheries Science & Aquaculture*, *29*(4), 706–720.

Musinguzi, L., Efitre, J., Odongkara, K., Ogutu-Ohwayo, R., Muyodi, F., Natugonza, V., ... Naigaga, S. (2016). Fishers' perceptions of climate change, impacts on their livelihoods and adaptation strategies in environmental change hotspots: A case of Lake Wamala, Uganda. *Environment, Development and Sustainability*, *18*(4), 1255–1273.

Nunes, J. M. C., Sampaio, Y., & Barros, F. (2011). How much can the shrimp industry afford to reduce environmental impacts? *Environment and Development Economics*, *16*(1), 91–113.

Nurdin, S., et al. (2021). Developing sustainable small-scale fisheries livelihoods in Indonesia: Trends, enabling and constraining factors, and future opportunities. *ResearchGate*.

Obi, C. I. (2010). Oil extraction, dispossession, resistance, and conflict in Nigeria's oil-rich Niger Delta. *Canadian Journal of Development Studies*, *30*(1–2), 219–236.

Ogbonna, D. N., et al. (2023). Towards a framework for mangrove restoration and conservation in Nigeria. *Science of the Total Environment*, *899*, 165644. https://doi.org/10.1016/j.scitotenv.2023.165644

Olaoye, O. J., & Ojebiyi, W. G. (2018). Marine fisheries in Nigeria: A review. *Marine ecology-Biotic and abiotic interactions*.

Olatunji, A. E., & Olah, O. M. (2012). *The socio-economic status of artisanal fishers in Cross River*. Nigeria: Cross River State.

Pinkerton, E. (2017). Hegemony and resistance: Disturbing patterns and hopeful signs in the impact of neoliberal policies on small-scale fisheries around the world. *Marine Policy*, *80*, 1–9.

Poggie, J. J., Pollnac, R. B., & Weeks, E. (1995). Cross-cultural aspects of fisheries development in the Gulf of Thailand. *Ocean & Coastal Management*, *28*(1-3), 103–133.

Pomeroy, R. S., Garces, L., Pido, M., & Silvestre, G. (2010). Ecosystem-based fisheries management in small-scale tropical marine fisheries: Emerging models of governance arrangements in the Philippines. *Marine Policy*, *34*(2), 298–308.

Primavera, J. H. (2006). Overcoming the impacts of aquaculture on the coastal zone. *Ocean & Coastal Management*, *49*(9–10), 531–545.

Richards, D. R., & Friess, D. A. (2016). Rates and drivers of mangrove deforestation in Southeast Asia, 2000–2012. *Proceedings of the National Academy of Sciences*, *113*(2), 344–349.

Rouf, M. A., & Jensen, K. R. (2001). Some fisheries aspects of the Sundarbans mangrove forest of Bangladesh. *Estuaries*, *24*(2), 423–430.

Salas, S., Chuenpagdee, R., Seijo, J. C., & Charles, A. (2007). Challenges in the assessment and management of small-scale fisheries in Latin America and the Caribbean. *Fisheries Research*, *87*(1), 5–16.

Sarker, S. K., Reeve, R., Thompson, J., Paul, N. K., & Matthiopoulos, J. (2018). Are we failing to protect threatened mangroves in the Sundarbans world heritage ecosystem? *Remote Sensing*, *10*(9), 1441.

Shaffril, H. A. M., Samah, A. A., & D'Silva, J. L. (2017). Climate change: Social adaptation strategies for fishermen. *Marine Policy*, *81*, 256–261.

Silva, I. R., et al. (2024). Challenges to mangroves of the semiarid equatorial coast of Brazil in the Anthropocene. *Cambridge Prisms: Coastal Futures, 2*. https://doi.org/10.1017/cft.2024.16

Siry, H. Y. (2011). In search of appropriate approaches to coastal zone management in Indonesia. *Ocean & Coastal Management, 54*(6), 469–477.

Smit, B., & Wandel, J. (2006). Adaptation, adaptive capacity and vulnerability. *Global Environmental Change, 16*(3), 282–292.

Solomon, S., Qin, D., Manning, M., Chen, Z., Marquis, M., Averyt, K., ... Miller, H. (2007). IPCC fourth assessment report (AR4). *Climate Change, 374*.

Spalding, A. K. (2023). An introduction to marine studies. In A. K. Spalding & D. O. Suman (Eds.), *Oceans and society: An introduction to marine studies* (pp. 3–13). New York: Routledge.

Spalding, M., et al. (2010). *World atlas of mangroves*. London: Earthscan.

Sudmeier-Rieux, K., Arce-Mojica, T., Boehmer, H. J., Doswald, N., Emerton, L., Friess, D. A., ... & Walz, Y. (2021). Scientific evidence for ecosystem-based disaster risk reduction. *Nature Sustainability, 4*(9), 803–810.

Teh, L. S., Teh, L. C., & Sumaila, U. R. (2011). Quantifying the overlooked socio-economic contribution of small-scale fisheries in Sabah. *Malaysia. Fisheries Research, 110*(3), 450–458.

Thiammueang, D., Chuenpagdee, R., & Juntarashote, K. (2012). The "crab bank" project: Lessons from the voluntary fishery conservation initiative in Phetchaburi Province. *Thailand. Agriculture and Natural Resources, 46*(3), 427–439.

Thompson, P. M., & Islam, M. A. (2010). Environmental impacts and resource sustainability in Bangladesh with reference to fisheries and wetlands. *Progress in Environmental Microbiology, 1*, 205–244.

Thorburn, C. C. (2000). Changing customary marine resource management practice and institutions: The case of sasi lola in the Kei Islands, Indonesia. *World Development, 28*(8), 1461–1479.

Tindall, C., & Holvoet, K. (2008). From community-based to community-driven development in the Niger Delta. *Community Development Journal, 43*(4), 444–455.

Truong, N. X., Vassdal, T., Ngoc, Q. T. K., Anh, N. T. K., & Thuy, P. T. T. (2011). Technical efficiency of gillnet fishery in Da Nang, Vietnam: Application of stochastic production frontier. *Fish for the People, 9*(1), 26–39.

UNEP. (2019). *Climate change and mangroves in Africa*. Kenya: Nairobi.

Vardi, T. (2023). The role of mangroves in coastal protection and biodiversity. *AquaWorld*. https://aquaworld.org.ng/the-role-of-mangroves-in-coastal-protection-and-biodiversity/

Walters, B. B., Rönnbäck, P., Kovacs, J. M., Crona, B., Hussain, S. A., Badola, R., ... Dahdouh-Guebas, F. (2008). Ethnobiology, socio-economics and management of mangrove forests: A review. *Aquatic Botany, 89*(2), 220–236.

World Bank. (2023). New fisheries initiative will benefit over a million people in the Philippines. *Press Release*. https://www.worldbank.org/en/news/press-release/2023/05/31/new-fisheries-initiative-will-benefit-over-a-million-people-in-the-phi

Zabbey, N. (2009). *Impacts of extractive industries on the biodiversity of the Niger Delta*. A consultancy report for the UNDP/GEF Niger Delta Biodiversity Project.

Zabbey, N., Sam, K., & Onyebuchi, A. T. (2010). Remediation of contaminated lands in the Niger Delta, Nigeria: Prospects and challenges. *Scientific Research Essays, 5*(21), 3163–3168.

Zhang, W., Graf, H. F., Leung, Y., & Herzog, M. (2012). Different El Niño types and tropical cyclone landfall in East Asia. *Journal of Climate, 25*(19), 6510–6523.

Zu Ermgassen, P. S., et al. (2020). Fishers who rely on mangroves: Modelling and mapping the global intensity of mangrove-associated fisheries. *Estuarine, Coastal and Shelf Science*, *247*, 106975. https://doi.org/10.1016/j.ecss.2020.106975

Zu Ermgassen, P. S., et al. (2021). Fishers who rely on mangroves: Modelling and mapping the global intensity of mangrove-associated fisheries. *Estuarine, Coastal and Shelf Science*, *248*, 107159. https://doi.org/10.1016/j.ecss.2020.107159

7 Sailing Toward Sustainability: *Final Reflections and the Road Ahead*

7.1 INTRODUCTION

The word "Sundarbans" is perceived differently in the two territories. In Bangladesh, the Sundarbans refers to a natural region, specifically Ramsar Site No. 560, covering an area of 6,017 km^2. In India, this word denotes the 5,400 km^2 expanse of the Sundarbans Biosphere Reserve (SBR) and Ramsar Site No. 2370, which covers 4,230 km^2. Despite attempts to synchronize the legal frameworks of Bangladesh and India, traditional diplomacy would probably prevail, particularly concerning border conflicts, natural areas, and the management of fishery resources (Ramesh et al., 2017; Sarker et al., 2022). The uniqueness of the Sundarbans lies in the evolving transboundary waterscape that intermingles subject and object, victim and victimized (Mollick et al., 2021). The main challenges for the Sundarbans artisanal fishery are limited capacity, corruption, and fishing rights that are unevenly distributed. These issues led to a crisis of overlapping bureaucracy between artisanal fishers and fishing rights officials (Marcinko et al., 2021). These rights holders have no historical claim to the maritime resource. Consequently, comprehensive and multifarious transformative change can be realized through the analysis of the socioeconomic system (SES) (Yazdi & Neyshabouri, 2014) and substantial reconfiguration of governance institutions via the Institutional Analysis and Development framework to fulfill the Sustainable Development Goals (Mishra et al., 2024; Sene-Harper et al., 2019; Campos-Silva et al., 2021; Sowman & Raemaekers, 2018). Hardin (1968) argues that the real tragedy is not in the commons but in the commoditization of resources, when natural assets are treated as commodities to be traded, ignoring their ecological value and societal significance. When global entities restrict the responsibility and governance of local communities, policies that weaken social cohesiveness, support the existing power structures, and extend economic inequality will lead to human and environmental consequences (Utete et al., 2019). Effective governance relies on reliable information (Santos et al., 2020), and access to information is uneven among coastal resource users in the Sundarbans (Satumanatpan & Pollnac, 2020). Therefore, the amalgamation of advanced local ecological knowledge (LEK) and conventional scientific methodologies, including cultural models, will enhance the understanding of global change impacts and encourage initiatives to promote adaptive responses and facilitate stakeholder negotiations (Barua & Barua, 2023; Barua et al., 2023). Achieving sustainability requires a shared understanding among stakeholders in collaborative decision-making processes (Sowman & Raemaekers, 2018; Tull et al., 2019). It is imperative to

DOI: 10.1201/9781003729990-7

recognize and leverage existing "bright spots" where successful applications have yielded improved social and ecological outcomes. By collaborating with fishers, researchers, practitioners, and policymakers, we can promote sustainable and effective adaptation, facilitate social learning processes, and augment environmental stewardship (Silva et al., 2019). A hybrid methodology must include elements of spatial and temporal resilience evaluation to preserve system dynamics (Satumanatpan & Pollnac, 2020). It is important to recognize local experts and knowledge holders, raising their voices in civic, scientific, and political discussions to create effective policies. Many restrictions, especially those related to protected areas, restricted seasons for finfish, the ban on collecting prawn seed, and the rules against dewatering and blocking khals, need better sharing of information to improve compliance. Effective mass awareness programs on the essential trade-offs inherent in sustainable fisheries management and social discourse have been implemented in the Indian Sundarbans (Said & Chuenpagdee, 2019; Salik et al., 2019; Quiros et al., 2018). This outreach should include an educational component to inform fishermen about the connection between their harvesting techniques, fish abundance, and bycatch mortality (Philcox et al., 2010). Roberts et al. (2024) argued that the current idea of "sustainable fishing," which has been largely embraced since the mid-twentieth century, is outdated and too simple. This approach mainly focuses on maintaining catch quantities within set limits, ignoring the environmental or social impacts of different fishing methods. They question whether all fishing gear impact ecosystems and sustainability in the same way. They highlight that existing norms overlook broader issues, such as the safety and rights of fishing crews. This shows that the existing framework overlooks critical human and ecological factors. To optimize human advantages from fishery while preserving its sustainability, many models have been created to assess the maximum sustainable yield (MSY), maximum economic yield (MEY), or the principal ecosystem-based fisheries management (EBFM) strategy. However, these models fail to account for ecosystem complexity, the trade-offs between ecological and economic factors, and the uncertainties linked to ecological processes due to lack of knowledge (Grafton et al., 2007; Roberts, 1997; Skern-Mauritzen et al., 2016). The user-friendly model FISHDA (Fishing Industries' Support in Handling Decisions Application) can integrate ecological, spatial, and social methodologies to assess the condition and sustainability of small-scale fisheries (SSFs) (Garmendia et al., 2010) using minimal or easily obtainable data. This approach is particularly beneficial in data-deficient regions like Sundarbans (Licuanan et al., 2008; Muallil et al., 2014).

The concept of next-generation fisheries is more than short-term sustainability; it demands radical transformation; it is about long-term resilience, flexibility, and ecological stability. In contrast to traditional fisheries management methods that only aim to avoid overfishing, durable fisheries integrate economic security, ecosystem health, and community well-being to establish a fishing industry that has flourished for centuries. To ensure durable fisheries, we must go beyond conventional conservation methods and embrace cutting-edge innovations, technological advancements, and community-led solutions. The fishing industry of the future will not only extract from nature—it will regenerate ecosystems, create wealth, and build resilience against climate change. A comprehensive policy framework is necessary to enhance the livelihoods and sustainability of fishers in the Sundarbans. This section delineates essential components of a proposed policy framework aimed at improving the conditions of Sundarbans fishers (Figure 7.1). As Robert Swan warns, "The greatest threat to our planet is the belief that someone else will save it." This perception is especially

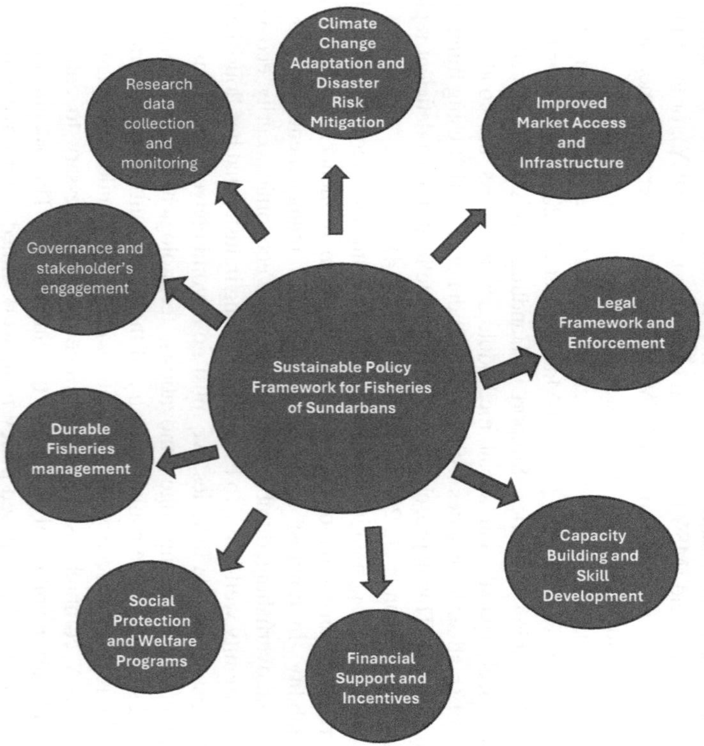

Category	Key Innovations
Smart Fisheries	• AI-Powered Stock Assessments: Uses satellite & sonar for real-time fish tracking. • Smart Fishing Nets: Acoustic sensors reduce bycatch. • Blockchain Traceability: Ensures compliance with sustainability standards.
Marine Protected Areas (MPAs) 2.0	• Dynamic MPAs: Expand/contract using AI-based fish migration data. • Rotational Marine Reserves: Temporary closures for fish stock replenishment.
Catch Limits & Precision Fishing	• DNA Barcoding: Portable scanners to verify species and quotas. • E-Vessel Monitoring (e-VMS): Solar-powered boat trackers prevent illegal fishing.
Alternative Livelihoods & Blue Economy	• Seaweed & Algae Aquaculture: Generates income & reduces overfishing. • Biofloc Technology: Zero-waste fish farming. • Blue Carbon Credit Market: Monetizes mangrove rehabilitation.
Climate-Resilient Fisheries	• Floating Fishing Villages: Adapt to rising sea levels. • Hybrid Solar Cold Storage: Reduces post-harvest losses.
Early Warning & Climate Forecasting	• AI Climate Maps: Predicts storms, salinity shifts, & fish migration. • Fisher-Led Climate Data: Boats with sensors enhance forecasting.
Mangrove Restoration & Resilience	• Mangrove-Based Shrimp Farming: Restores coastal habitats. • Seed-Ball Drones: Fast-track reforestation efforts.
Fisher Training & Skill Development	• E-Fishing Schools: Digital courses on sustainable fishing. • Hologram Learning: AI-powered remote education.
Women's Leadership in Fisheries	• Women-Led Fish Processing: Branding/export of Sundarbans seafood. • Blockchain Microfinance: AI-based credit for female entrepreneurs.
Financial Innovations	• Fisheries Credit Score: Enables low-cost loans. • Climate-Linked Insurance: AI-triggered payouts for extreme weather. • Cash-for-Conservation: Rewards for reducing bycatch & reporting violations.
Eco-Incentives	• Mangrove Bonds: Investors fund coastal resilience. • AI-Powered E-Auctions: Fair pricing for fishers.
Next-Gen Market & Trade	• Fisher-to-Consumer (F2C) Platforms: Direct seafood sales. • Hydrogel Smart Packaging: Extends seafood freshness.
Legal & Policy Innovations	• AI Licensing: Auto-renewal based on compliance. • E-Courts for Disputes: Faster conflict resolution. • Indo-Bangladesh Digital Fisheries Treaty: AI-governed cross-border management.

FIGURE 7.1 Next-generation policy framework for sustainable small-scale fisheries in the Sundarbans.

relevant for the Sundarbans, where deferring responsibility to distant institutions or future interventions risks irreversible ecological and social loss. Sustainable fisheries governance must therefore be collective, locally grounded, and forward-looking—recognizing that safeguarding the Sundarbans depends on immediate, shared action by communities, policymakers, scientists, and society at large.

7.2 STEERING THE FUTURE: STRATEGIC RECOMMENDATIONS FOR NEXT-GENERATION SUSTAINABLE SMALL-SCALE FISHERIES

Promising ways to tackle the various issues challenging marine ecosystems and coastal communities are offered by transforming fisheries systems through sustainable practices and technological innovation. The strategies, with the right institutional frameworks and technologies, environmental preservation, and economic growth, can protect the environment while also boosting the economy. Governments, industries, academic institutions, and fishing communities need to collaborate to successfully implement these innovations successfully. To fully realize the potential of technological advancement, investments in infrastructure, capacity building, and institutional development are necessary. A collaborative problem-solving case study on fishery management, rooted in a transdisciplinary approach, situates the important need for problem-solving between stakeholders. According to Bradley et al. (2019), feedback data is an essential part of effective fishery data systems, allowing fishers and managers to collect, have access to, and be granted access to fisheries data as they collaborate to achieve a previously agreed upon goal. A new approach to fisheries data systems will promote innovation to increase data coverage, accuracy, and resolution. It will also reduce costs and allow for adaptive, responsive, near real-time management decision-making to improve fisheries outcomes.

The ecological significance, social complexity, and climate vulnerability of the Sundarbans region make it a perfect place to test these innovations. Other coastal areas around the world can learn from the experiences gained here. By working together and staying committed to sustainability and innovation, we can achieve the goal of technology-driven, sustainable fisheries that benefit both people and the environment. A road map for adapting fisheries systems to the opportunities and challenges of the twenty-first century is provided in the following section.

7.2.1 Smart Fisheries: Using AI and IoT to Promote Sustainable Harvesting

7.2.1.1 AI-Powered Stock Evaluation and Tracking

The use of artificial intelligence marks a shift from traditional survey methods to real-time, data-driven monitoring in fisheries stock assessment. To evaluate fish populations and their distribution patterns accurately, machine learning algorithms can analyze large amounts of sonar data, satellite images, and environmental information (Mahesh, 2020; Rather et al., 2024). In fisheries management, combining Remote Sensing, Global Imaging System, and Potential Fishing Zone technologies represents a significant change (Sedyaaw et al., 2025). AI-enabled satellite monitoring systems can identify illegal fishing activities, track fishing vessel movements, and observe changes in marine ecosystems (Cheng et al., 2023). These systems offer extensive coverage of fishing areas by using optical sensors and synthetic aperture radar (SAR) to locate vessels even under overcasting conditions. Fish stock movements, influenced by oceanographic conditions, water temperature, and seasonal patterns, can be predicted by the integration of machine learning models. Despite numerous studies and publications on the oceanographic applications of SAR images, new technologies necessitate more advanced research in this field. In Table 7.1, key parameters and innovations in digital fisheries monitoring using SAR have been presented. Moreover, many of these applications have not reached enough maturity to be applied in operational practices, and further studies are necessary in these fields (Asiyabi et al., 2023).

AI-enhanced sonar and acoustic monitoring technologies are more accurate than conventional techniques at identifying different fish species and estimating population sizes. By identifying species, age groups, and even individual fish behavior patterns, deep learning networks trained on acoustic signatures can facilitate more accurate stock assessments and focused conservation initiatives (Yassir et al., 2023). Although AI-adapted electronic fishing is still new to fisheries, practical applications to improve the selectivity of fishing gear may not be immediately apparent. AI models are dependent on the quality of the training data, and imagery is still currently lacking (Abangan et al., 2023).

7.2.1.2 Smart Fishing Equipment Powered by IoT

An important step in reducing environmental impact while maintaining fishing efficiency is creating smart fishing gear. Smart nets that use acoustic sensors and automated release systems can detect and release nontarget species and juvenile fish before they are transported on fishing vessels. Smart nets equipped with acoustic sensors and automated release mechanisms can significantly reduce bycatch. Simple cod-end modifications such as changes in mesh size and/or orientation have been among the most popular and effective gear modifications to improve size selectivity of target and bycatch species. When cod-end size selection did not suffice, exclusion grids were introduced to either exclude unwanted bycatch species or improve size selection of the target species. More research efforts have focused on testing gear changes that aim to improve the selectivity of passive gears (Lucchetti et al., 2023). Dynamic monitoring of fishing vessels using deep learning can ensure their safety and improve the efficiency of managing them at the port. To make dynamic predictions of fishery resources, AI algorithms can more accurately predict the gathering areas and fishing intensity of fishing vessels, providing critical information support for marine ecological protection and spatial planning management (Cheng et al., 2023). WorldFish's AI-powered tool SmartCatch has won the AI for Planet Prize at the 2025 AI for Good Global Summit, hosted by the United Nations in Geneva. Designed for low-resource settings, SmartCatch could help small-scale fishers identify and record their catches with a simple mobile phone photo. The tool uses artificial intelligence to detect species, size, and weight. This supports faster and more accurate data collection, even in areas without reliable internet.

TABLE 7.1
Key Parameters and Innovations in SAR Systems for Different Oceanographic Applications

Parameter	Key Advantage	Limitation	Best Applications	Innovations
Ocean surface winds	Significant coverage, uniform global observation	Inability to obtain an accurate speed for very low or high wind speed	Climate modeling, weather forecasting	AI fusion with scatterometer and buoy data
Ocean surface current	Detects mesoscale circulation and eddies	Requires extra calibration	Shipping, search and rescue	Combine tracking observations, buoy drifting, and remote sensing to validate
Sea level	Detects coastal changes and storm surge effects	Limited accuracy in open ocean	Identify risk of coastal flooding	Pair with altimetry and tide gauges
Ocean internal waves	Maps large-scale internal waves and mixing	Potential misidentification due to different signals	Submarine navigation, fisheries	AI-based filtering for misclassification
Aquatic vegetation/ coastlines	Tracks flooding and vegetation cover	Complex interpretation of signals due to mixed signals	Wetland management, ecosystem health	Fusion of SAR measurement with optical sensors
Ship detection	Tracks ships during nighttime and inclement weather	False positives due to sea clutter and noise	Maritime security, illegal fishing	Neural networks to classify ship sizes
Oil spills	Detects surface slicks in real time	Misidentification of natural films	Environmental monitoring and compliance	Polarimetric SAR and spectral signatures
Ocean tides	Captures tidal oscillation	Limited in rough sea state	Coastal erosion studies, tidal observation	Multitemporal SAR stacking
Lakes, reservoirs, and small water bodies	Maps small inland waterbodies	Resolution issues with small lakes	Water resources management	High-resolution SAR and UAV data integration
Bathymetry	Infers the seabed from wave refraction	Shallow water only (<30 m)	Coastal navigation safety	AI and LiDAR bathymetry integration

Source: Modified after Asiyabi et al. (2023).

These Internet of Things (IoT)-enabled nets make decisions about keeping catches in real time by using species recognition algorithms, pressure sensors, and underwater cameras. The net automatically opens specific escape panels to allow undersized fish or protected species to escape. Innovations include self-operating underwater vehicles with cameras and small sensors, aerial drones, computer simulations of fishing gear design, environmental DNA (eDNA) tests, computer visualizations, and deep learning. The successful application of such tools and methods promises to improve our understanding of factors that influence capture, escape, and stress of caught species (Poisson et al., 2022). In addition to lessening the ecological impact, this technology assists fishermen in adhering to rules and avoiding fines. Incorporating GPS tracking and communication systems into fishing gear enables precise location monitoring. Also, data on fishing effort, catch composition, and environmental conditions can be collected. This data helps with management strategies and improves stock assessments. To track fishing activity in real time, these systems use GPS tracking, automatic identification systems (AIS), and vessel detection algorithms (Kroodsma et al., 2018). Advanced e-VMS systems can recognize patterns in fishing behavior, locate boats operating in prohibited zones, and automatically notify law enforcement. To differentiate between various fishing activities and identify possible violations before they occur, machine learning algorithms examine the movement patterns of vessels.

7.2.1.3 Blockchain-Powered Systems for Traceability

The utilization of blockchain technology in the fishing industry has been extensively studied and implemented to address issues such as illegal fishing and carbon emissions control. There are substantial difficulties involved in implementing blockchain using the abundant data available in the fisheries supply chain. Such difficulties include handling large amounts of data (which could take the form of photos or videos) that could be used for tracking products over their entire life cycle, and as cross-border trade and market expansion continue to grow, additional complexities are also created. Additionally, blockchain's storage capacity limitations present hurdles in fully accommodating and comprehensively storing detailed supply data from a complex and expanding supply chain (Alwi et al., 2024). The use of blockchain technology in seafood supply chains promises a level of accountability and transparency that has never been possible. A blockchain-based traceability system can be used to prevent illegal, unreported, or unregulated (IUU) fishing; it can also guarantee that harvested fish comply with the requirements for sustainability certificates (Tsolakis et al., 2020). Each fish or batch of seafood receives a unique digital identity, which is verified using blockchain technology and stored with blockchain records. This unique identity contains information on the catch location, fishing methods, vessel identification, and handling protocols. This immutable record enables verification of sustainability while also tracking the product from catch to consumer and preventing fraudulent labeling. The blockchain system can also automatically pay fishers, based on delivery requirements or specifications for quality and sustainability compliance, using smart contracts. This automated process ensures that sustainable methods applied by fishers are compensated while minimizing transaction costs. Blockchain technology can enhance transparency and sustainability in the seafood industry (Tsolakis et al., 2020). In fisheries, blockchain has the potential to curb illegal activities, improve supply chain coordination, boost operational efficiency, support sustainable practices, and help identify market trends (Hastig & Sodhi, 2020; Marttila et al., 2019; Shamsuzzoha et al., 2024). In Table 7.2, different computational applications have been presented, which can be used in SSF management.

TABLE 7.2
Computational Applications (Machine Learning and Data Mining Tools) in Small-Scale Fisheries Management

Domain	Computational Methods Used	Applications	Key References
Fisheries Resource Management	• Object-oriented modeling and multiagent simulations • Time series decomposition + ARIMA • General regression neural networks • Self-organizing feature maps • A priori algorithm • Convolutional Neural Network (CNN) • Tree-based models, random forest, eXtreme Gradient Boosting, Light Gradient Boosting Machine, and categorical boosting driven by RFECV	• Simulation of ecological, economic, and social dynamics in fisheries • Identification of biological associations in nature • Fish abundance estimation from catch data • Ecosystem monitoring and management	Bousquet et al. 1994; Su et al. (2008); Papageorgiou (2009); Taka et al. 2017); Xu et al. (2024); Selvaraj et al. 2020; Palmer et al. 2009
Fish Catch Monitoring	• Support Vector Machine (SVM) • Random Forest, Deep learning (CNN, LSTM) • Bayesian networks • Open-source computer vision models	• Prediction of fish catches • Identifying fishing ground suitability • Real-time image-based fish detection • Biomass estimation from acoustic data	Vincent et al. 2025; Silva et al. 2022
Fishing Technology	• Decision trees • Rule-based expert systems • Logistic regression • Artificial neural networks	• Fishing gear efficiency analysis • Reducing bycatch • Optimizing fishing strategies	Haque and Al Jufaili 2026; Jaree et al. 2008; Kim and Lee 2020

(Continued)

TABLE 7.2 (Continued)

Domain	Computational Methods Used	Applications	Key References
Environment Monitoring	• Wavelet transform + neural networks • Fuzzy logic systems • Genetic algorithm • Hybrid neural networks	• Monitoring water quality and fish habitats • Prediction of harmful algal blooms • Environmental impact assessments	Siddique 2013; Huang et al. 2018; Mohd et al. 2025
Socioeconomics	• Multiple linear regression • K-means clustering • Decision trees	• Analysis of fishing community livelihoods • Economic resilience modeling • Fisheries market prediction	Khan et al. 2025; Reza et al. 2025; Yoder et al. 2021

Source: Modified after Gladju et al. (2022).

7.2.2 Precision Fisheries Management and Adaptable Marine Protected Areas

7.2.2.1 Adaptive MPA Management

Fish migration patterns and the dynamic nature of marine ecosystems are frequently overlooked by traditional marine protected areas (MPAs) with set borders. Well-managed MPAs can boost conservation within no-take zones and benefit nearby fisheries through spillover. There are principally two different forms of spillover: ecological spillover (movement of fish biomass out of MPAs) and fishery spillover (the portion of fish biomass that can be fished according to regulations). Hydro-acoustics and tagging studies demonstrate evidence of ecological spillover, while experimental catch data are necessary for assessing fishery spillover, specifically the spillover that supports local fisheries the most (Di Lorenzo et al., 2016). For many wide-ranging species, dynamic area-based management tools (ABMTs), including mobile protected areas, may be the practical option. By utilizing AI and satellite data in the enactment of dynamic MPAs, it makes it feasible to dynamically alter protected area boundaries in response to ecological conditions and species distribution patterns (Maxwell et al., 2020).

Using satellite tracking of tagged fish, oceanographic data, and AI-powered predictive models, it is possible to identify critical habitats and migration corridors that vary seasonally or in response to environmental alterations. Dynamic MPAs can be expanded or contracted to include these critical spaces to optimize protection without imposing limitations on fishing communities (Chassot et al., 2011). The ongoing tracking of vessel activities via solar-powered electronic vessel monitoring systems (e-VMS) allows for enforcement of fishing regulations and MPA boundaries.

The concept of rotational marine reserves presents another option for an MPA closure rather than a permanent one, where fishing is prohibited in different places on a rotating basis. This regime allows fish populations to reset and provides fishing opportunities close to the protected site. AI algorithms can be developed to optimize the rotations based on recovery rates, timing of spawning events, and species life cycle protocols. MPAs are often a common link between fisheries management and biodiversity conservation and present stakeholders with different, fluid objectives, and expectations. This presents a system often necessitating differing management practices and attempts at compromises. As a result, these trade-offs can generate conflict among resource users and stakeholders at the local level, and even spill over into higher political levels. The adoption of Social-Ecological Network Analysis (SENA) as a flexible method for confronting the complex relationships at the human-nature interface in contextually demonstrable ways can be proposed. If MPAs are to continue to be viewed and used as economically, socially, and politically viable mechanisms for managing the ocean, it remains necessary to demonstrate that they are consistently effective in achieving their aims. However, in scientific practice, comprehensive evaluation of MPA effectiveness presents challenges, and this critical process is often overlooked (Halpern et al., 2010; Kriegl et al., 2021).

7.2.2.2 Species Monitoring and DNA Barcoding

Portable DNA scanners are used at fishing ports to rapidly identify protected species and verify compliance with quotas. DNA barcoding technology can identify species from very small tissue samples within minutes, providing law enforcement personnel with information about the species composition of catches (Rasmussen & Morrissey, 2021).

These generators produce a large, robust, genetic database for tracking fish populations, monitoring illegal catches, and determining whether conservation is effective. By linking blockchain traceability systems with DNA data, illegal and unreported catches can be circulated through the legal supply chain, while providing additional confirmation of sustainable fishing practices. A significant positive relationship was observed between the eDNA concentration and fish species abundance. Seasonal changes in fish community structure were estimated using eDNA concentrations, which may reveal the activity seasons of different fish (Wu et al., 2023).

7.2.3 Integration of the Blue Economy and Alternative Sources of Income

7.2.3.1 Aquaculture of Seaweed and Algae

The expansion of seaweed aquaculture has numerous benefits for coastal communities, including carbon sequestration, multiple revenue streams, and reduced pressure on wild capture fisheries. Specifically, for families engaged in fishing, seaweed aquaculture provides year-round income options, given the limited input needed, and it can be integrated with existing fishing practices (Duarte et al., 2021).

Large-scale seaweed aquaculture provides the potential to utilize valuable biomass for food, feed, and industrial uses while also taking up a considerable amount of CO_2 from seawater and mitigating ocean acidification. Combining seaweed aquaculture with fish aquaculture leads to polyculture systems, improves water quality, and increases total production.

Recently, machine learning modeling, based on artificial neural networks (ANN), and optimization based on Genetic Algorithms (GA) have been incorporated to improve seedling production to consider its implications for commercial aquaculture (Dineshkumar et al., 2024). Technological innovations in seaweed aquaculture, such as automated seeding systems, sensors to monitor underwater and harvesting devices, can reduce labor costs and increase productivity. The environmental adaptability of AI-driven growth models enhances maximum yield, predict harvest times, and improve planting schedules (Aung et al., 2025) by optimizing for those variables in real time. Enhancing the conservation of genetic diversity could be achieved through the development of disease-resistant macroalgae and the establishment of germplasm banks. Cooperation among scientists, researchers, policymakers, and industry stakeholders is essential for advancing the potential of seaweed cultivation to support sustainability and resilience (Khan et al. 2025).

7.2.3.2 Using Biofloc Technology (BFT) to Promote Sustainable Aquaculture

BFT is a game-changing technology that can help solve interlinked global issues such as food security, pollution reduction, and use of sustainable resources, providing an expandable framework for the future of aquaculture and beyond (Li et al., 2025). BFT is an advance in aquaculture and describes the process of turning fish waste into fish feed through the help of microbial communities. This method not only improves production efficiency and decreases production costs but also improves environmental impacts due to the zero-waste system (Emerenciano et al., 2021).

By converting the toxic ammonia from fish waste into protein-rich microbial biomass that acts as a natural feed, biofloc is able to maintain near-optimal water quality. This biological system makes aquaculture more sustainable and profitable as it also eliminates water exchange and reliance on external feed. Biofloc systems can be closely monitored using new systems that have sensors for dissolved oxygen, ammonia, nitrite, pH, and other parameters. The convergence of AI and IoT with biofloc aquaculture represents a significant step toward a more automated, cost-effective, and ecologically sustainable fish farming paradigm (Alghamdi and Haraz, 2025).

7.2.3.3 Programs for Blue Carbon Credits

Blue carbon credit markets allow ecosystem restoration while providing coastal communities new income sources. Resources related to the blue carbon market project investment indicate that mangroves are some of the most effective natural climate solutions for creating climate-resilient coastal communities (Karpowicz et al., 2024).

Projects related to carbon markets based on mangroves (MbCMPs) involving conservation, protection, and reforestation efforts have been increasingly developed and implemented in recent years and have gained increased prominence in the global natural climate solution/blue carbon markets. According to McLeod et al. (2011), mangrove forests store up to four times as much carbon per hectare as terrestrial forests, making them one of the planet's most efficient carbon sinks. Local communities can directly benefit financially from carbon credits generated by community-based mangrove restoration initiatives, which are sold on global markets. Satellite monitoring paired with AI-powered verification systems ensure precise measurements of carbon sequestration, while also preventing duplication of credits. Silvofishery systems (Fithor, 2025) provide several benefits when utilizing both aquaculture and mangrove restoration, such as carbon sequestration, habitat restoration, coastal protection, and sustainable seafood production.

7.2.4 Disaster Preparedness and Climate Resilience

7.2.4.1 Infrastructure That Adapts to Climate Change

Sea level rise, intensified storms, and modified precipitation patterns represent grave challenges for coastal fishing communities because of a changing climate. Building climate-resilient infrastructure is necessary for continuing fishing activities and protecting coastal communities. Long-term solutions for coastal infrastructure include harbors that can adjust to shifting sea levels and floating fishing platforms. Contemporary sea defense approaches, a hybrid of traditional and natural solutions, provide novel and sustainable approaches for increasing resilience and decreasing the vulnerability of coastal zones. While previous studies have predominantly focused on the ecological benefits of nature-based and hybrid engineering solutions, there remains a scarcity of data and evidence regarding wave–structure interactions at these sea defenses (Xu et al., 2024). Buildings are designed with new engineering materials that can withstand extreme weather while still functioning. The use of different materials to identify how material roughness influences overtopping and whether a sustainable material can be utilized for practical implications is being explored. The use of real vegetation is being examined to study the effects of flexible coral reefs and underwater plants on wave attenuation and pollutant dispersion near coastal areas. Different structural configurations with different kinds of saltwater-tolerant vegetation were evaluated for their capacity to mitigate overtopping, dissipate wave energy, and investigate influences on the longevity of the structures. Additional tests were performed on slope angles in search of optimal shapes or angles, which can reflect wave energy with different breakwater configurations. Sediment transport was evaluated by testing an angled, hit-and-miss concrete base (Rubinato et al., 2024).

Modular designs support expansion or relocation as environmental conditions change. The emergence of hybrid solar and wind systems is changing the energy landscape by providing communities with reliable electricity access and building a more sustainable future (Nwagu et al., 2025). Renewable energy systems with hybrid solar panels, wind turbines, and battery storage provide reliable power supply for fishing communities and processing facilities. Such hybrid systems provide energy security during extreme weather disruption to normal electricity supply and provide a better energy reliability than fossil fuel power.

7.2.4.2 Climate Adaptation Driven by AI

Climate change projection models enhanced by artificial intelligence provide fishing communities with valuable information for adaptation planning. The models use oceanographic data, weather patterns, and ecological indicators to predict the location of fish, storm surges, and salinity levels. AI-generated climate adaptation maps help fishers adjust their strategies to the anticipated environmental changes. These tools can alert fishers to dangerous weather events, provide alternative fishing methods, and provide information about the best fishing sites. Fishermen can support climate data collection and improve localized weather forecasts using community-based monitoring programs employing inexpensive environmental sensors. This collaborative method promotes the learning of sustainable local practices while enhancing regional climate modeling consistency.

7.2.4.3 Restoration of Mangroves as Natural Climate Insurance

Mangrove ecosystems give biodiversity and fisheries productivity and naturally protect against erosion, storm surges, and sea level rise. Systematic restoration programs using drone technology and seed-balls will accelerate mangrove recovery in degraded areas (Worthington et al., 2020).

Drone-based methods of reforestation use uncrewed aerial vehicles to deliver mangrove seed materials throughout precisely mapped restoration areas in study sites. Compared to planting by hand, this method is more cost-effective and efficient, particularly in hard-to-reach places. With the aid of GPS-guided drones, the placing of seed is coordinated, ensuring the requirements of the species and the conditions of the site are accommodated (Stamatopoulos et al., 2024).

7.2.5 Capacity Building and Future Skills Development

7.2.5.1 E-learning Platforms and Digital Literacy

As the fisheries sector advances to more technology-laden systems, there is a need to conduct comprehensive capacity-building programs for fishing communities. Fishing communities in remote coastal areas can receive training on digital literacy, access to markets, and other sustainable fishing practices through e-learning platforms via the web and mobile (Savo et al., 2021).

Fishermen can enhance their learning at their own pace through programming that employs interactive online courses on sustainable fishing gear, fish stock assessment, and basic business skills. Programs can accommodate a diversity of educational backgrounds in communities through visual learning and multilingual delivery methods.

Emerging technology such as virtual and augmented reality technologies can provide engaging training on complex processes such as equipment maintenance, quality assurance, and fish handling. Such technologies allow for experience-based learning to occur without the cost and risk associated with traditional learning methods. Case examples include China's "Smart Fishery," Indonesia's "Aruna," and India's "eFishery," which exemplify how technology can extend profitability, reduce waste, and improve sustainability in fisheries. However, technology alone does not guarantee benefits for those reliant on SSFs, as significant adoption barriers—beyond infrastructure and cost—persist. Resistance to innovation due to perceived value, image, risk, and tradition remains a considerable challenge (Alhassan et al., 2025; Purwanto et al., 2021) (Table 7.3). Simple interfaces and minimal data needs help drive mobile app adoption in fisheries. Although many apps exist globally, further innovation is possible (Calderwood, 2022). Practical skills, such as using e-logbooks, mobile apps, GPS, and market platforms, should be offered in local languages and customized to meet the needs of the community (Alhassan., 2025).

TABLE 7.3
Innovation in Small-Scale Fisheries: Challenges and Solutions

Dimension	Key Aspects	Cases	Challenges	Solutions
Digital technologies	GPS navigation, automated feeding, VMS, mobile Apps, GIS, AI, IoT, Blockchain, Big Data Analytics	Indonesia's Aruna, India's eFishery, China's Smart Fishery, Southeast Asia adoption	Limited infrastructure, poor internet in rural areas, high gadget costs, and digital illiteracy	Expand rural connectivity, subsidize devices, and offer digital training
Economic impacts	Reduced costs, optimized fuel, better labor efficiency, reduced postharvest losses, and higher prices	China's AI harvest prediction, European VMS, India's mKRISHI app	Unequal access → digital divide, small-scale fishers unable to invest	Cooperative platforms, public–private partnerships, targeted subsidies
Sustainability gains	Improved traceability, compliance with regulations, minimized overfishing, and optimized harvest timing	Southeast Asian blockchain traceability, Big Data for long-term trends	Low awareness of sustainability benefits, resistance to changing practices	Awareness campaigns, ecological stewardship programs, policy incentives
Barriers to adoption	Value (unclear benefit vs cost), image (distrust of agencies/NGOs), risk (financial uncertainty), tradition (loss of old practices)	Reported in Indonesia, India, and rural African fisheries	Fear of losing control, low education, language barriers, and lack of trust in digital systems	Innovation Resistance Theory (IRT), culturally adapted training, and trust-building
Policy and governance	Regulation, investment, and compliance with international standards	EU fisheries monitoring, Asian government pilots	Weak regulatory frameworks, insufficient funding	Clear policies, international collaboration, pilot-to-scale transitions

Source: Modified after AlHassan et al. (2025).

7.2.5.2 Holographic Instruction and Outreach in Rural Areas

Holographic projections and other cutting-edge visualization technologies can directly deliver professional knowledge to isolated fishing communities. AI-enabled holographic teachers can provide tailored instruction regarding fish processing, sustainable aquaculture, and business development in native languages and cultural contexts. Holographic-displayed mobile education units can reach even the most remote communities that do not have access to formal educational materials.

With three-dimensional visualizations, they can also communicate complex concepts like ecosystem management, climate adaptation, and modern fishing practices.

Through the integration of real-time communication mechanisms, remote communities can ask questions, interface with suitable experts, and receive customized technical assistance. This approach ensures access to contemporary information and practices for all communities, regardless of geographical distance.

7.2.6 Innovations in Finance and Green Incentive Initiatives

7.2.6.1 Weather-Indexed Insurance

Weather-indexed insurance products promote sustainable practices while protecting fishing communities from climate-related losses. When predetermined weather variations are exceeded, insurance offerings can leverage satellite information and/or weather station data to trigger automated payments (Jensen & Barrett, 2017).

Parametric insurance policies allow for a quick recovery from disasters, since the policies do not require an evaluation of damages and offer speedier claims processing. Smart contracts, based on blockchain technologies, can decrease administrative costs and enhance trust in the insurance process through automatic payments based on meteorological data. Including sustainable fishing practices into insurance contracts contributes to reducing environmental risk and can encourage conservation behaviors. Fishermen who employ low-impact fishing gear, participate in assessments of stock, and are following regulations are offered discounted premiums and/or preferential policy terms. Collaboration between the insurance industry, ocean research community, and marine users must be increased to help blue food industries effectively manage risk over the coming decades in many countries (Hobday et al., 2025).

7.2.6.2 Reimbursement for Ecosystem Services

Payments for ecosystem services programs (PESPs) are increasingly being adopted globally to enhance sustainability outcomes. There are also hundreds of studies yearly on various aspects of PESPs. Still, research on their contributions to the sustainability of communities and the ecosystems they depend upon at the global scale is rare (Le et al., 2024). Financial rewards can be available through cash-for-conservation programs for actions such as habitat restoration, bycatch reduction, or reporting of illegal fishing by fishermen (Börner et al., 2017).

Mangrove restoration bonds provide innovative funding options for projects aimed at protecting the coast. In return for returns based on the delivery of ecosystem services, such as storm protection, carbon sequestration, and fisheries enhancement, investors contribute up-front funds for restoration projects. Accurate evaluation of conservation results is ensured by the measurement and validation of ecosystem services using satellite monitoring and AI analytics. Participating communities can receive conservation incentives efficiently and transparently by blockchain-based payment systems.

7.2.6.3 Gender-Inclusive Business Development

Women are key to fisheries value chains, particularly in postharvest marketing and processing. Facilitating women-led businesses and cooperatives can expand the sector's productivity overall and improve income distribution (Harper et al., 2020). Women-led fish processing businesses can focus on value-added products (including smoked seafood, organic dried fish, or eco-certified crab products destined for export markets). Through direct support, such as technical training, help to access equipment, and market linkage, these firms can work their way into higher-value market segments. Certification programs and brand development allow women-run businesses to distinguish their products and charge more for them. Products from locally owned businesses have a market edge when they carry an organic certification, fair trade label, or sustainable label.

In fishing communities, there is an ongoing credit challenge—women business owners continue to face issues with accessing credit. Fee-free, quick loan approvals and automated credit scoring through blockchain-sourced microfinance platforms can provide credit accessibility solutions (Chen & Bellavitis, 2020). Alternative data sources, including social media, mobile usage, and community activities, supported by AI-based credit scoring systems, can assess creditworthiness for individuals lacking formal financial histories. Providing more accurate and unbiased lending, AI credit scoring reduces bias and discrimination. Smart contracts, an integrated aspect of blockchain technology, automates loan issuance and repayment and consequently reduces overhead. In addition to streamlining repayment, the risk of default is reduced by autopayments based on catch records or sales data.

7.2.7 Next-Generation Trade and Market Access

7.2.7.1 Innovation in Supply Chains and Digital Fish Markets

Traditional fish marketing systems often include multiple intermediaries that can significantly benefit from producers. AI-powered electronic auction platforms create competition among buyers to ensure fair pricing and reduce exploitation from middlemen (Love et al., 2021).

Real-time price data and market intelligence give fishers more confidence in when and where to sell their catch. Mobile applications can provide access to multiple markets at the same time, leading to increased competition and enhanced returns to producers.

Fish-to-consumer (F2C) e-commerce platforms create a direct marketing strategy allowing fishers to bypass intermediaries and sell directly to restaurants, retailers, and customers. They benefit fishers by increasing profit potential and simultaneously providing direct access to information regarding traceability, sustainability certifications, and quality assurance for consumers. Nevertheless, cultural insight is necessary to learn which strategy to pursue, how to partner with partners in the chain, understand consumer behavior, prioritize marketing channels, website design, type of product, and packaging. Therefore, the appropriate target markets should be fully considered and discussed prior to market entry into markets (Voldnes et al., 2021).

7.2.7.2 Cold Chain Management

Postharvest losses in tropical fisheries can reach over 30% of total catch because of inadequate cold storage and transportation. Regulatory and demand-driven intelligent inventory management and AI-powered cold chain optimization systems can model demand patterns, optimize routing, and potentially mitigate cold chain wastage (Ruiz-Garcia et al., 2009). An RFID smart tag developed for food traceability/monitoring that can measure and report temperature and relative humidity recorded is one example. Cold storage allocations can be improved through predictive analytics to anticipate demand that can be applied to market conditions allocation. "Developed RFID smart tag will help improve the competitiveness of fish/food companies, facilitate their logistic management, and enhance consumer trust in fish/food logistics" (Abad et al., 2009). Machine learning algorithms evaluate a profile of historical sales, seasonal demand/availability, and market conditions to optimize distribution to reduce waste. Innovative bio-based packaging and refrigeration systems assisted by hydrogen can improve product shelf-life while reducing environmental costs. Smart sensors monitor real-time information on temperature, humidity, and gas concentrations to make sure the product will be maintained above or below ideal storage conditions of the product (Song et al., 2025).

7.2.7.3 Global Certification and Eco-labeling Branding

Eco-labeling facilitates sustainable fisheries by market-based means. SSFs certified by the Marine Stewardship Council (MSC) can cooperate to save on costs by forming assessment units or mutually paying for an assessment with neighboring fisheries targeting the same stock. When price premiums are lacking, direct marketing is crucial. To increase its impact among SSFs, the MSC must seek price premiums in new markets (Wakamatsu & Wakamatsu, 2017). Blockchain-based eco-labeling systems allow consumers to access verified information on the sustainability of seafood, its source, and its production processes. A geographically specific seafood certification system enables blockchain and AI verification to provide local products with a premium market designation. For fishing communities, value-added product development opens new markets. Innovations using materials that would otherwise be discarded, such as fish leather, marine collagen supplements, and bioplastics made from seafood waste, can fetch high prices in global markets. To access international markets, compliance with social, environmental, and food safety laws will be required. Digital certification systems offer credible confirmation of sustainability claims, while simplifying compliance procedures. Certification schemes can create significant value within the fisheries value chain by improving the relationship between fisheries stakeholders, as well as reforming the structure of SSFs (de Melo et al., 2024).

7.2.8 Innovation in Law and Policy for Digital Fisheries Governance

7.2.8.1 AI-Powered Compliance and Licensing

Digital fishing permits can enhance regulatory efficiency and reduce administrative burden by renewing automatically and easily with compliance information. As Soto et al. (2018) explain, AI-powered compliance monitoring systems track fishing activities and automatically modify permit conditions based on performance information. E-courts for fisheries disputes facilitate easy and efficient resolution of conflicts related to quotas, licenses, or protected areas. Online dispute resolution will increase access to justice for remote communities while lowering costs and time delays. Real-time monitoring and enforcement systems enable adaptive management responses to changing conditions. Automatic notifications and interventions can help ensure that compliance issues are addressed in a timely manner and violations are addressed before they occur.

7.2.8.2 Collaboration in Transboundary Fisheries

Transboundary fisheries primarily address the resources, fleets, trade, and governance. Resolving the "boundary paradox" reflects the multilevel dynamics of reconceptualizing, modifying, and reimagining boundaries (Song et al., 2017). Transboundary SSFs imply coordinated management responses. In addition to reducing conflicts, digital fisheries treaties with centralized rules and AI-automated coordination can enhance conservation strategies. For fishermen forced to migrate due to environmental changes that have rendered historical fishing sites useless, climate migration safety nets can aid. Cooperative efforts to develop alternative livelihoods enable communities to adapt to changing conditions while maintaining social and cultural connections. Shared monitoring and enforcement mechanisms improve coverage, benefiting each country financially. Collaborative patrols, shared databases, and collective legal frameworks all work together to enhance conservation and management efforts.

7.3 CONCLUSION

SSFs often have specific challenges because they rely heavily on local ecosystems and have limited access to modern technologies. The adoption of the Voluntary Guidelines for Securing Sustainable Small-Scale Fisheries in the Context of Food Security and Poverty Eradication (SSF Guidelines) in 2014 established a new platform within international discussions, promoting the rights of SSF communities while bringing to light the indifference, inequity, and injustices faced by them. These guidelines are especially significant as SSFs have often been neglected in policy discussions, despite their considerable social and economic impact. However, the effective application of the SSF Guidelines necessitates continuous efforts from researchers across various fields, working closely with government entities, civil society organizations, and fishers. Engaging all pertinent expertise, particularly from legal professionals and policymakers, in an active and collaborative manner can significantly advance the implementation of the SSF Guidelines (Nakamura et al., 2021). The diversity of coastal transitions has led to varying levels of success for small-scale fishers engaging in them; the broader context of intensifying commodification, widespread environmental degradation, and coastal squeeze together comprise a setting of heightened pressure on livelihood activities that compound people's vulnerability (Fabinyi et al., 2022). Involving local communities in regulatory design and implementation not only builds trust but also strengthens compliance with sustainability and safety standards, ensuring that technological advancements do not leave small-scale fishers behind. Infrastructure, training, and institutional development must be heavily invested in for the successful implementation of technological innovations in fisheries. By this we can manage risks and build local capacity. These should begin with pilot projects and gradually scale up effective interventions using phased implementation strategies (FAO, 2024). Partnerships between the public and private sectors can help gather necessary funds while ensuring the fishing communities' benefit. Combining private sector innovation and investment, with government support for research and development will promote technology adoption. Community ownership and participation are key to ensuring sustainable operation and maintenance of technological systems. Successful technology transfer initiatives must include training programs, technical assistance, and the growth of local institutions. In a study, MacKeracher et al. (2019) found that capacity building is one part of livelihood development, and that achieving long-term adoption of innovations might require continued training, support to fishers for access to materials, and initiatives to develop alternative relationships with traders. Adoption and knowledge sharing were not significantly related to the multiscale characteristics examined (age, gender, education, resource dependence, village population size, market access).

Potential effects on the environment and society must be considered when introducing new technologies. The manufacturing, deployment, and disposal stages of technology should all be included in environmental impact assessments. Social impact assessment (SIA) is a well-established framework for examining the expected consequences of a planned intervention, such as a new policy, program, or technological development, on the well-being of a community (Esteves et al., 2012; Freudenburg, 1986; Rainock et al., 2018). Evaluations of social impacts should examine how they affect community cohesion, employment, and income distribution. To protect vulnerable populations and ensure that technological progress supports inclusive development, mitigation measures may be necessary. In contrast, socioeconomic impact assessment (SEIA) assesses the social and economic effects of new technologies, proposals, or projects. While definitions vary, SEIAs systematically identify and weigh socioeconomic costs and benefits, aiming to minimize negative impacts and enhance positive outcomes. They also serve as tools for scientific advice to policymakers (Rodrigues & Rituerto, 2022; Spök, 2010). Unexpected consequences can be addressed while optimizing benefits through adaptive management techniques that track results and modify interventions based on experience. Stakeholder input and regular evaluation ensure that technological advancements meet environmental goals and community needs. In recent times, Social-Ecological Network Analysis (SENA) has emerged as an effective approach for understanding the interconnectedness of humans and ecosystems. By emphasizing the influence of relationships among system elements on processes and outcomes, SENA has contributed to a deeper scientific comprehension of how coordination and cooperation among resource users impact their effectiveness, while providing researchers with a practical method to measure various aspects of socioecological system (SES) vulnerability and adaptation (Frawley et al., 2024). In many developing countries, the success determinants in SSFs are still in the infancy stage. Most of the reported success stories come from developed countries where SSFs are not very prominent (Rahman et al., 2019; Uddin et al., 2022). To accomplish positive, sustainable outcomes, transboundary efforts must be developed by an effective multilayered, collaborative, and participatory process, with dedicated resources for policy formulation and consensus building. Implementation of neoliberal policies for the artisanal fisheries resources in Sundarbans can prevent marine resource overexploitation by strengthening property rights and reducing fishing effort. "Neoliberalism individualizes responsibility" (Davis & Aldieri, 2021). Fineman's notion of vulnerability emphasizes the universality of vulnerability and argues that a "responsive state" produces "resilient" subjects to neoliberal pressures. Additionally, Fineman's resilience rhetoric may reinforce neoliberal self-management thinking. On the contrary, Butler's concept of precarity highlights specific social factors, rather than universal vulnerabilities, that weaken certain individuals. As we face neoliberal influences, vocabulary of precarity offers a more effective challenge than vulnerability, highlighting social resistance rather than individual or institutional resilience (Davis & Aldieri, 2021). However, as the Sundarbans small-scale fishers reside in rural areas, neoliberalism consolidated wealth and power into urban and industrial fishing operations and away from customarily governed open-access fishing on which small-scale fishers depend (Sannigrahi et al., 2020; Steckler et al., 2022). Thus, reanalysis and transformation on the roles of both government and public on a diverse array of public–private partnerships are of prime importance to stop this "ocean grabbing" and to strengthen and support the individuality, cultural ethics, and artisanal indigenous fishing practices of this unique coastal livelihood in a sustainable manner. As in this "tide country, transformation is the rule of life: rivers stray from week to week, and islands are made and unmade in days. In other places forests take centuries, even millennia, to regenerate; but mangroves can recolonize a denuded island in ten to fifteen years. Could it be the very rhythms of the earth were quickened here so that they unfolded at an accelerated pace?" (Ghosh, 2005).

BIBLIOGRAPHY

Abad, E., Palacio, F., Nuin, M., De Zarate, A. G., Juarros, A., Gómez, J. M., & Marco, S. (2009). RFID smart tag for traceability and cold chain monitoring of foods: Demonstration in an intercontinental fresh fish logistic chain. *Journal of Food Engineering*, *93*(4), 394–399.

Abangan, A. S., Kopp, D., & Faillettaz, R. (2023). Artificial intelligence for fish behavior recognition may unlock fishing gear selectivity. *Frontiers in Marine Science*, *10*, 1010761.

Alghamdi, M., & Haraz, Y. G. (2025). Smart biofloc systems: Leveraging artificial intelligence (AI) and internet of things (IoT) for sustainable aquaculture practices. *PRO*, *13*(7), 2204.

Alhassan, M. D. (2025). ICT regulation and sustainable development goals: The mediating role of ICT access and use in a global context. *Information Development*, *02666669251371204*.

Alwi, A., Sasongko, N. A., Suryana, Y., & Subagyo, H. (2024). Blockchain and big data integration design for traceability and carbon footprint management in the fishery supply chain. *Egyptian Informatics Journal*, *26*, 100481.

Asiyabi, R. M., Ghorbanian, A., Tameh, S. N., Amani, M., Jin, S., & Mohammadzadeh, A. (2023). Synthetic aperture radar (SAR) for ocean: A review. *IEEE Journal of Selected Topics in Applied Earth Observations and Remote Sensing*, *16*, 9106–9138.

Aung, T., Abdul Razak, R., Rahiman Bin, M., & Nor, A. (2025). Artificial intelligence methods used in various aquaculture applications: A systematic literature review. *Journal of the World Aquaculture Society*, *56*(1), e13107.

Barua, P. (2021). Coping practices of coastal fishermen in response to climate change for southern coastal belt of Bangladesh. *Social Values and Society*, *3*(2), 74–80.

Barua, P., & Barua, M. (2023). Indigenous knowledge practices by farmers in response to salinity intrusion in coastal Bangladesh. *IUP Journal of Knowledge Management*, *21*(3), 50–70.

Barua, P., Mitra, A., Mitra, A., & Eslamian, S. (2023). Resilience of farmers in response to salinity intrusion problem in agricultural fields of coastal region of Bangladesh. In A. Alam (Ed.), *Climate change, agriculture and society*. Cham: Springer.

Blasiak, R., Spijkers, J., Tokunaga, K., Pittman, J., Yagi, N., & Österblom, H. (2019). Climate change and marine fisheries: Least developed countries top global index of vulnerability. *PLoS One*, *14*(6), e0179632.

Börner, J., Baylis, K., Corbera, E., Ezzine-de-Blas, D., Honey-Rosés, J., Persson, U. M., & Wunder, S. (2017). The effectiveness of payments for environmental services. *World Development*, *96*, 359–374.

Bousquet, F., Cambier, C., & Morand, P. (1994). Distributed artificial intelligence and object-oriented modelling of a fishery. *Mathematical and Computer Modelling*, *20*(8), 97–107.

Bradley, D., Merrifield, M., Miller, K. M., Lomonico, S., Wilson, J. R., & Gleason, M. G. (2019). Opportunities to improve fisheries management through innovative technology and advanced data systems. *Fish and Fisheries*, *20*(3), 564–583.

Calderwood, J. (2022). Smartphone application use in commercial wild capture fisheries. *Reviews in Fish Biology and Fisheries*, *32*(4), 1063–1083.

Campos-Silva, J. V., Peres, C. A., Hawes, J. E., Haugaasen, T., Freitas, C. T., Ladle, R. J., & Lopes, P. F. (2021). Sustainable-use protected areas catalyze enhanced livelihoods in rural Amazonia. *Proceedings of the National Academy of Sciences*, *118*(40), e2105480118.

Chassot, E., Bonhommeau, S., Reygondeau, G., Nieto, K., Polovina, J. J., Huret, M., . . . & Demarcq, H. (2011). Satellite remote sensing for an ecosystem approach to fisheries management. *ICES Journal of Marine Science*, *68*(4), 651–666.

Chen, M. A., & Bellavitis, C. (2020). Blockchain disruption and decentralized finance: The rise of decentralized business models. *Journal of Business Venturing Insights*, *13*, e00151.

Cheng, X., Zhang, F., Chen, X., & Wang, J. (2023). Application of artificial intelligence in the study of fishing vessel behavior. *Fishes*, *8*(10), 516.

Christensen, V., Coll, M., Piroddi, C., Steenbeek, J., Buszowski, J., & Pauly, D. (2021). The global ocean is an ecosystem: Simulating marine life and fisheries. *Global Ecology and Biogeography*, *23*(4), 429–442.

Davis, B. P., & Aldieri, E. (2021). Precarity and resistance: A critique of Martha Fineman's vulnerability theory. *Hypatia*, *36*(2), 321–337.

de Melo, M. T. G., da Silva Barros, J. M., Ribeiro, A. R. B., Lima, T. L. D. A., & Sobral, M. F. F. (2024). The role of certifications and eco-labels in fisheries: A systematic literature review of their benefits and challenges. *Environmental Sciences Europe*, *36*(1), 200.

Di Lorenzo, M., Claudet, J., & Guidetti, P. (2016). Spillover from marine protected areas to adjacent fisheries has an ecological and a fishery component. *Journal for Nature Conservation*, *32*, 62–66.

Dineshkumar, R., Kazi, M. A., & Mantri, V. A. (2024). Application of artificial neural network and genetic algorithm modeling for in vitro regeneration of seaweed seedling production. In *Plant cell culture protocols* (pp. 99–107). New York: Springer US.

Duarte, C. M., Bruhn, A., & Krause-Jensen, D. (2021). A seaweed aquaculture imperative to meet global sustainability targets. *Nature Sustainability*, *5*, 185–193.

Emerenciano, M. G., Martínez-Córdova, L. R., Martínez-Porchas, M., & Miranda-Baeza, A. (2021). Biofloc technology (BFT): A tool for water quality management in aquaculture. *Water Quality Management*, 91–109.

Esteves, A. M., Franks, D., & Vanclay, F. (2012). Social impact assessment: The state of the art. *Impact Assessment and Project Appraisal*, *30*(1), 34–42.

Fabinyi, M., Belton, B., Dressler, W. H., Knudsen, M., Adhuri, D. S., Aziz, A. A., . . . Vandergeest, P. (2022). Coastal transitions: Small-scale fisheries, livelihoods, and maritime zone developments in Southeast Asia. *Journal of Rural Studies*, *91*, 184–194.

FAO. (2020). *The state of world fisheries and aquaculture 2020: Sustainability in action*. Food and Agriculture Organization of the United Nations.

FAO. (2024). The State of World Fisheries and Aquaculture 2024 – Blue Transformation in action.

Fithor, A. (2025). Innovative silvofishery model in restored mangrove forests: A 10-year assessment. *Heliyon*, *11*(2).

Frawley, T. H., González-Mon, B., Nenadovic, M., Gladstone, F., Nomura, K., Zepeda-Domínguez, J. A., & Basurto, X. (2024). Self-governance mediates small-scale fishing strategies, vulnerability and adaptive response. *Global Environmental Change*, *84*, 102805.

Freudenburg, W. R. (1986). Social impact assessment. *Annual Review of Sociology*, *12*(1), 451–478.

Gallagher, A. J., Hammerschlag, N., Shiffman, D. S., & Giery, S. T. (2022). Evolved for extinction: The cost and conservation implications of specialization in hammerhead sharks. *Bioscience*, *64*(7), 619–629.

Garmendia, E., Gamboa, G., Franco, J., Garmendia, J. M., Liria, P., & Olazabal, M. (2010). Social multi-criteria evaluation as a decision support tool for integrated coastal zone management. *Ocean & Coastal Management*, *53*(7), 385–403.

Ghosh, A. (2005). *The hungry tide*. Houghton Mifflin.

Gladju, J., Kamalam, B. S., & Kanagaraj, A. (2022). Applications of data mining and machine learning framework in aquaculture and fisheries: A review. *Smart Agricultural Technology*, *2*, 100061.

Grafton, R. Q., Kompas, T., & Hilborn, R. W. (2007). Economics of overexploitation revisited. *Science*, *318*(5856), 1601–1601.

Halpern, B. S., Lester, S. E., & McLeod, K. L. (2010). Placing marine protected areas onto the ecosystem-based management seascape. *Proceedings of the National Academy of Sciences*, *107*(43), 18312–18317.

Haque, S. A., & Al Jufaili, S. M. (2026). Applications of artificial intelligence in fisheries: From data to decisions. *Big Data and Cognitive Computing*, *10*(1), 19.

Hardin, G. (1968). The tragedy of the commons. *Science*. https://doi.org/10.1126/science.162.3859.1243

Harper, S., Zeller, D., Hauzer, M., Pauly, D., & Sumaila, U. R. (2020). Women and fisheries: Contribution to food security and local economies. *Marine Policy*, *39*, 56–63.

Hastig, G. M., & Sodhi, M. S. (2020). Blockchain for supply chain traceability: Business requirements and critical success factors. *Production and Operations Management*, *29*(4), 935–954.

Hobday, A. J., Little, L. R., Watson, J. R., & Spillman, C. M. (2025). Parametric insurance for climate adaptation in fisheries and aquaculture. *Reviews in Fish Biology and Fisheries*, *35*(1), 175–185.

Huang, M., Tian, D., Liu, H., Zhang, C., Yi, X., Cai, J., . . . Ying, G. (2018). A hybrid fuzzy wavelet neural network model with self-adapted fuzzy c-means clustering and genetic algorithm for water quality prediction in Rivers. *Complexity*, *2018*(1), 8241342.

Islam, M. M., Begum, P., Begum, A., & Herbeck, J. (2024). When hazards become disasters: Coastal fishing communities in Bangladesh. *Environmental Hazards*, *45*(3), 533–549.

Jarre, A., Paterson, B., Moloney, C. L., Miller, D. C., Field, J. G., & Starfield, A. M. (2008). Knowledge-based systems as decision support tools in an ecosystem approach to fisheries: Comparing a fuzzy-logic and a rule-based approach. *Progress in Oceanography*, *79*(2-4), 390–400.

Jensen, N., & Barrett, C. (2017). Agricultural index insurance for development. *Applied Economic Perspectives and Policy*, *39*(2), 199–219.

Karpowicz, D. A., Mohan, M., Watt, M. S., Montenegro, J. F., King, S. A., Selvam, P. P., ... Ewane, E. B. (2024). Mangrove-based carbon market projects: 15 considerations for engaging and supporting local communities. *Diversity*, *16*(9), 574.

Khan, M. B., Shafiq, N., & Zada, N. S. (2025, October). Investigating the influence of sea weed. In *Proceedings of the international conference on smart cities—Volume 1: ICSC 2024, September 10–11, Kota Kinabalu, Malaysia* (p. 95). Springer Nature.

Khan, S., Das, S. G., Kalra, S., Chaudhary, N., & Singh, G. (2025, February). Factors influencing ai index across regions, political regimes, and development indicators: A random forest and K-mean clustering approach. In *2025 International conference on technology enabled economic changes (InTech)* (pp. 672–677). IEEE.

Kim, K. I., & Lee, K. M. (2020). Convolutional neural network-based gear type identification from automatic identification system trajectory data. *Applied Sciences*, *10*(11), 4010.

Kriegl, M., Elías Ilosvay, X. E., von Dorrien, C., & Oesterwind, D. (2021). Marine protected areas: At the crossroads of nature conservation and fisheries management. *Frontiers in Marine Science*, *8*, 676264.

Kroodsma, D. A., Mayorga, J., Hochberg, T., Miller, N. A., Boerder, K., Ferretti, F., & Worm, B. (2018). Tracking the global footprint of fisheries. *Science*, *359*(6378), 904–908.

Le, T. A. T., Vodden, K., Wu, J., Bullock, R., & Sabau, G. (2024). Benefits and risks from payments for ecosystem services programs across the globe. *Frontiers in Environmental Science*, *12*, 1419821.

Li, C., Ge, Z., Dai, L., & Chen, Y. (2025). Integrated application of biofloc technology in aquaculture: A review. *Water*, *17*(14), 2107.

Licuanan, W. Y., Mamauag, S. S., Gonzales, R. O. M., & Aliño, P. M. (2008). The minimum sizes of fish sanctuaries and fishing effort reductions needed to achieve sustainable coastal fisheries in Calauag and Tayabas bays. *Philippine Agricultural Scientist*, *91*(1), 51–59.

Love, D. C., Allison, E. H., Asche, F., Belton, B., Cottrell, R. S., Froehlich, H. E., & Zhang, W. (2021). Emerging COVID-19 impacts, responses, and lessons for building resilience in the seafood system. *Global Food Security*, *28*, 100494.

Lucchetti, A., Melli, V., & Brčić, J. (2023). Innovations in fishing technology aimed at achieving sustainable fishing. *Frontiers in Marine Science*, *10*, 1310318.

MacKeracher, T., Foale, S. J., Gurney, G. G., & Purcell, S. W. (2019). Adoption and diffusion of technical capacity-building innovations by small-scale artisanal fishers in Fiji. *Ecology and Society*, *24*(2).

Mahesh, B. (2020). Machine learning algorithms—A review. *International Journal of Science and Research*, *9*, 381–386.

Marcinko, C. L., Nicholls, R. J., Daw, T. M., Hazra, S., Hutton, C. W., Hill, C. T., Clarke, D., Harfoot, A., Basu, O., & Das, I. (2021). The development of a framework for the integrated assessment of SDG trade-offs in the Sundarban biosphere reserve. *Water*, *13*, 528.

Marttila, J., Nousiainen, M., Sheppard, B., Malka, M., & Karjalainen, R. (2019). *Tracey–your traceability and trade data companion*. TX–Technololy Exploration Oy.

Maxwell, S. M., Hazen, E. L., Lewison, R. L., Dunn, D. C., Bailey, H., Bograd, S. J., & Costa, D. P. (2020). Dynamic Ocean management: Defining and conceptualizing real-time management of the ocean. *Marine Policy*, *58*, 42–50.

Mcleod, E., Chmura, G. L., Bouillon, S., Salm, R., Björk, M., Duarte, C. M., & Silliman, B. R. (2011). A blueprint for blue carbon: Toward an improved understanding of the role of vegetated coastal habitats in sequestering CO_2. *Frontiers in Ecology and the Environment*, *9*(10), 552–560.

Mishra, M., de Lucena, A. J., & Maharaj, B. (2024). *Climate change and regional socio-economic systems in the global south*. Springer.

Mohd, N. N. M. S. N., Kamal, A. A. Z., Hussin, A. A. A., Annas, A. H., Mohammad-Noor, N., & Razali, R. M. (2025). Optimizing radial basis function networks for harmful algal bloom prediction: A hybrid machine learning approach. *International Journal of Electrical and Computer Engineering (IJECE)*, *15*(6), 5647–5654.

Mollick, A. S., Roy, M., Khan, N. I., Islam, W., Sadath, N., & Nath, T. K. (2021). Assessing good governance in protected areas (PA) co-management: A case study of the Sundarbans mangrove forests of Bangladesh.

Muallil, R. N., Mamauag, S. S., Cabral, R. B., Celeste-Dizon, E. O., & Aliño, P. M. (2014). Status, trends and challenges in the sustainability of small-scale fisheries in the Philippines: Insights from FISHDA (fishing industries' support in handling decisions application) model. *Marine Policy*, *44*, 212–221.

Nakamura, J., Chuenpagdee, R., & El Halimi, M. (2021). Unpacking legal and policy frameworks: A step ahead for implementing the small-scale fisheries guidelines. *Marine Policy*, *129*, 104568.

Nwagu, C. N., Ujah, C. O., Kallon, D. V., & Aigbodion, V. S. (2025). Integrating solar and wind energy into the electricity grid for improved power accessibility. *Unconventional Resources*, *5*, 100129.

Palmer, M., Quetglas, A., Guijarro, B., Moranta, J., Ordines, F., & Massutí, E. (2009). Performance of artificial neural networks and discriminant analysis in predicting fishing tactics from multispecific fisheries. *Canadian Journal of Fisheries and Aquatic Sciences*, *66*(2), 224–237.

Papageorgiou, L. G. (2009). Supply chain optimisation for the process industries: Advances and opportunities. *Computers & Chemical Engineering*, *33*(12), 1931–1938.

Philcox, N., Knowler, D., & Haider, W. (2010). Eliciting stakeholder preferences: An application of qualitative and quantitative methods to shrimp aquaculture in the Indian Sundarbans. *Ocean & Coastal Management*, *53*(3), 123–134.

Poisson, F., Budan, P., Coudray, S., Gilman, E., Kojima, T., Musyl, M., & Takagi, T. (2022). New technologies to improve bycatch mitigation in industrial tuna fisheries. *Fish and Fisheries*, *23*(3), 545–563.

Purwanto, E., Sjarief, R., & Anwar, C. (2021). The effect of the innovation resistance factors on the digital fishery platform acceptance in developing country. *Review of International Geographical Education Online*, *11*(5), 507–518.

Quiros, T. E. A. L., Beck, M. W., Araw, A., Croll, D. A., & Tershy, B. (2018). Small-scale seagrass fisheries can reduce social vulnerability: A comparative case study. *Ocean and Coastal Management*, *157*, 56–67.

Rahman, M. M., Khan, M. N. I., Hoque, A. F., & Ahmed, I. (2019). Carbon stock in the Sundarban mangrove forest: Spatial variations in vegetation types and salinity zones. *Wetland and Ecology Management*, *23*(2), 269–283.

Rainock, M., Everett, D., Pack, A., Dahlin, E. C., & Mattson, C. A. (2018). The social impacts of products: A review. *Impact Assessment and Project Appraisal*, *36*(3), 230–241.

Ramesh, R., Purvaja, R., Krishnan, P., Lakshmi, A., Abhilash, K., & Kingsley, P. W. (2017). Conservation of coastal wetlands: An appraisal of the policy and legal framework in south Asian nations. In *Wetland science* (pp. 515–544). Springer.

Rasmussen, R. S., & Morrissey, M. T. (2021). DNA-based methods for the identification of commercial fish and seafood species. *Comprehensive Reviews in Food Science and Food Safety*, *7*(3), 280–295.

Rather, M. A., Ahmad, I., Shah, A., Hajam, Y. A., Amin, A., Khursheed, S., . . . Rasool, S. (2024). Exploring opportunities of artificial intelligence in aquaculture to meet increasing food demand. *Food Chemistry: X*, *22*, 101309.

Reza, S. A., Rahman, M. K., Hossain, M. S., Rabbi, M. N. S., Mozumder, A. Q., Safat, S., . . . Ahad, M. A. (2025). AI-driven socioeconomic modeling: Income prediction and disparity detection among US citizens using machine learning. *Advances in Consumer Research*, *2*(4).

Roberts, C. M. (1997). Ecological advice for the global fisheries crisis. *Trends in Ecology and Evolution*, *12*, 35–38.

Roberts, S., Jacquet, J., Majluf, P., & Hayek, M. N. (2024). Feeding global aquaculture. *Science. Advances*, *10*(42), eadn9698.

Rodrigues, R., & Rituerto, M. D. (2022). Socio-economic impact assessments for new and emerging technologies. *Journal of Responsible Technology*, *9*, 100019.

Rubinato, M., Rosario Ortiz, F. L., Piazza Meireles, R., Martins, R., & Sun, P. N. (2024). Are we keeping coastal waters safe and clean after heavy rainfall events? An overview and update. *ACS ES&T Water*, *4*(11), 4664–4667.

Ruiz-Garcia, L., Lunadei, L., Barreiro, P., & Robla, I. (2009). A review of wireless sensor technologies and applications in agriculture and food industry: State of the art and current trends. *Sensors*, *9*(6), 4728–4750.

Said, A., & Chuenpagdee, R. (2019). Aligning the sustainable development goals to the small-scale fisheries guidelines: A case for EU fisheries governance. *Marine Policy*, *107*, 103599.

Salik, K. M., Jahangir, S., Zahdi, W. u. Z., & Hasson, S. U. (2019). Climate change vulnerability and adaptation options for the coastal communities of Pakistan. *Ocean and Coastal Management*, *112*(2), 61–73.

Sannigrahi, S., Zhang, Q., Pilla, F., Joshi, P. K., Basu, B., Keesstra, S., Roy, P. S., Wang, Y., Sutton, P. C., Chakraborti, S., Paul, S. K., Sen, S., Campos-Vargas, C., Binks, S., Balan, S., & Baghel, P. (2020). Responses of ecosystem services to natural and anthropogenic forcings: A spatial regression based assessment in the world's largest mangrove ecosystem. *Science of Total Environment*, *715*(2), 189–210.

Santos, R. E., Pinto-Coelho, R. M., Drumond, M. A., Fonseca, R., & Zanchi, F. B. (2020). Damming Amazon rivers: Environmental impacts of hydroelectric dams on Brazil's Madeira River according to local fishers' perception. *Ambio*, *49*, 1612–1628.

Sarker, S., Huda, A., Niloy, M., & Chowdhury, G. W. (2022). Trophic transfer of microplastics in the aquatic ecosystem of Sundarbans mangrove forest, Bangladesh. *Science of Total Environment*, *838*(2), 345–370.

Satumanatpan, S., & Pollnac, R. (2020). Resilience of small-scale fishers to declining fisheries in the Gulf of Thailand. *Coastal Management*, *48*, 1–22.

Savo, V., Lepofsky, D., Benner, J. P., Kohfeld, K. E., Bailey, J., & Lertzman, K. (2021). Observations of climate change among subsistence-oriented communities around the world. *Nature Climate Change*, *6*(5), 462–473.

Sedyaaw, P., Pathan, D. I., Bhaladhare, D. R., Kamble, P. A., Kawade, S. S., & Sharangdhar, A. S. (2025). *Overfishing and marine biodiversity loss. In vanishing biodiversity: Global threats to flora and fauna* (pp. 337–352).

Selvaraj, J. J., Arunachalam, V., Coronado-Franco, K. V., Romero-Orjuela, L. V., & Ramírez-Yara, Y. N. (2020). Time-series modeling of fishery landings in the Colombian Pacific Ocean using an ARIMA model. *Regional Studies in Marine Science*, *39*, 101477.

Sene-Harper, A., Matarrita-Cascante, D., & Larson, L. R. (2019). Leveraging local livelihood strategies to support conservation and development in West Africa. *Environment and Development*, *29*(3), 16–28.

Shamsuzzoha, A., Marttila, J., & Helo, P. (2024). Blockchain-enabled traceability system for the sustainable seafood industry. *Technology Analysis & Strategic Management, 36*(11), 3891–3905.

Siddique, N. (2013). *Intelligent control: A hybrid approach based on fuzzy logic, neural networks and genetic algorithms* (Vol. 517). Springer.

Silva, C. N., Dainys, J., Simmons, S., Vienožinskis, V., & Audzijonyte, A. (2022). A scalable open-source framework for machine learning-based image collection, annotation and classification: A case study for automatic fish species identification. *Sustainability, 14*(21), 14324.

Silva, M. R. O., Pennino, M. G., & Lopes, P. F. M. (2019). Social-ecological trends: Managing the vulnerability of coastal fishing communities. *Journal of Ecological Indicator, 45*(3), 55–75.

Skern-Mauritzen, M., Ottersen, G., Handegard, N. O., Huse, G., Dingsør, G. E., Stenseth, N. C., & Kjesbu, O. S. (2016). Ecosystem processes are rarely included in tactical fisheries management. *Fish and Fisheries, 17*(1), 165–175.

Smith, H., & Basurto, X. (2019). Defining small-scale fisheries and examining the role of science in shaping perceptions of who and what counts: A systematic review. *Frontiers in Marine Science, 30*(3), 55–65.

Song, A. M., Scholtens, J., Stephen, J., Bavinck, M., & Chuenpagdee, R. (2017). Transboundary research in fisheries. *Marine Policy, 76*, 8–18.

Song, X., Zhang, X., Dong, G., Ding, H., Cui, X., Han, Y., . . . Wang, L. (2025). AI in food industry automation: Applications and challenges. *Frontiers in Sustainable Food Systems, 9*, 1575430.

Soto, D., Ross, L. G., Handisyde, N., Bueno, P. B., Beveridge, M. C., Dabbadie, L., . . . & Wainberg, A. A. (2018). *Climate change and aquaculture: Potential impacts, adaptation and mitigation*. FAO Fisheries and Aquaculture Technical Paper No. 530.

Sowman, M., & Raemaekers, S. (2018). Socio-ecological vulnerability assessment in coastal communities in the BCLME region. *Journal of Marine Systems, 188*, 160–171.

Spök, A. (2010). *Assessing socio-economic impacts of GMOs: Issues to consider for policy development; final report*. Bundesministerium für Gesundheit, Sekt. II.

Stamatopoulos, I., Le, T. C., & Daver, F. (2024). UAV-assisted seeding and monitoring of reforestation sites: A review. *Australian Forestry, 87*(2), 90–98.

Steckler, M. S., Oryan, B., & Wilson, C. A. (2022). Synthesis of the distribution of subsidence of the lower Ganges-Brahmaputra Delta. *Bangladesh. Reviews of Earth Sciences, 70*(3), 45–65.

Stock, C. A., Alexander, M. A., Bond, N. A., Brander, K. M., Cheung, W. W., Curchitser, E. N., & Werner, F. E. (2021). On the use of IPCC-class models to assess the impact of climate on living marine resources. *Progress in Oceanography, 88*(1-4), 1–27.

Su, N. J., Yeh, S. Z., Sun, C. L., Punt, A. E., Chen, Y., & Wang, S. P. (2008). Standardizing catch and effort data of the Taiwanese distant-water longline fishery in the western and Central Pacific Ocean for bigeye tuna. *Thunnus obesus. Fisheries Research, 90*(1–3), 235–246.

Taka, H., Sasaki, T., & Wada, M. (2017, December). Supporting system for fishery resource management utilizing convolutional neural network. In *2017 20th international symposium on wireless personal multimedia communications (WPMC)* (pp. 442–447). IEEE.

Thompson, B. S., Bladon, A. J., Fahad, Z. H., Mohsanin, S., & Koldewey, H. J. (2016). Evaluation of the ecological effectiveness and social appropriateness of fishing regulations in the Bangladesh Sundarbans using a new multi-disciplinary assessment framework. *Fisheries Research, 183*, 410–423.

Tsolakis, N., Niedenzu, D., Simonetto, M., Dora, M., & Kumar, M. (2020). Supply network design to address United Nations sustainable development goals: A case study of blockchain implementation in Thai fish industry. *Journal of Business Research*, *131*, 495–519.

Tull, M., Metcalf, S. J., & Gray, H. (2019). The economic and social impacts of environmental change on fishing towns and coastal communities: A historical case study of Geraldton, Western Australia. *ICES Journal of Marine Sciences*, *73*, 1437–1446.

Uddin, M. S., Steveninck, E. R., Stuip, M., & Shah, M. A. R. (2022). Economic valuation of provisioning and cultural services of a protected mangrove ecosystem: A case study on Sundarbans reserve Forest, Bangladesh. *Ecosystem Service*, *5*(2), 88–93.

Utete, B., Phiri, C., Mlambo, S. S., Muboko, N., & Fregene, B. T. (2019). Vulnerability of fisherfolks and their perceptions towards climate change and its impacts on their livelihoods in a peri-urban lake system in Zimbabwe. *Environmental Development and Sustainability*, *21*, 917–934.

Vincent, A. N., Sakthidasan, K., Sunhaloo, M. S., Clement, J. C., Laloo, N., & Bagurubumwe, U. (2025). Comparative analysis of RNN, LSTM and CNN algorithms for marine data prediction. *Journal of Coastal Conservation*, *29*(6), 73.

Voldnes, G., Pleym, I. E., Ageeva, T., Alm, S., Nyrud, T., & Rosnes, J. T. (2021). E-commerce of seafood–a review of existing research. *Journal of International Food & Agribusiness Marketing*, *33*(1), 3–35.

Wakamatsu, M., & Wakamatsu, H. (2017). The certification of small-scale fisheries. *Marine Policy*, *77*, 97–103.

Worthington, T., Zu Ermgassen, P., Friess, D., Krauss, K., Lovelock, C., Thorley, J., & Spalding, M. (2020). A global biophysical typology of mangroves and its relevance for ecosystem structure and deforestation. *Scientific Reports*, *10*(1), 1–11.

Wu, L., Wu, Q., Inagawa, T., Okitsu, J., Sakamoto, S., & Minamoto, T. (2023). Estimating the spawning activity of fish species using nuclear and mitochondrial environmental DNA concentrations and their ratios. *Freshwater Biology*, *68*(1), 103–114.

Xu, S., Wang, J., Chen, X., & Zhu, J. (2024). Identifying optimal variables for machine-learning-based fish distribution modeling. *Canadian Journal of Fisheries and Aquatic Sciences*, *81*(6), 687–698. https://doi.org/10.1139/cjfas-2023-0197

Yassir, A., Andaloussi, S. J., Ouchetto, O., Mamza, K., & Serghini, M. (2023). Acoustic fish species identification using deep learning and machine learning algorithms: A systematic review. *Fisheries Research*, *266*, 106790.

Yazdi, J., & Neyshabouri, S. S. (2014). Identifying low impact development strategies for flood mitigation using a fuzzy-probabilistic approach. *Environmental Modelling & Software*, *60*, 31–44.

Yoder Clark, A., Blumenfeld, N., Lal, E., Darbari, S., Northwood, S., & Wadpey, A. (2021). Using K-means cluster analysis and decision trees to highlight significant factors leading to homelessness. *Mathematics*, *9*(17), 2045.

Annexure

A Visual Narrative of Everyday Life in Sundarbans' Small-Scale Fisheries

Photo credit: Anisa Mitra, Prabal Barua, Sitesh Manna

FIGURE A.1 (a) A fleet of traditional fishing boats anchored on the shore, ready for their next venture into the waters. The boats, equipped with fishing gear and colorful decorations, represent the livelihood of local fishermen dependent on the rich aquatic resources. (b) Fishermen and traders actively engaged in unloading freshly caught seafood from fishing boats. The scene captures the vibrant fish trade, with crates filled with ice and fish being prepared for transport to markets, highlighting the economic significance of the fishing industry in coastal communities. (c) Traditional fishing in a muddy riverbank, where fishermen use large nets to catch fish. The scene reflects the reliance on local waterways for livelihood, showcasing a sustainable, community-driven approach to fishing in a natural setting. (d) Fisherman navigates his wooden boat through a vast river, surrounded by lush, green mangroves. This image highlights the traditional and independent fishing practices in riverine communities, emphasizing the deep connection between nature and livelihood.

FIGURE A.2 (a) A young girl sits beside a vast spread of fish drying on green nets under the open sky, near a coastal area. This image portrays the traditional method of fish preservation, an essential livelihood activity for many coastal communities, while also hinting at the daily realities faced by those involved in the fishing industry. (b) Children play near large black nets spread out on the sandy ground, filled with drying fish. This scene highlights the traditional fish drying process, a crucial part of the fishing economy in coastal villages. In the background, villagers go about their daily routines, reflecting the close connection between livelihood and community life. (c) A busy fishing harbor with multiple wooden boats anchored and fishermen actively working. The cloudy sky and choppy waters suggest a coastal setting where traditional fishing practices are still prevalent, forming an essential part of the local economy. (d) A young boy holding a large fishing net in front of a traditional wooden hut. The rural setting, dry sandy ground, and handmade fishing equipment highlight the subsistence fishing lifestyle, where even children contribute to daily livelihood activities.

FIGURE A.3 (a) Fishermen and their families working along the riverbank, drying and preparing fishing nets near wooden boats. The peaceful riverside setting, combined with traditional fishing practices, reflects a close connection to nature and a livelihood dependent on water. (b) A basket filled with freshly caught fish at a small-scale fishery site. (c) Fishermen working in a small-scale fishery, with blue barrels used for storage or transportation of fish, set along a dirt path near a fishing site. (d) Expansive river scene with boats scattered across the water.

FIGURE A.4 (a) Women engage in the labor-intensive process of sorting and drying fish in an outdoor setting. Large piles of fish are being prepared, with some already packed into plastic bags. The rural backdrop, including bamboo huts and electrical infrastructure, highlights the intersection of traditional livelihoods and modern development. (b) Father and son, engage in traditional fishing using a net in a narrow, shallow waterway. The surrounding landscape is lush with tall grasses, indicating a rural setting where fishing serves as a vital source of livelihood. (c) Children and young fishermen by the riverbank, engaged in fishing with a traditional net. This scene captures a simple, rural lifestyle where fishing is both a livelihood and a means of sustenance. (d) A bustling fish market scene with workers managing crates of freshly caught fish. The wet and cluttered surroundings highlight the fast-paced environment of the seafood trade in a local marketplace.

FIGURE A.5 (a) A freshly caught assortment of fish in a basket, showcasing various species of marine and freshwater fish, possibly from a local fishing community. The variety of fish sizes and colors indicates a diverse aquatic ecosystem. (b) Fishermen sorting and preserving freshly caught shrimp with ice on a wooden boat. This image highlights the traditional fishing practices and the importance of maintaining seafood freshness for market distribution. (c) Fishermen sorting through a fresh catch of small fish on a wooden boat, showcasing their daily work in the fishing industry near a river or coastal area. (d) Fishermen navigate a wooden boat along a calm river, carrying traditional bamboo fishing traps, likely used for crab or fish catching in a mangrove region.

FIGURE A.6 (a) Fishermen work together to haul in a large catch from their net, showcasing the teamwork and effort involved in traditional fishing practices. Their livelihood depends on the bounty of the waters. (b) A family by the riverside sorts and cleans their fresh catch, with children eagerly participating, reflecting the deep-rooted connection between fishing and community life. (c) A young boy proudly displays his catch from a handmade crab trap along the muddy banks of a mangrove-lined river, showcasing traditional fishing practices in coastal communities. (d) A local fish vendor inspects the fresh catch as fishermen sort and clean their haul on a boat docked by the riverside, depicting the daily trade and livelihood of riverine communities.

(a) (b)

FIGURE A.7 (a) A fisherman and his companion navigate their boat along a calm river, carrying supplies under a makeshift shelter. The scene reflects the resilience and simplicity of riverine life. (b) A woman (Meendhara) skillfully sorting the Meens (Prawn fry) through her fishing net the muddy riverbank. Her dedication highlights the daily struggles and resilience of coastal livelihoods.

Index

Pages in *italics* refer to figures, pages in **bold** refer to tables, and pages followed by "n" refer to notes.